New Trends in Statistical Physics

Festschrift in Honor of Leopoldo García-Colín's 80th Birthday

New Trends in Statistical Physics

Festschrift in Honor of Leopoldo García-Colín's 80th Birthday

Alfredo Macias
Universidad Autónoma Metropolitana-Iztapalapa, Mexico

Leonardo Dagdug
Universidad Autónoma Metropolitana-Iztapalapa, Mexico

World Scientific

NEW JERSEY • LONDON • SINGAPORE • BEIJING • SHANGHAI • HONG KONG • TAIPEI • CHENNAI

Published by

World Scientific Publishing Co. Pte. Ltd.
5 Toh Tuck Link, Singapore 596224
USA office: 27 Warren Street, Suite 401-402, Hackensack, NJ 07601
UK office: 57 Shelton Street, Covent Garden, London WC2H 9HE

British Library Cataloguing-in-Publication Data
A catalogue record for this book is available from the British Library.

NEW TRENDS IN STATISTICAL PHYSICS
Festschrift in Honor of Leopoldo García-Colín's 80th Birthday

ISBN-13 978-981-4307-53-6
ISBN-10 981-4307-53-X

Printed in Singapore.

Preface

Friends, colleagues, and former students of Leopoldo García-Colín contributed to this volume to celebrate and express to him our respect, admiration, and affection on the occasion of his 80th birthday. We want to celebrate, with this book, his scientific and scholarly achievements, the inspirational quality of his teaching, his graciousness as a colleague, his thoughtful guidance of graduated students, his service to the department, the university and the physics community at large, his open, courteous, easy accessibility to anyone needing his counsel or expertise.

Leopoldo García-Colín has worked in many different fields of statistical physics, and has applied it to biological physics, solid state physics, relativity and cosmology. This book contains original and peer-reviewed articles from his friends, colleagues, and former students, and represents a unique collection of papers on statistical physics and related subjects.

We would like to express our gratitude to everyone who contributed to this homage volume.

This endeavor could not have been realized without the financial support of Dr. José Lema–Labadie, President of the Universidad Autónoma Metropolitana, and Dr. Oscar Monroy–Hermosillo, Rector of the campus Iztapalapa of the same university, we thank very grateful them for sponsoring this international and multidisciplinary endeavor.

We hope that this volume will serve to foster the impressive growth of statistical physics in our region, and additionally to reinforce the already existing strong ties between Mexican scientists and their colleagues in other parts of the world.

Alfredo Macías and Leonardo Dagdug

Leopoldo S. García–Colín

Leopoldo Garca–Colin Scherer was born in Mexico City on 27th November, 1930. He received his B.Sc. in Chemistry from the National University of México (UNAM) in 1953. His thesis dealt with the thermodynamic properties of D_2 and HD as part of a project whose leader was one of Mexicos's foremost scientist, Alejandro Medina. He obtained his Ph.D. at the University of Maryland in 1959 under Elliott Montroll's supervision. There, he began his dedicated career as a teacher and researcher and started an important research program on the broad area of Statistical Mechanics, Kinetic Theory, Irreversible Thermodynamics, Critical Phenomena, Chemical Physics, and other related subjects. Prof. Garcia–Colin has been recognized as the founder of Statistical Physics research in Mexico.

Upon his return to Mexico in 1960 he faced the very hard duty of constructing totally new research groups in those fields of Sciences cultivated by him. This was an exhausting task, demanding time, effort, and an insight in the development of high level Educative Institutions, and Government Laboratories with a solid scientific basis. The first task was to optimize the scientific formation of scientists, mixing the selection and preparation of students, with a program of graduate studies outside México in selected places abroad, including universities of the United States, England, Holland, Germany, Belgium, etc. The role of García–Colín was essential through his personal scientific relations, since through their disposition of accepting the supervision of post–graduated students they contributed to the enhancement of Statistical Mechanics in Mexico.

His research interests encompass many areas of Physics and Chemical–Physics, like statistical physics of non-equilibrium systems, non–linear irreversible thermodynamics, kinetic theory of gases and plasmas, phase transitions, non–linear hydrodynamics, and glass transition. He is author of more than 240 research papers, 60 on popular science, 17 books dealing with thermodynamics, statistical mechanics, non–equilibrium phenomena, and quantum mechanics. His textbook on Classical Thermodynamics is well known in Spanish speaking countries.

Along many years, beside the fields previously mentioned, his scientific interests have spurred important incursions into protein folding and transport theory in Astrophysics and Cosmology, Air Pollution Problems, Science Policy, and Educational Research. He has maintained a continuous correspondence with many scientists and travels frequent all over the world. He is an active member of various prestigious International Societies, like the Third World Academy of Sciences, for instance, since 1988.

He has been an enthusiastic supporter for book editing, and conferences organization, such as the well known Mexican Meeting on Mathematical and Experimental Physics, which started in 2001 and is already in its fourth edition.

He was founder and Professor of Physics of the Escuela Superior de Física y Matemáticas at the National Polytechnical Institute (IPN) (1960–1963). Subsequently, he went to the Universidad Autónoma de Puebla (1964–1966), and then to the Sciences Faculty of the UNAM (1967–1984). He was researcher at the National Institute for Nuclear Research (ININ) (1966–1967), and thereafter he became Head of the Processes Basic Research Section at Mexican Petroleum Institute (1967–1974). In 1974, he started the Department of Physics and Chemistry at UAM–Iztapalapa as Head (1974–1978). Since 1988 he is National Researcher, level III, and now, additionally, National Emeritus Researcher and Distinguished Professor at UAM–Iztapalpa.

Among his extra–academic activities Professor García–Colín was vice–president and later President of the Mexican Physical Society. He is an active member of the American Physical Society, of the American Association for Physics Teachers, of the Mexican Academy of Sciences, of the Mexican

Chemical Society, and of the American Association for the Advancement of Science, among others. His work has been recognized through several honors and awards, among them the Physics Award form the University of Maryland (1956–1957), the Prize in Exact Sciences of awarded by the now Mexican Academy of Sciences (1965), and the Merit Medal from the Universidad Autónoma de Puebla (1965). He held the van der Waals Chair at the University of Amsterdam, Holland (1976), and received the National Prize of Sciences and Arts from the mexican government (1988), to mention the significative ones.

Professor Leopoldo García–Colín became member of El Colegio Nacional in 1977. His opening talk: Modern Ideas on Liquid–Gas Transition, was answered by Professor Marcos Moshinsky.

His scientific and teaching career has been also recognized by degrees of Doctor Honoris Causa by Universidad Iberoamericana (1991), by the University of Puebla (1995), and by the National University of México (2006). In 2007 Professor García–Colín became Emeritus Professor at UAM–Iztapalapa.

Contents

PART 1

Statistical Physics and Related Subjects

Chapter 1

Canonical and Grand–Canonical Ensembles for Trapped Bose Gases

Abel Camacho, Alfredo Macías

Departamento de Física, Universidad Autónoma Metropolitana–Iztapalapa, Apartado Postal 55–534, C.P. 09340, México, D.F., México

acq@xanum.uam.mx; amac@xanum.uam.mx

Abel Camacho–Galván

DEP–Facultad de Ingeniería

Universidad Nacional Autónoma de México.

In the experimental devices used for Bose–Einstein condensation the gas is trapped resorting to magneto–optical methods. In other words, the concept of volume of the corresponding container (as usually found in any text book on statistical mechanics) becomes in these experiments a rather vague parameter. In the present work we deduce the concepts of volume and pressure (as a function of the variables associated to the trap) that allow us to define the corresponding ensembles. Additionally, the issue of the fluctuations in the number of particles is addressed.

Dedicated to Leopoldo García–Colín on the occasion of his 80th birthday.

Contents

1.1. Introduction

The concept of Bose–Einstein condensate (BEC) dates back to the year of 1924 with a work by S. N. Bose in which the statistics of the quanta of light was analyzed.[1] A. Einstein pursued the idea of Bose predicting the occurrence of a phase transition in a gas of non–interacting atoms.[2] The discovery of Bose–Einstein condensation (BEC) with the help of traps[3] has spurred an enormous amount on the theoretical and experimental realms associated to this topic. Among the issues addressed we may find, for instance, the mathematical problems related to BEC,[4] some theoretical and heuristic aspects,[5,6] or even its possible use as tools in precision tests for gravitational physics.[7,8] The analysis of the corresponding thermodynamical properties is also a relevant aspect of BEC.[9]

The experimental design related to the current technology in connection with BEC entails the absence of a container (in the sense considered in any text book on Statistical Mechanics), a fact that requires a careful thermodynamical and statistical–mechanical analysis. Indeed, the text books always consider the case of a gas enclosed in a container whose volume is V. Under these circumstances the definition of the thermodynamical parameter of volume shows no problem at all, and in consequence its conjugate thermodynamical variable, pressure, is also well defined.

Experimentally, the gas is trapped resorting to magneto–optical methods.[10] Bearing this fact in mind we may wonder what is the meaning of volume here since the container is in this case absent. In the present work it will be proved that we may define the concept of volume as a function of the parameters of the trapping potential. In addition, the corresponding conjugate thermodynamical variable will be derived, showing that these two parameters fulfill the standard laws of thermodynamics.

Finally, the issue of fluctuations in the number of particles will be addressed and in this manner we will deduce the conditions defining those cases in which the canonical and grand–canonical ensembles formalisms are equivalent. This point has a relevant role in the context of the comparison between theoretical predictions and experiment. Indeed, in the most general sense the canonical and grand–canonical ensembles, even for the simplest situation (a gas conformed by non–interacting particles enclosed within a container of volume V), are not equivalent. This can be understood recalling the equivalence between them is a function of the isothermal compressibility, i.e., it is proportional to the square root of it.[11] The crucial remark in this context is that even for this idealized situation the isothermal compressibility diverges if the temperatures tends to the condensation temperature, i.e., the equivalency between the canonical and grand–canonical ensembles breaks down even for this very simple case. Clearly, the comparison between theory and experiment requires a careful analysis of this point.

1.2. Ensemble for a Bose Gas Trapped by a Harmonic Oscillator Potential

The current technology resorts to experimental devices in which the atom clouds are confined with the help of laser trapping or magnetic traps.[12] In addition, for the case of alkali atoms some of the available confining potentials can be approximated by a three–dimensional harmonic oscillator.[5] In other words, our trapping potential can, mathematically, be approximated by

$$V(x,y,z) = \frac{m}{2}\Big[\omega_x^2 x^2 + \omega_y^2 y^2 + \omega_z^2 z^2\Big]. \tag{1.1}$$

The frequencies of our harmonic oscillators along the coordinate–axes are denoted by ω_l, where $l = x, y, z$, while m is the mass of the corresponding atoms. In general the frequencies of our harmonic oscillators along the coordinate–axes are different. The interaction among the particles of the cloud is here not taken into account.

We now deduce the density of states per energy unit. In this deduction the continuum approximation is employed and the number of states ($G(\epsilon)$) with energy equal or less that a certain value, say ϵ, is the volume of the first octant bounded by the plane $\epsilon = \epsilon_x + \epsilon_y + \epsilon_z$, where ϵ_l, $l = x, y, z$, is the energy associated to the harmonic oscillator along the l–axis,[10]

$$G(\epsilon) = \frac{\epsilon^3}{6(\hbar\omega_p)^3}. \tag{1.2}$$

The density is provided by ($\frac{dG}{d\epsilon} = \Omega(\epsilon)$)

$$\Omega(\epsilon) = \frac{\epsilon^2}{2(\hbar\omega_p)^3}, \tag{1.3}$$

$$\omega_p = \Big(\omega_x\omega_y\omega_z\Big)^{1/3}. \tag{1.4}$$

The internal energy (U) and the number of particles in the excited states (N_e) are given by[10]

$$U = 3\kappa T\Big(\frac{\kappa T}{\hbar\omega_p}\Big)^3 g_4(z), \tag{1.5}$$

$$N_e = g_3(z)\Big(\frac{\kappa T}{\hbar\omega_p}\Big)^3. \tag{1.6}$$

In these last two expressions κ denotes Boltzmann constant and $g_l(z)$ are the so–called Bose functions.[13] Additionally, z is the fugacity, which is a function of the chemical potential (μ) and the temperature (T)

$$z = \exp\Big(\mu/\kappa T\Big). \tag{1.7}$$

1.3. Volume and Pressure for a Bose Gas Trapped by a Harmonic Oscillator Potential

For the sake of clarity from now on we will assume an isotropic harmonic oscillator, i.e., $\omega_x = \omega_y = \omega_z$. It is readily seen that for the case of a gas trapped by a potential with the form of a harmonic oscillator, at least at this point, the concept of volume has a dubious meaning. Nevertheless, we must state that our ensemble is defined resorting to three physical parameters, namely, T, μ, and ω. The case in which the gas is within a container has also three parameters defining the ensemble, T, μ, and V, i.e., both ensembles are defined with the same number of independent variables. This last remark implies that the number of assumptions, needed to relate the statistical–mechanical calculations with thermodynamics, are in both situations equal.

The idea is now to define the concept of volume, and therefore of pressure, and to prove that these definitions obey the laws of thermodynamics. Our proposal is that for a gas trapped by a harmonic oscillator potential the concept volume, consistent with the laws of thermodynamics, is provided by

$$V = \Big(\frac{\hbar}{m\omega}\Big)^{3/2}. \tag{1.8}$$

This definition shall be no surprise. Indeed, for a one–dimensional harmonic oscillator there is a *natural* distance parameter defined by the constants of the problem (mass of the particles, Planck's constant, and frequency of the oscillator), i..e, $l = \sqrt{\frac{\hbar}{m\omega}}$.[14] Of course, this argument does not prove that this definition of volume does not violate the laws of thermodynamics. Let us now carry out this proof.

Define the intensive variable conjugate to V as Ξ. Considering the internal energy (U) as function of the extensive variables, namely, V, N, and S we find[15]

$$U = TS + \mu N - \Xi V. \tag{1.9}$$

The Second Law of Thermodynamics

$$dU = TdS + \mu dN - \Xi dV, \tag{1.10}$$

allows us to deduce Ξ. Indeed, from this law we find, keeping T and N constant

$$\Xi = -T\Big(\frac{\partial S}{\partial V}\Big)_{T,N} + \Big(\frac{\partial U}{\partial V}\Big)_{T,N}. \tag{1.11}$$

Additionally, the Second Law of Thermodynamics renders the so–called Gibbs–Duhem equation for this system[15]

$$SdT + Nd\mu - Vd\Xi = 0. \tag{1.12}$$

This last expression can be contemplated as a condition to be satisfied by our parameters V and Ξ. In other words, Gibbs–Duhem is a criterion that allows us to check the compatibility of our proposal with the laws of thermodynamics.

Clearly, if we find the entropy, S, we, immediately, deduce Ξ. At this point we address the issue of the entropy. The axioms of statistical mechanics allow us, resorting to the method of averages over ensembles, to deduce mechanical thermodynamical properties like pressure, internal energy, etc, Nevertheless, concepts like entropy are obtained resorting to the comparison between averages over the ensemble and thermodynamical expressions.[16] Our deduction of entropy will be done in this spirit.

Without any exaggeration we may clearly state that in the realm of statistical mechanics the concept of partition function is one of the core ideas.[13,16] From it all the thermodynamics of the involved system can be elicited. For a gas trapped without a *container* but resorting to a harmonic oscillator potential we may consider an ensemble in which its elements may interchange energy and particles. Clearly, if we consider one of these elements, then its corresponding Schrödinger equation will have a set of eigenenergies which will depend upon the number of particles contained in the aforementioned element of the ensemble. We will denote this set by $\{E_{N,j}(\omega)\}$. Let us explain our notation. $E_{N,j}(\omega)$ means that we are considering an element of the ensemble with N particles and with an energy equal to the j–th energy level associated to the corresponding Hamiltonian operator. Clearly, the energy levels shall depend upon the involved frequency of the harmonic potential, a fact explicitly pointed out in our notation. For our ensemble its corresponding partition function reads.[13]

$$Z(\beta, \gamma, \{E_{Nj}(\omega)\}) = \sum_{N=0}^{\infty} \sum_{j} \exp\Big(-\beta E_{Nj}(\omega)\Big) \exp\Big(-\gamma N\Big). \quad (1.13)$$

Here the parameters β and γ are Lagrange multipliers associated to the constraints of this ensemble ,[13] i.e., the energy and number of particles related to the whole ensemble are constant, Clearly, the energy and number of particles in each element of the ensemble are not constant. Define now

$$f(\beta, \gamma, \{E_{Nj}(\omega)\}) = \ln Z. \quad (1.14)$$

Therefore,

$$df = -Ud\beta - <N> d\gamma - \beta \sum_{N=0}^{\infty} \sum_{j} P_{Nj} dE_{Nj}. \quad (1.15)$$

U is the internal energy, $<N>$ the number of particles in thermodynamical equilibrium, and P_{Nj} the probability of finding an element of this ensemble with N particles and in the j–th. energy eigenstate.[13]

Then

$$d\big(f + U\beta + \gamma <N>\big) = \beta dU + \gamma d <N> - \beta \sum_{N=0}^{\infty} \sum_{j} P_{Nj} dE_{Nj}. \quad (1.16)$$

Comparing this last expression against the Second Law of Thermodynamics ($TdS = dU - dW_m - \mu d <N>$) we *postulate* (as is also done with the case of a gas in a container[13,16]) that β is proportional to $1/T$ and

$\gamma = -\mu\beta$, notice that dW_m denotes the reversible mechanical work. Consider now the following change, $\omega \to \omega + d\omega$, this modification in the frequency will impinge upon the energy eigenvalues, i.e., $E_{Nj} \to E_{Nj} + dE_{Nj}$ since they emerge from the N-particles Schrödinger equation. The motion equation is time–invariant[14] and therefore this contribution to the change in energy ($\sum_{N=0}^{\infty}\sum_j P_{Nj}dE_{Nj}$) has to be, inexorably, associated to reversible work. It is readily seen that these energy eigenvalues cannot hinge upon β or γ. We may understand this last statement from two different, though equivalent, perspectives. Indeed, mathematically β and γ are Lagrange multiplies introduced in the formalism in order to take into account some conserved quantities associated to the ensemble. We may rephrase this stating that they are, from the mathematical point of view, disconnected with the N-particles Schrödinger equation, and in consequence the energy eigenvalues cannot depend on β or γ. From a physical point of view we have identified β and γ as functions of the temperature. It is readily seen that the N-particles Schrödinger equation has no relation to the concept of temperature, i.e., E_{Nj} cannot depend upon β or γ.

These last arguments allow us to identify

$$-\sum_{N=0}^{\infty}\sum_j P_{Nj}dE_{Nj} = \Xi dV. \tag{1.17}$$

In addition, we may also deduce the entropy

$$S = k\ln Z + U/T - \mu < N > /T. \tag{1.18}$$

We may cast the bosonic partition function in terms of the energies of single–particle (ϵ) and not in terms of the eigenenergies of N–particles (E_{Nj}) as before[13,16]

$$Z = \Pi_\epsilon\Big[1 - z\exp(-\beta\epsilon)\Big]^{-1}. \tag{1.19}$$

The continuum approximation renders

$$\kappa\ln Z = U/18T, \tag{1.20}$$

$$S = \frac{19U}{18T} - \mu < N > /T. \tag{1.21}$$

Consider $T > T_c$, where T_c is the condensation temperature, then (from now on we denote the total number of particles as N and forget the notation $< N >$)

$$N = g_3(z)\Big(\frac{\kappa T}{\hbar\omega}\Big)^3. \tag{1.22}$$

This expression entails

$$\Big(\frac{\partial\mu}{\partial\omega}\Big)_{T,N} = \frac{3}{\omega}\kappa T\frac{g_3(z)}{g_2(z)}, \tag{1.23}$$

and with this partial derivative we may now calculate Ξ.

$$\Big(\frac{\partial U}{\partial V}\Big)_{T,N} = \frac{6N\kappa T}{V}\Big(\frac{g_4(z)}{g_3(z)} - \frac{g_3(z)}{g_2(z)}\Big), \tag{1.24}$$

$$\Big(\frac{\partial S}{\partial V}\Big)_{T,N} = \frac{N\kappa}{3V}\Big(19\frac{g_4(z)}{g_3(z)} - 13\frac{g_3(z)}{g_2(z)}\Big), \tag{1.25}$$

$$\Xi = \frac{N\kappa T}{3V}\Big(5\frac{g_3(z)}{g_2(z)} - \frac{g_4(z)}{g_3(z)}\Big). \tag{1.26}$$

This last expression plays the role of pressure for the case of a gas trapped by an isotropic harmonic oscillator potential.

Let us now consider the expression for some thermodynamic potentials, in particular we address the issue of its behavior in the neighborhood of the condensation temperature.

The pressure has been provided in the last expression. Below the condensation temperature (since the fugacity is equal to 1, i.e., $z = 1$) we have that

$$\Xi = \frac{N_e\kappa T}{3V}\Big(5\frac{\zeta(3)}{\zeta(2)} - \frac{\zeta(4)}{\zeta(3)}\Big). \tag{1.27}$$

In this last expression $\zeta(\nu)$ denotes Riemann's Zeta function. Since particles in the ground state do not contribute significantly to the pressure in this last expression the parameter N_e is related to particles not in the ground state. From our previous results, see expressions (1.6) and (1.(8), we have that

$$\Xi = (\kappa T)^4 \frac{\zeta(3)}{3}\left(\frac{m}{\hbar^3\omega}\right)^{3/2}\left(5\frac{\zeta(3)}{\zeta(2)} - \frac{\zeta(4)}{\zeta(3)}\right). \tag{1.28}$$

Concerning the internal energy, below the condensation temperature we have, according to (1.5)

$$U = 3\left(\kappa T\right)^4 \frac{\zeta(4)}{\left(\hbar\omega\right)^3}. \tag{1.29}$$

These last two expressions allow us to cast pressure as a function of the energy density ($u = U/V$)

$$\Xi = \frac{\zeta(3)}{9\zeta(4)}\left(5\frac{\zeta(3)}{\zeta(2)} - \frac{\zeta(4)}{\zeta(3)}\right)u \tag{1.30}$$

This last expression has an interesting side. Indeed, notice that according to expression (1.5) and (1.26) above the condensation temperature pressure and energy density are not related by an expression of the form $\Xi = \gamma u$, where γ is a constant. Mathematically this can be understood noting that energy depends upon $g_4(z)$, whereas Ξ is a function of $g_i(z)$, where $i = 4, 3, 2$. But below the condensation temperature pressure and energy density are proportional to each other. In this context the behavior is more conventional. Let us explain this last statement. For a bosonic gas enclosed in a container of volume V and where the corresponding particles do not interact among them it can be proved that if the relation between energy and momentum of single particle has the structure $\epsilon = p^s$, and there are n–spacelike coordinates, then the pressure, P, and energy density, u, fulfill the relation $P = \frac{s}{n}u$.[13] Bearing this result in mind we can define s for this case (assuming $n = 3$) with the condition

$$s = \frac{\zeta(3)}{3\zeta(4)}\left(5\frac{\zeta(3)}{\zeta(2)} - \frac{\zeta(4)}{\zeta(3)}\right). \tag{1.31}$$

1.4. Fluctuation in the Number of Particles

The role of fluctuations has always been a fundamental aspect in this context.[17,18] An issue that becomes relevant in this context is the equivalence between the *canonical* and *grand–canonical* ensembles. This aspect can be formulated in terms of the relative fluctuations, which are a function of the standard deviation and the average number of particles in thermodynamical equilibrium.[13,16]

The standard deviation in the number of particles reads[13]

$$\sigma_N^2 = \kappa T\Big(\frac{\partial N}{\partial \mu}\Big)_{T,V}. \tag{1.32}$$

A simple calculation leads us to

$$\Big(\frac{\partial \mu}{\partial N}\Big)_{T,V} = \Big(\frac{V}{N}\Big)^2\Big(\frac{\partial \Xi}{\partial V}\Big)_{T,N}, \tag{1.33}$$

$$\Big(\frac{\partial \Xi}{\partial V}\Big)_{T,N} = \frac{N\kappa T}{3V^2}\Big(10\frac{g_1(z)g_3^2(z)}{g_2^3(z)} - \frac{g_4(z)}{g_3(z)} - 13\frac{g_3(z)}{g_2(z)}\Big). \tag{1.34}$$

As mentioned before, the parameter defining the equivalence or non–equivalence between these two ensembles is the relative fluctuations in the number of particles, i.e., σ_N^2/N^2. For this particular case

$$\sigma_N^2/N^2 = \frac{3}{N}\frac{g_3(z)g_2^3(z)}{\Big(10g_1(z)g_3^2(z) - g_4(z)g_2^3(z) - 13g_3^2(z)g_2^2(z)\Big)}. \tag{1.35}$$

1.5. Conclusions

In the present work we have considered the case of a Bose gas trapped by an isotropic harmonic oscillator, and the concepts of *volume* and *pressure* have been deduced in such a way that its compatibility with the laws of thermodynamics is guaranteed.

The idea of volume for this case is based upon the only distance parameter that can be defined with the parameters related to the case, namely, mass of the particle, Planck's constant, and the frequency of the oscillator, i.e., $V = \left(\frac{\hbar}{m\omega}\right)^{3/2}$.

In this context the concept of pressure emerges from the introduction of this volume parameter into the Second Law of Thermodynamics and afterwards a process at constant T and N is considered. In this way the fulfillment of the laws of thermodynamics is guaranteed.

Finally, the issue of the relative fluctuations has been addressed and the criterion that explains when these two ensembles are equivalent (we have called them *canonical* and *grand–canonical*) has been deduced. Notice that our results tell us they are not equivalent if

$$10g_1(z)g_3^2(z) \rightarrow g_4(z)g_2^3(z) + 13g_3^2(z)g_2^2(z). \tag{1.36}$$

This expression defines, implicitly, a temperature (here denoted by $T = \hat{T}$) at which our equivalence is broken down. In other words, in general these two ensembles are not equivalent. This feature is no surprise since it appears even in the case of a Bose gas trapped by a container.[13] In addition notice that (1.35) behaves, as a function of the number of particles N in the usual form.[13] In this sense the behavior is the usual one, it goes $\sigma_N^2/N^2 \sim 1/N$. In relation with the dependence upon T, let us define

$$10g_1(\hat{z})g_3^2(\hat{z}) = g_4(\hat{z})g_2^3(\hat{z}) + 13g_3^2(\hat{z})g_2^2(\hat{z}), \quad z(T = \hat{T}) = \hat{z}. \tag{1.37}$$

Then if $T \rightarrow \hat{T}$ our result for the relative fluctuations tell us that $\sigma_N^2/N^2 \rightarrow \infty$.

Let us now calculate the specific heat keeping the parameters of the trapping potential constant, i.e., at V constant. It can be easily shown that if $C_\omega = \left(\frac{\partial U}{\partial \omega}\right)_{T,N}$, then

$$C_\omega = 3\kappa\left(\frac{\kappa T}{\hbar\omega}\right)^3\left(4g_4(z) - 3\frac{g_3^2(z)}{g_2(z)}\right), \quad T > T_c, \tag{1.38}$$

$$C_\omega = 3\kappa\left(\frac{\kappa T}{\hbar\omega}\right)^3\zeta(4), \quad T < T_c. \tag{1.39}$$

Clearly, the specific heat has a discontinuity at $T = T_c$. Indeed, the substraction of the corresponding limits reads:

$$\Delta C_\omega = 9\kappa N\frac{\zeta(3)}{\zeta(2)}. \tag{1.40}$$

We recover a result already known.[10]

Acknowledgments

This research was supported by CONACyT Grants 48404–F and 47000-F.

References

1. S. N. Bose, *Plancks Gesetz und Lichtquantenhypothese*, Z. für Phys. 26, 178–181 (1924)
2. A. Einstein, *Quantentheorie des einatomigen idealen Gases*, Sitzber. Kgl. Preuss. Akad. Wiss. 3, 261–267, (1924)
3. M. R. Anderson et al, *Observation of Bose–Einstein condensation in an dilute atomic vapor*, Science **269**, 198–201, (1995).
4. E. H. Lieb, R. Seiringer, J. P. Solovej, and J. Yngvason, "The Mathematics of the Bose Gas and its Condensation", Birkhaeuser–Verlag, Berlin, 2005.
5. F. Dalfovo, L. P. Pitaevski, and S. Stringari, *Theory of Bose–Einstein condensation in trapped gases*, Rev. Mod. Phys. **71**, 463–512, (1999).
6. S. Stenholm, *Heurisitc field theory of Bose–Einstein condensates*, Phys. Rep. **363**, 173–217, (2002).
7. D. Colladay and P. MacDonald, *Bose–Einstein condensates as a probe for Lorentz violation*, Phys. Rev. **D73**, 105006, (2006).
8. E. Castellanos and A. Camacho, *Critical points in a relativistic gas induced by the quantum structure of spacetime*, Gen. Rel. Grav. **41**, 2677–2685, (2009).
9. B. Klünder and A. Pelster, *Systematic Semiclassical Expansion for Harmonically Trapped Ideal Bose Gases*, Eur. Phy. J. B**68**, 457-465, (2009).
10. C. J. Pethick and H. Smith, "Bose–Einstein Condensation in Dilute Gases", Cambridge University Press, Cambridge, 2006.
11. D. A. McQuarrie, "Statistical Mechanics", Harper and Row, New York, 1976.
12. A. J. Legget, *Bose–Einstein condensates in the alkali gases: Some fundamental concepts* Rev. Mod. Phys. 73, 307–354, (2001).
13. R. K. Phatria, "Statistical Mechanics", Butterworth–Heineman, Oxford, 1996.
14. J. J. Sakurai, "Modern Quantum Mechanics", Addison–Wesley Publishing Company, New York, 1994.
15. L. García-Colín Scherer, "Introducción a la Termodinámica Clásica", Trillas, México, 2005.
16. L. García-Colín Scherer, "Introducción a la Física Estadística", El Colegio Nacional, México, 2005.
17. D. M. Greenberger, N. Erez, M. O. Scully, A. A. Svidzinsky, M. S. Zubairy, "Planck, Photon Statistics, and Bose-Einstein Condensation", in Progress in Optics, Vol. **50**, edited by E. Wolf, Elsevier, Amsterdam, 275-330, 2007.
18. V. V. Kocharovsky et al,"Fluctuations in Ideal and Interacting Bose-Einstein Condensates: From the laser phase transition analogy to squeezed states and Bogoliubov quasiparticles", Advances in Atomic, Molecular and Optical Physics Vol. **53**, 291–369, 2006.

Chapter 2

Shockwaves and Local Hydrodynamics; Failure of the Navier-Stokes Equations

Wm. G. Hoover and Carol G. Hoover

Ruby Valley Research Institute
Highway Contract 60, Box 598
Ruby Valley, Nevada 89833
hooverwilliam@yahoo.com

Shockwaves provide a useful and rewarding route to the nonequilibrium properties of simple fluids far from equilibrium. For simplicity, we study a strong shockwave in a dense two-dimensional fluid. Here, our study of nonlinear transport properties makes plain the connection between the observed local hydrodynamic variables (like the various gradients and fluxes) and the chosen recipes for defining (or "measuring") those variables. The *range* over which nonlocal hydrodynamic averages are computed turns out to be much more significant than are the other details of the averaging algorithms. The results show clearly the incompatibility of microscopic time-reversible cause-and-effect dynamics with macroscopic instantaneously-irreversible models like the Navier-Stokes equations.

Contents

2.1. Introduction

Leopoldo García-Colín has studied nonequilibrium fluids throughout his research career. In celebrating his Eightieth Birthday we conform here to his chosen field of study. Though Leo's approach is typically quite general, looking for improvements on linear transport theory, he has studied particular problems too. A specially interesting and thought-provoking study, with Mel Green, of the nonuniqueness of bulk viscosity,[1] emphasised the general problem of finding appropriate definitions for state variables far from equilibrium. The magnitude of the bulk viscosity gives the additional viscous pressure due to the compression *rate*. The pressure difference evidently depends upon the underlying definition of the equilibrium reference pressure. The reference pressure itself in turn depends upon the choice between temperature and energy in *defining* the reference state. In the end, the same physics results, as it always must; the valuable lesson is that many different languages can be used to describe the underlying physics. There is the tantalizing possibility that some one approach is better than others.

In fact, temperature itself can have *many* definitions away from equilibrium.[2] Away from equilibrium the *thermodynamic* temperature would depend upon defining a nonequilibrium entropy – and there is good evidence that there is no such entropy. This is because nonequilibrium distribution functions are typically fractal, rather than smooth.[3] The *kinetic* temperature, a measure of the velocity fluctuation, becomes a *tensor* away from equilibrium.[2,4] At low density this temperature is the same as the pressure tensor, $P = \rho T$. For dense fluids the potential energy introduces *nonlocality*, complicating the definition of constitutive averages. The simplest of the many *configurational* temperatures[5–7] depends on force fluctuations, and so likewise has tensor properties. Because configurational temperature can be negative[8] and because thermodynamic temperature is undefined away from equilibrium, we focus our attention on kinetic temperature here.

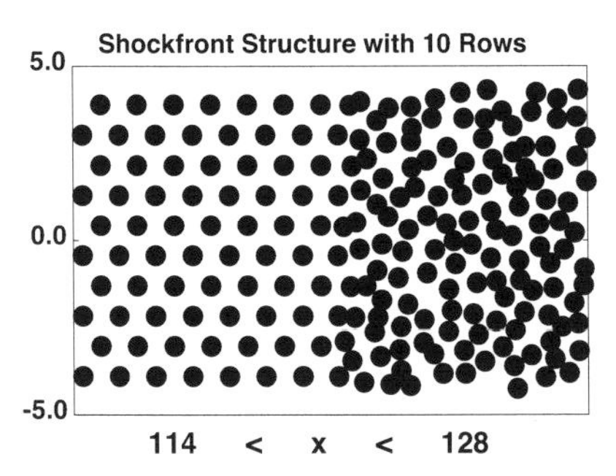

Fig. 2.1. Closeup of a strong shockwave. The cold stress-free solid on the left moves to the right at twice the speed of the hot fluid, which exits to the right. The boundaries in the vertical y direction are periodic. The pair potential is $\phi(r < 1) = (10/\pi)(1 - r)^3$. The overall density change is $\sqrt{4/3} \to 2\sqrt{4/3}$ and $u_s = 2u_p = 1.93$. The system height is $10\sqrt{3/4} \simeq 8.66$.

Shockwaves are irreversible transition regions linking a "cold" and a "hot" state.[8–11] Such a shock region contains nonequilibrium gradients in density, velocity, and energy. The irreversible change from cold to hot takes place in just a few free paths in a time of just a few collision times.[11] The localized nature of shockwaves makes them ideal for computer simulation. Their gross one-dimensional nature, illustrated in Figure 1, makes it possible to compute *local* averages in a region of width h. Because h is necessarily small it is evident that the average values depend on it. Thus the average temperature depends upon both the underlying definition of temperature and additionally on the details of the local averaging.

In this work we begin by describing molecular dynamics simulations, for a strong, nominally stationary and one-dimensional shockwave, in a two-dimensional fluid. Next, we discuss the Navier-Stokes description of such a wave and then set out to compare the two approaches, focusing on the definition of local hydrodynamic variables. A close look at the momentum and heat fluxes shows clear evidence for the incompatibility of the microscopic and macroscopic constitutive relations.

2.2. The Microscopic Model System and a Continuum Analog

We consider structureless particles of unit mass in two space dimensions interacting with the short-ranged purely-repulsive pair potential,

$$\phi(r < 1) = (10/\pi)(1 - r)^3 .$$

As shown in Figure 1, particles enter into the system from the left, moving at the shock velocity u_s. Likewise, particles exit at the right with a lower mean speed, $u_s - u_p = u_s/2$, where u_p is the "Particle" or "piston" velocity. The velocity ratio of two which we choose throughout is consistent with twofold compression. We carried out series of simulations, all with a length of 250 and the shock near the system center, with system widths of from 10 to 160 rows. Figure 1 shows a closeup of the center of such a 10-row flow for the narrowest system width, $10\sqrt{3/4} \simeq 8.66$.

To analyze the results from molecular dynamics one and two-dimensional average values of the density, energy, pressure, heat flux and the like were computed using the one- and two-dimensional forms of Lucy's weight function:[12,13]

$$w^{1D}(r < 1) = (5/4h)(1 - r)^3(1 + 3r) \; ; \; r \equiv |x|/h .$$

$$w^{2D}(r < 1) = (5/\pi h^2)(1-r)^3(1+3r) \; ; \; r \equiv \sqrt{x^2+y^2}/h \; .$$

The averages are not significantly different to those computed with Hardy's more cumbersome approach.[14]

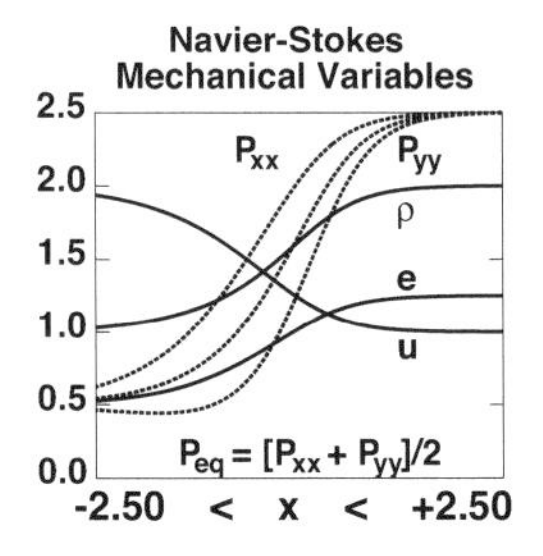

Fig. 2.2. The nonzero pressure-tensor components from the Navier-Stokes equations solution, P_{xx} and P_{yy}, as well as their average, $P = (\rho^2/2) + \rho T$, are shown as dashed lines along with the velocity, energy, and density profiles. In this simple example it is assumed that the bulk viscosity vanishes so that the average of the longitudinal and transverse components is equal to the equilibrium pressure.

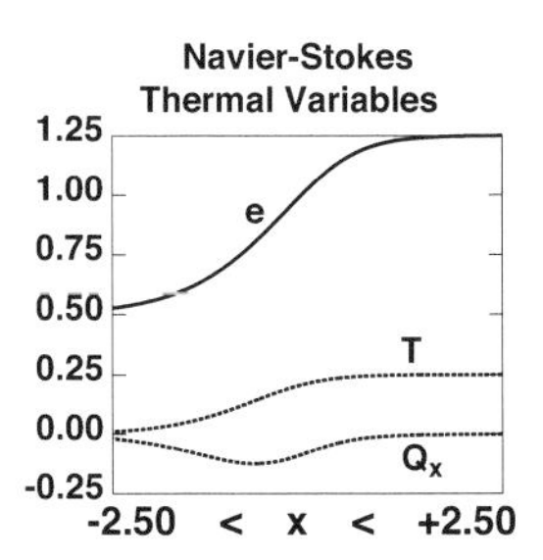

Fig. 2.3. The energy, (scalar) temperature, and heat flux vector from the Navier-Stokes equations are shown here. The heat conductivity and shear viscosity coefficients were assumed equal to unity in the underlying calculation.

A preview of the one-dimensional averages results from molecular dynamics' simplest continuum analog, a solution of the stationary Navier-Stokes equations. For simplicity, in the Navier-Stokes analog we use the constitutive relations for the van-der-Waals-like model with shear viscosity and heat conductivity of unity:

$$P = \rho e = (\rho^2/2) + \rho T \; ; \; e = (\rho/2) + T$$

$$(P_{xx} - P_{yy})/2 = -du/dx \; ; \; (P_{xx} + P_{yy})/2 = P \; ;$$

$$Q_x = -dT/dx \ .$$

This model is similar to our microscopic simulation model, but has a nonzero initial pressure and energy. A set of self-consistent cold and hot boundary conditions for the Navier-Stokes velocity, pressure, energy, and scalar temperature is as follows:

$$u : [2 \to 1] \ ; \ \rho : [1 \to 2] \ ; P : \ [1/2 \to 5/2] \ ; e : [1/2 \to 5/4] \ ; T : [0 \to 1/4] \ .$$

These boundary conditions satisfy conservation of mass, momentum, and energy. The (constant) mass, momentum, and energy fluxes *throughout* the shockwave (not just at the boundaries) are:

$$\rho u = 2 \ ; \ P_{xx} + \rho u^2 = 5/2 \ ; \ \rho u[e + (P_{xx}/\rho) + (u^2/2)] + Q_x = 6 \ .$$

The most noteworthy feature of the numerical solution is the slight decrease of P_{yy} below the equilibrium value on the cold side of the shock. Figure 2 shows the mechanical variables and Figure 3 the thermal variables near the center of the shock as computed from the Navier-Stokes equations.[11] A serious shortcoming of the Navier-Stokes equations is their failure to distinguish the longitudinal and transverse temperatures.

2.3. Averaged Results from Molecular Dynamics

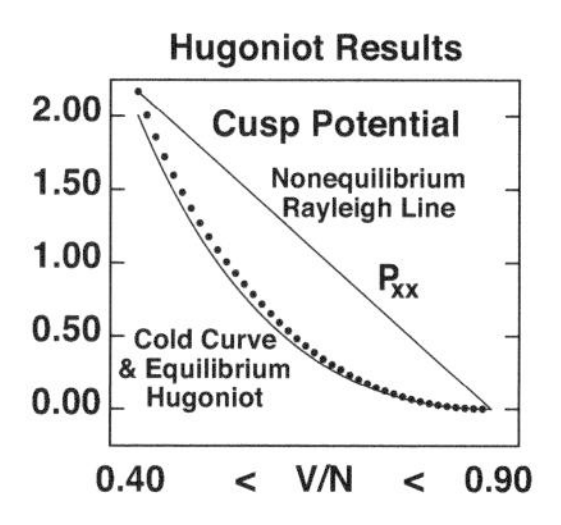

Fig. 2.4. The longitudinal pressure tensor component P_{xx} varies linearly with volume, and follows the Rayleigh line. The cold curve corresponds to the pressure of a perfect static triangular lattice. The equilibrium Hugoniot, indicated by dots, corresponds to thermodynamic equilibrium states accessible from the initial cold state by shockwave compression. For P_{yy} see Figure 7.

One-dimensional averages reproduce the linear dependence of P_{xx} on the volume very well. That linear dependence is the "Rayleigh Line", shown

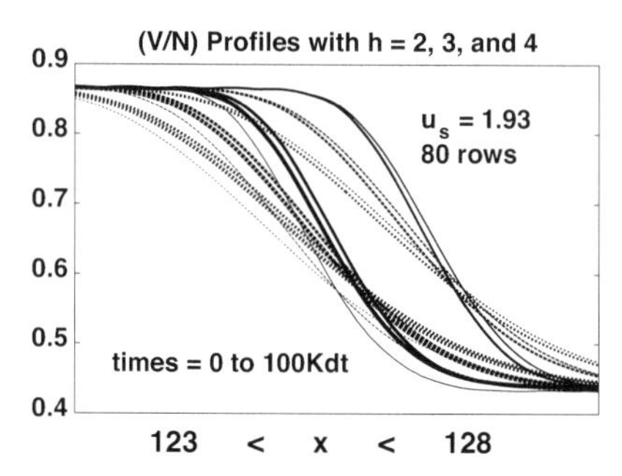

Fig. 2.5. Shockwave profiles at five equally-spaced times. The five line widths correspond to times of 0, 25,000dt, 50,0000dt, 75,000dt, and 100,000dt. The steepest profiles correspond to $h = 2$. Results for $h = 3$ and $h = 4$ are also shown. The fourth-order Runge-Kutta timestep, here and throughout, is given by $dt = 0.02/u_s \simeq 0.01$.

in Figure 4. The "cold curve" in that Figure is the calculated pressure for a cold triangular lattice:

$$P_{\rm cold}V = 3NrF(r);\ r = \sqrt{(V/V_0)}\ ;\ V_0 = \sqrt{3/4}N\ ;\ F(r) = (30/\pi)(1-r)^2\ .$$

That pressure lies a bit below the Hugoniot curve (the locus of all equilibrium states which can be reached by shocking the initial state). The Hugoniot pressure at each volume was generated by trial-and-error isothermal (isokinetic) molecular dynamics runs, leading to the temperatures satisfying the Hugoniot relation:

$$E_{\rm hot} - E_{\rm cold} = +\Delta V[P_{\rm hot} + P_{\rm cold}]/2\ ;\ \Delta V = V_{\rm cold} - V_{\rm hot}\ .$$

Figure 5 shows typical one-dimensional snapshots of the shockwave profile, $V(x)$. The averages shown in the Figure were computed at 5 equally-spaced times, separated by 25,000 timesteps. The fluctuating motion of the shockwave, of order unity in 100,000 timesteps, corresponds to fluctuations in the averaged shock velocity of order 0.001.

The apparent shockwidth is sensitive to the range of the weighting function h. $h = 2$ is evidently *too small*, as it leads to discernable wiggles in the profile. The wider profiles found for $h = 3$ and $h = 4$ indicate that the constitutive relation describing the shockwave must depend explicitly on h. That is, h must be chosen sufficiently large to avoid unreasonable wiggles, but must also be sufficiently small to capture and localize the changes occurring within the shockwave.

Two-dimensional averages are no more difficult to evaluate. The density at a two-dimensional gridpoint, for instance, can be evaluated by summing the contributions of a few dozen nearby particles:

$$\rho_r \equiv \sum_j w_{rj}^{2D} = \sum_j w^{2D}(|r - r_j|) .$$

Such sums are automatically continuous functions of the gridpoint location r. They necessarily have continuous first and second derivatives too, provided that the weight function has two continuous derivatives, as does Lucy's weight function.[12,13] Linear interpolation in a sufficiently-fine grid can then provide *contours* of macroscopic variables. Figure 6 is an illustration, and shows the contour of average density at 10 equally-spaced times. The boundary value of u_s for that Figure was chosen as 1.92 rather than the shock velocity of 1.93. Thus the shockfront moves slowly to the left in the Figure, with an apparent picture-frame velocity of -0.01.

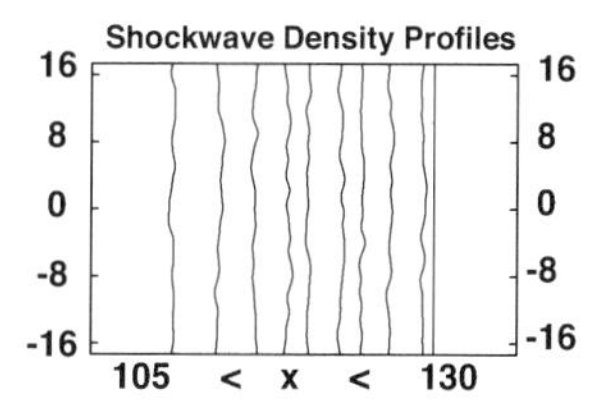

Fig. 2.6. Ten average-density contours (corresponding to the shockfront position) at equally-spaced times for a 40-row system. The amplitude of the fluctuations, of order unity, is similar to the range of the weight function h. The total timespan is 225,000 timesteps. Because the entrance velocity, at $x = 0$, is $u_s = 1.92$, rather than 1.93, the shockfront moves slowly toward the left, with a picture-frame velocity of about -0.01.

Figure 7 shows the nonequilibrium equation of state within the shockwave, the variation of the pressure tensor components P_{xx} and P_{yy} with the specific volume, $(V/N) \equiv (1/\rho)$. P_{xx} is insensitive to the smoothing length h (as is required by the momentum conservation condition defining the Rayleigh line) while P_{yy} shows a slight dependence on h. This lack of sensitivity of the pressure tensor suggests that nonequilibrium formulations of the equation of state within the shock can be successful.

The conventional Newton and Fourier constitutive relations require that gradients be examined too. *Gradients* can be evaluated directly from sums including ∇w. Consider the gradient of the velocity at a gridpoint r as an example (where w_{rj} is the weight function for the distance separating the

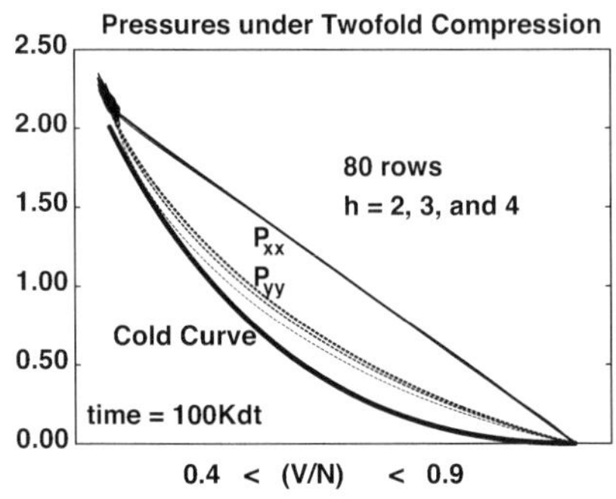

Fig. 2.7. Typical snapshot of the dependence of P_{xx} and P_{yy} on the volume (V/N). The line width increases with the range $h = 2$, 3, and 4. The range-dependence of P_{xx} is too small to be seen here while it is possible to see a small increase in P_{yy} with increasing h.

gridpoint from Particle j):

$$(\rho\nabla \cdot v + v \cdot \nabla\rho)_r \equiv \nabla \cdot (\rho v) \equiv \nabla_r \cdot \sum_j w_{rj} v_j = \sum_j v_j \cdot \nabla_r w_{rj} \ .$$

Using the identity,

$$\rho = \sum_j w_{rj} \ ,$$

gives

$$(\rho\nabla \cdot v)_r = \sum_j (v_j - v_r) \cdot \nabla_r w_{rj} \ .$$

The tensor temperature gradient can be evaluated in the same way:

$$(\rho\nabla \cdot T)_r = \sum_j (T_j - T_r) \cdot \nabla_r w_{rj} \ .$$

Figure 8 compares the velocity gradients as calculated using three values of h to the pressure tensor using the same three values. We see that the velocity gradient is much more sensitive than is the stress to h, suggesting a sensitive dependence of the Newtonian viscous constitutive relation on the range of the weight function. The data in the Figure indicate a shear viscosity of the order of unity. Gass' Enskog-theory viscosity[15] confirms this estimate.

Figure 9 shows the temperature gradients. There are two of these for each h because the longitudinal and transverse temperatures differ. Again the magnitudes of the gradients are relatively sensitive to h while the maximum in the nonequilibrium flux Q_x is less so. Again the heat conductivity from the data is of the order of Gass' Enskog-theory estimate.

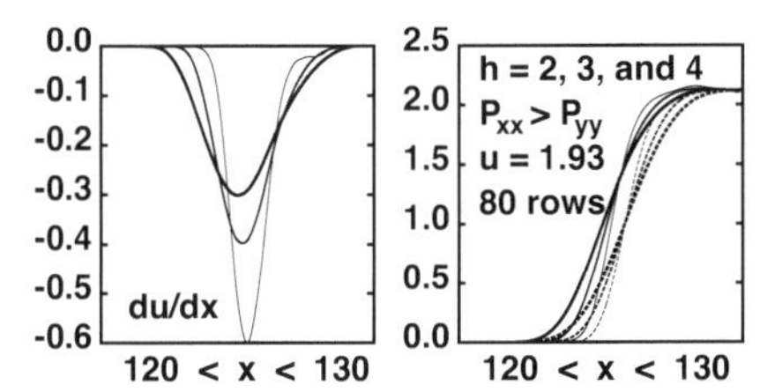

Fig. 2.8. The velocity gradient, for three values of h, is much more sensitive than is the shear stress, $(P_{yy} - P_{xx})/2$, to h. The gradient extrema, at 125.35, 125.18, and 125.01 precede the shear stress extrema at 125.67, 125.61, and 125.57 for $h = 2$, 3, and 4.

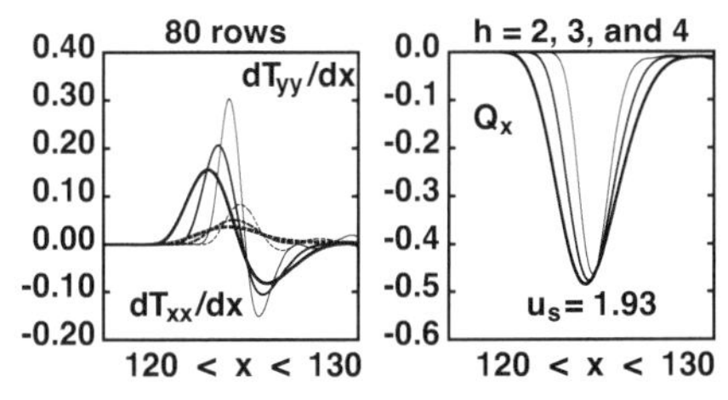

Fig. 2.9. The temperature gradients (the dashed lines correspond to the transverse temperature) for three values of h correspond to the heat fluxes found with the same h values. The pronounced maximum in T_{xx} indicates a violation of Fourier's Law, as the heat flux does not show a corresponding change of sign. The maxima in dT_{xx}/dx occur at distances 124.93, 124.49, and 124.09, significantly leading the flux maxima at 125.44, 125.29, and 125.15 for $h = 2$, 3, and 4.

2.4. Conclusion: Failure of Navier-Stokes Equations

Results from earlier shockwave simulations[9,11,16] have uniformly been described as showing "good" or "fairly good" agreement with continuum predictions. Examining the more nearly accurate profiles made possible with improved averaging techniques shows that the agreement is actually limited, and in a qualitative way. A more detailed look at the data shown in Figures 8 and 9 reveals a consistent "fly in the ointment" pattern: the largest fluxes are *not* located at the largest values of the gradients. The fluxes lag behind the gradients by a (relaxation) time of order unity. This shows that no simple instantaneous relationship links the fluxes to the gradients. *In molecular dynamics the instantaneous stress cannot be proportional to the instantaneous strain rate.*

The reason for this apparent contradiction of linear transport theory is plain enough: the underlying molecular dynamics is time-reversible, so that pressure is necessarily an *even* function of velocity and time. This same symmetry must be true also of any spatially-averaged instantaneous pressure. Because there is no possibility to find an instantaneous irreversible constitutive relation with time-reversible molecular dynamics, it is apparent that any attempt to "explain" local molecular dynamics averages through irreversible macroscopic constitutive relations is doomed to failure.

There is of course no real difficulty in carrying out the instantaneous averages, in one or two or three space dimensions, for today's molecular dynamics simulations. On the other hand, the gap between the microscopic and macroscopic pictures becomes an unbridgable chasm when the detailed spatiotemporal contradictions between the two approaches are considered.

2.5. Prospects

The prospect of understanding shockwaves in gases has stimulated studies of dilute gases, based on the Boltzmann equation.[17–20] Leo has been a driving force for this work. Though the analysis is highly complex[19,20] it has become apparent that the Boltzmann equation is itself nicely consistent with corresponding solutions using molecular dynamics,[17,18] up to a Mach number $M = u_s/c_{\rm cold}$ of 134. The applicability of the Burnett equations, which include all second-order contributions of the gradients to the fluxes, is still in doubt for strong shockwaves in dilute gases.[17,18]

Dense fluids will require a new approach. Local averages must be defined. Longitudinal and transverse temperatures must be treated separately. The causal timelag between the forces (velocity and temperature gradients) and the resulting momentum and heat fluxes must be included in the modelling. Although these challenges are enormous, today's fast computers place the responsibility for successfully meeting them squarely on physicists' imaginations. The excuse that the problem is too hard to tackle is no longer valid. We can look forward to many more contributions from Leo, his coworkers, and those inspired and stimulated by his work.

Acknowledgments

We thank Paco Uribe, Michel Mareschal, Leopoldo García-Colín, and Brad Holian for their comments on an earlier draft of this work. Paco Uribe kindly exhumed and resurrected the Burnett solutions of References 19 and 20 and established that there is indeed a timelag between the Burnett forces and fluxes in dilute-gas shockwaves for hard-sphere Mach numbers of both 2 and 134. This seems to contradict the lower part of Figure 2 in Reference 17.

References

1. L. S. García-Colín and M. S. Green, "Definition of Temperature in the Kinetic Theory of Dense Gases", Physical Review **150**, 153-158 (1966).
2. W. G. Hoover and C. G. Hoover, "Nonlinear Stresses and Temperatures in Transient Adiabatic and Shear Flows *via* Nonequilibrium Molecular Dynamics", Physical Review E **79**, 046705 (2009).
3. Wm. G. Hoover, "Liouville's Theorems, Gibbs' Entropy, and Multifractal Distributions for Nonequilibrium Steady States", Journal of Chemical Physics **109**, 4164-4170 (1998).
4. W. G. Hoover, C. G. Hoover, and M. N. Bannerman, "Single-Speed Molecular Dynamics of Hard Parallel Squares and Cubes" Journal of Statistical Physics **136**, 715-732 (2009).
5. O. G. Jepps, PhD Dissertation, Australian National University, Canberra, 2001.
6. E. Braga and K. P. Travis, "A Configurational Temperature Nosé-Hoover Thermostat", Journal of Chemical Physics **123**, 134101 (2005).
7. Wm. G. Hoover and C. G. Hoover, "Nonequilibrium Temperature and Thermometry in Heat-Conducting ϕ^4 Models", Physical Review E **77**, 041104 (2008).
8. Wm. G. Hoover and C. G. Hoover, "Tensor Temperature and Shockwave Stability in a Strong Two-Dimensional Shockwave", Physical Review E **80**, 011128 (2009).
9. B. L. Holian, W. G. Hoover, B. Moran, and G. K. Straub, "Shockwave Structure *via* Nonequilibrium Molecular Dynamics and Navier-Stokes Continuum Mechanics", Physical Review A **22**, 2798-2808 (1980).
10. H. M. Mott-Smith, "The Solution of the Boltzmann Equation for a Shockwave", Physical Review **82**, 885-892 (1951).
11. W. G. Hoover, "Structure of a Shockwave Front in a Liquid", Physical Review Letters **42**, 1531-1534 (1979).
12. L. B. Lucy, "A Numerical Approach to the Testing of the Fission Hypothesis", The Astronomical Journal **82**, 1013-1024 (1977).
13. Wm. G. Hoover, *Smooth Particle Applied Mechanics; the State of the Art* (World Scientific, Singapore, 2006).
14. R. J. Hardy, "Formulas for Determining Local Properties in Molecular Dynamics Simulations: Shockwaves", Journal of Chemical Physics **76**, 622-628 (1982).
15. D. M. Gass, "Enskog Theory for a Rigid Disk Fluid", Journal of Chemical Physics **54**, 1898-1902 (1971).

16. R. E. Duff, W. H. Gust, E. B. Royce, M. Ross, A. C. Mitchell, R. N Keeler, and W. G. Hoover, "Shockwave Studies in Condensed Media", in *Behavior of Dense Media under High Dynamic Pressures*, Proceedings of the 1967 Paris Conference, pages 397-406 (Gordon and Breach, New York, 1968).
17. E. Salomons and M. Mareschal, "Usefulness of the Burnett Description of Strong Shock Waves", Physical Review Letters **69**, 269-272 (1992).
18. B. L. Holian, C. W. Patterson, M. Mareschal, and E. Salomons, "Modeling Shockwaves in an Ideal Gas: Going Beyond the Navier-Stokes Level", Physical Review E **47**, R24-R27 (1993).
19. F. J. Uribe, R. M. Velasco, and L. S. García-Colín, "Burnett Description of Strong Shock Waves", Physical Review Letters **81**, 2044-2047 (1998).
20. F. J. Uribe, R. M. Velasco, L. S. García-Colín, and E. Díaz-Herrera, "Shockwave Profiles in the Burnett Approximation", Physical Review E **62**, 6648-6666 (2000).

Chapter 3

Nonequilibrium Stationary Solutions of Thermostated Boltzmann Equation in a Field

F. Bonetto

School of Mathematics, GaTech, Atlanta GA 30332

J.L. Lebowitz

Departments of Mathematics and Physics, Rutgers University, Piscataway NJ 08854

We consider a system of particles subjected to a uniform external force $\mathbf{E}$ and undergoing random collisions with "virtual" fixed obstacles, as in the Drude model of conductivity.[1] The system is maintained in a nonequilibrium stationary state by a Gaussian thermostat. In a suitable limit the system is described by a self consistent Boltzmann equation for the one particle distribution function f. We find that after a long time $f(\mathbf{v},t)$ approaches a stationary velocity distribution $f(\mathbf{v})$ which vanishes for large speeds, *i.e.* $f(\mathbf{v}) = 0$ for $|\mathbf{v}| > v_{max}(\mathbf{E})$, with $v_{max}(\mathbf{E}) \simeq |\mathbf{E}|^{-1}$ as $|\mathbf{E}| \to 0$. In that limit $f(\mathbf{v}) \simeq \exp(-c|\mathbf{v}|^3)$ for fixed $\mathbf{v}$, where c depends on mean free path of the particle. $f(\mathbf{v})$ is computed explicitly in one dimension.

Dedicated to Leopoldo García–Colín on the occasion of his 80th birthday.

Contents

3.1. Introduction

Nonequilibrium stationary states (NESS) in real systems must be maintained by interaction with (effectively infinite) external reservoirs. Since a complete microscopic description of such reservoirs is generally not feasible it is necessary to represent them by some type of modeling.[2]

However, unlike systems in equilibrium, which maintain themselves without external inputs and for which one can prove (when not inside a coexistence region of the phase diagram) that bulk behavior is independent of the nature of the boundary interactions, we do not know how different microscopic modelling of external inputs, affects the resulting NESS.

One particular type of modelling dynamics leading to NESS is via Gaussian thermostats, see.[3] Analytical results as well as computer simulations have shown that the stationary states produced by these dynamics behave in many cases in accord with those obtained from more realistic models.[4] This has led us to continue our study of the NESS in current carrying thermostated systems. In our previous work, see[3,5] we carried out extensive numerical and analytical investigations of the dependence of the current on the electric field for a model system consisting of N particles with unit mass, moving among a fixed periodic array of discs in a two dimensional square Λ with periodic boundary conditions, see Fig. 1. They are acted on by an external (electric) field $\mathbf{E}$ parallel to the x-axis and by a "Gaussian thermostat". (The discs are located so that there is a finite horizon, *i.e.* there is a maximum distance a particle can move before hitting a disc or obstacle).

The equations of motion describing the time evolution of the positions $\mathbf{q}_i$ and velocities $\mathbf{v}_i$, $i = 1, ..., N$, are:

$$\dot{\mathbf{q}}_i = \mathbf{v}_i \qquad \mathbf{q}_i = (q_{i,x}, q_{i,y}) \in \Lambda' \tag{3.1}$$

$$\dot{\mathbf{v}}_i = \mathbf{E} - \alpha(\mathbf{J}, U)\mathbf{v}_i + F_{obs}(\mathbf{q}_i) \tag{3.2}$$

where

$$\alpha(\mathbf{J}, U) = \frac{\mathbf{J} \cdot \mathbf{E}}{U}, \qquad \mathbf{J} = \frac{1}{N}\sum_{i=1}^{N} \mathbf{v}_i, \qquad U = \frac{1}{N}\sum_{i=1}^{N} \mathbf{v}_i^2 \tag{3.3}$$

Here $\Lambda' = \Lambda \backslash \mathcal{D}$, with $\mathcal{D}$ the region occupied by the discs (obstacles) and F_{obs} represents the elastic scattering which takes place at the surface of

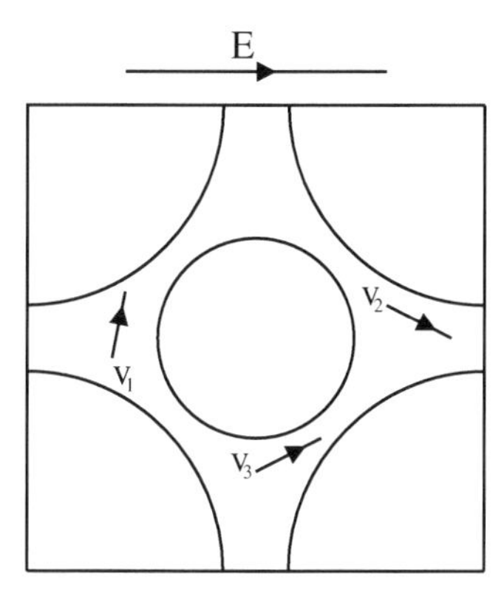

Fig. 3.1. General billiard structure with three particles shown.

the obstacles. The purpose of the Gaussian thermostat, represented by the term $\alpha(\mathbf{J}, U)\mathbf{v}$ in Eq. (3.1), is to maintain the total kinetic energy $1/2\sum_{i=1}^{N}\mathbf{v}_i^2$ constant, *i.e.* $U = v_0^2$. It also has the effect of making the flow Φ_t generated by Eq. (3.1) on the $(4N-1)$ dimensional energy surface non Hamiltonian when $\mathbf{E} \neq 0$. Another effect of the thermostat is to effectively couple all the particles in a mean field way, $\alpha(\mathbf{J}, U)$, depending only on the total momentum of the particles. Note that this is the only coupling between the particles in this system.

Our main interest is in the NESS of this model system. To get some analytical handle on the form of the NESS we also investigated numerically a model system in which the deterministic collisions with the obstacles are replaced by a stochastic process in which particle velocities get their orientations changed at random times, independently for each particle.[5] This yields, in the limit $N \to \infty$, a self consistent Boltzmann equation. In the spatially uniform case, the only one we shall consider here, the equation takes the form

$$\frac{\partial}{\partial t}f(\mathbf{v},t) + \frac{\partial}{\partial \mathbf{v}}\left[(\mathbf{E} - \mu\mathbf{v})f(\mathbf{v},t)\right] = -\lambda\int_{\mathbf{v}\cdot\mathbf{n}<0}\frac{\mathbf{v}'\cdot\mathbf{n}}{2}\left(f(\mathbf{v},t) - f(\mathbf{v}',t)\right)d\mathbf{n} \tag{3.4}$$

where $\mathbf{v}' = \mathbf{v} - 2\mathbf{n}(\mathbf{n}\cdot\mathbf{v})$, $\mathbf{n}$ is a unit vector and μ is determined self consistently by requiring that

$$\frac{d}{dt}\int \frac{\mathbf{v}^2}{2} f(\mathbf{v},t)d\mathbf{v} = -\mathbf{E}\cdot\langle\mathbf{v}\rangle + \mu\langle\mathbf{v}^2\rangle = 0 \tag{3.5}$$

or

$$\mu = \frac{\mathbf{E}\cdot\langle\mathbf{v}\rangle}{\langle\mathbf{v}^2\rangle} \tag{3.6}$$

This ensures that $\langle\mathbf{v}^2\rangle$ is independent of time, *i.e.* $\int \mathbf{v}^2 f(\mathbf{v},t)d\mathbf{v} = \int \mathbf{v}^2 f(\mathbf{v},0)d\mathbf{v}$. When $\mathbf{E} = 0$ also $\mu = 0$ and $f(\mathbf{v},t)$ will approach, as $t\to\infty$, the function $\bar{f}(|\mathbf{v}|,0) = \frac{1}{S(1)}\int f(\mathbf{v},0)d\mathbf{n}$, where $S(1)$ is the area of the unit sphere in n dimensions. There will thus not be a unique stationary solution of Eq. (3.4). The situation is different when $\mathbf{E}\neq 0$. In that case $\lim_{t\to\infty} f(\mathbf{v},t) = f(\mathbf{v})$ independent of $f(\mathbf{v},0)$. It is this NESS of Eq. (3.4) which we shall study here.

The outline of the rest of the paper is as follows. In section 2 we show that the stationary solution vanishes for $|\mathbf{v}| > |\mathbf{E}|/\mu$. In section 3 we obtain an exact expression for the steady state solution of this equation in one dimension. In section 4 we investigate the small field limit of this stationary distribution. Finally in section 5 we find the small field limit of the distribution in two dimensions.

3.2. Properties of Eq. (3.4) in Arbitrary Dimension

Here we investigate some general properties of Eq. (3.4) in the steady state. Observe that the right hand side of the equation preserves $\mathbf{v}^2$. On the other hand we have

$$(\mathbf{E}-\mu\mathbf{v})\cdot\mathbf{v}\leq 0 \qquad \text{if} \quad |\mathbf{v}| > \frac{|\mathbf{E}|}{\mu} \tag{3.7}$$

Thus if we define

$$F(V,t) = \int_{|\mathbf{v}|\leq V} f(\mathbf{v},t)d\mathbf{v} \tag{3.8}$$

we get that

$$\frac{d}{dt}F(V,t) = -\int_{|\mathbf{v}|=V} ([(\mathbf{E}\cdot\mathbf{v}) - \mu V^2]f(\mathbf{v},t)d\mathbf{v} \tag{3.9}$$

$$\leq -\mathcal{S}(V)(|\mathbf{E}| - \mu|V|)\sup_{|\mathbf{v}|=V} f(\mathbf{v},t) \tag{3.10}$$

where $\mathcal{S}(V)$ is the surface of the sphere of radius V. Since in the steady state we must have $\dot{F}(V,t) = 0$ it follows that $f(\mathbf{v}) = 0$ if $\mathbf{v} > |\mathbf{E}|/\mu$, where μ is now the stationary self consistent value. When $|\mathbf{E}| \to 0$, μ will decrease as $|\mathbf{E}|^2$, as long as $\lambda > 0$, and so the maximum speed will grow as $|\mathbf{E}|^{-1}$.

Observe that, as noted before, the limit for $\mathbf{E} \to 0$ is very different from that obtained when $\mathbf{E} = 0$ and $\mu = 0$. This behavior is due to the fact that the collision term preserves $|\mathbf{v}|^2$. In particular it would fail if one also adds a non linear Boltzmann particle-particle collisions term to the right hand side of Eq. (3.4). In that case the stationary distribution would be positive for all $\mathbf{v}$. We expect however that Eq. (3.4) would describe the behavior of the system for a long time when the effective rate of inter-particle collisions is small compared to the rate of collisions with the obstacles.

3.3. The Boltzmann Equation in One Dimension

We consider the one dimensional version of Eq. (3.4). This takes the form:

$$\frac{\partial}{\partial t} f(v,t) + \frac{\partial}{\partial v}\left[(E - \mu v) f(v,t)\right] = -\lambda |v| \left(f(v,t) - f(-v,t)\right) \tag{3.11}$$

with μ determined self consistently by the requirement that the average energy remain fixed, see Eq. (3.6). We look for a stationary solution of Eq. (3.4) in the form

$$f(v) = \phi(v) + E\psi(v) \tag{3.12}$$

with

$$\begin{aligned} \phi(v) &= \phi(-v) \\ \psi(-v) &= -\psi(-v) \end{aligned} \tag{3.13}$$

and ϕ and ψ defined for $v > 0$. The time independent Eq. (3.11) takes the form

$$\begin{aligned} \frac{\partial}{\partial v}\left[\mu v \phi(v) - E^2 \psi(v)\right] &= 0 \\ \frac{\partial}{\partial v}\left[\phi(v) - \mu v \psi(v)\right] &= -2\lambda v \psi(v) \end{aligned} \tag{3.14}$$

where μ now has its steady states value given by Eq. (3.6). Using the fact that ψ is odd the first equation implies that

$$\psi(v) = \frac{\mu v}{E^2}\phi(v). \tag{3.15}$$

Setting $\mu = \nu E^2$ we have $\psi(v) = \nu v \phi(v)$ and the second equation in Eq. (3.14) becomes

$$\frac{\partial}{\partial v}\left[(1-\nu^2 E^2 v^2)\phi(v)\right] = -2\lambda\nu v^2\phi(v) \tag{3.16}$$

Observe that since $f(v) \geq 0$ we have from the definition, Eq. (3.13), that $\phi(v) \geq 0$. It follows from Eq. (3.16) that, if $\phi(v) > 0$ for $v > \frac{1}{\nu E}$, then we will have $\lim_{v\to\infty}\phi = \infty$, so we must have

$$\phi(v) = 0 \qquad \text{for} \qquad v > \frac{1}{\nu E} \tag{3.17}$$

This is consistent with the general result in the previous section. Setting

$$\Phi(v) = (1-\nu^2E^2v^2)\phi(v) \tag{3.18}$$

Eq. (3.16) becomes

$$\frac{d}{dv}\log\Phi(v) = \frac{\Phi'(v)}{\Phi(v)} = -\frac{2\lambda\nu v^2}{1-\nu^2E^2v^2} = \frac{2\lambda}{\nu E^2} - \frac{2\lambda}{\nu E^2}\frac{1}{1-\nu^2E^2v^2} \tag{3.19}$$

This can be solved to give

$$\Phi(v) = C\exp\left(\frac{2\lambda}{\nu E^2}v + \frac{\lambda}{\nu^2E^3}\log\left(\frac{1-\nu E v}{1+\nu E v}\right)\right) \tag{3.20}$$

or

$$\phi(v) = \frac{C}{1-\nu^2E^2v^2}\left(\frac{1-\nu E v}{1+\nu E v}\right)^{\frac{\lambda}{\nu^2E^3}} e^{\frac{2\lambda}{\nu E^2}v} \tag{3.21}$$

for $v < \frac{1}{\nu E}$ and 0 otherwise. We finally get, using Eq. (3.13), that for $v \in \mathbb{R}$,

$$f(v) = \frac{C}{1-\nu E v}\left(\frac{1-\nu E|v|}{1+\nu E|v|}\right)^{\frac{\lambda}{\nu^2E^3}} e^{\frac{2\lambda}{\nu E^2}|v|} \tag{3.22}$$

for $|v| < \frac{1}{\nu E}$ and 0 otherwise. C is a normalization constant.

Observe that at $v = (\nu E)^{-1}$ the solution is singular if $\lambda < \nu^2E^3$. This means that the field is strong enough to push the velocity of the particle close to its limiting value before a collision happen[a]. Furthermore, when the field goes to 0 we do not obtain a Maxwellian distribution. To be sure, there is no reason why this should happen since we do not have any mechanism (such as inter-particle collisions) which would tend to bring the system to equilibrium, as one would expect to be the case in a realistic physical model.

It follows from Eq. (3.16) that

$$\int_0^{\frac{1}{\nu E}} v^2 \phi(v) dv = \frac{C}{2\lambda\nu} \tag{3.25}$$

so that the average kinetic energy in the steady state is $\langle v^2 \rangle = C/\lambda\nu = v_0^2$ and the average current is $\langle v \rangle = CE/\lambda\nu$.

3.4. Small Field Limit

Equation (3.22) can be rewritten in the form:

$$f(v) = \frac{C}{1-\nu E v} \exp\left[\frac{\lambda}{\nu^2 E^3}\left(\log\left(1-\nu E|v|\right) - \log\left(1+\nu E|v|\right)\right)\right] e^{\frac{2\lambda}{\nu E^2}|v|} \tag{3.26}$$

This expression is clearly analytic in E for $E < 1/\nu|v|$. In particular we can compute the zero order term by expanding $\log(1-\nu E|v|) - \log(1+\nu E|v|)$ in term of $\nu E|v|$. Clearly only terms odd in E will appear. The term linear in E cancel out the exponent $\frac{2\lambda}{\nu E^2}|v|$. The term proportional to E^3 gives, once multiplied by $\lambda/\nu^2 E^3$, a term $2/3\lambda\nu|v|^3$. This gives us:

$$f(v) = C(1+E\nu v)\exp\left(-\frac{2}{3}\lambda\nu v^3\right) + o(E) \tag{3.27}$$

---

[a]Note that if we let $\lambda \to 0$ than the singularity becomes non integrable and:

$$\lim_{\lambda\to 0} \phi(v) = \frac{1}{2}\delta\left(|v| - \frac{1}{\nu E}\right) \tag{3.23}$$

so that

$$f(v) = \delta\left(v - \frac{1}{\nu E}\right) \tag{3.24}$$

with $\langle v^2 \rangle = (1/\nu E)^2 = v_0^2$ specified a priori, so that $\nu = 1/Ev_0$ or $\mu = E/v_0$. We also have $\langle v \rangle = v_0$. This is exactly what happens in the N-particles d-dimensional thermostated case without collisions.All particles end up moving parallel to the field with the same speed independent of the strength of $|\mathbf{E}| > 0$.

where the term linear in E comes from the prefactor $1/(1-\nu E v)$. Here ν is the limiting value of ν which, given the (initial) kinetic energy $v_0^2/2$, is given by

$$\nu = \frac{K}{v_0^3 \lambda} + O(E^2) \tag{3.28}$$

with

$$K = \frac{3}{8} \frac{2^{\frac{1}{2}} 3^{\frac{3}{4}} \Gamma\left(\frac{2}{3}\right)^{\frac{3}{2}}}{\pi^{\frac{3}{2}}} \tag{3.29}$$

We can also obtain the behavior of $f(v)$ when E tends to 0 by setting $E = 0$ in Eq. (3.16). This gives:

$$\frac{\partial}{\partial v}\phi(v) = -2\lambda \nu v^2 \phi(v) \tag{3.30}$$

whose solution is found to be

$$\phi(v) = Ce^{-\frac{2}{3}\lambda \nu v^3} \tag{3.31}$$

This agrees with the result in Eq. (3.27) when $E \to 0$.

3.5. The General Case for Small E

In the two dimensional case the stationary version of Eq. (3.4) takes the form

$$\frac{\partial}{\partial \mathbf{v}}\left[(\mathbf{E} - \nu|\mathbf{E}|^2\mathbf{v})f(\mathbf{v})\right] = -\lambda \int_{\mathbf{v}\cdot\mathbf{n}<0} \frac{\mathbf{v}'\cdot\mathbf{n}}{2}\left(f(\mathbf{v}) - f(\mathbf{v}')\right) d\mathbf{n} \tag{3.32}$$

where $\mathbf{v}' = \mathbf{v} - 2\mathbf{n}(\mathbf{n}\cdot\mathbf{v})$, $\mathbf{n}$ is a unit vector and we have set $\mu = \nu|\mathbf{E}|^2$.

We do not have an exact solution of this equation for arbitrary field strength. We can however still find the small $|\mathbf{E}|$ behavior of $f(\mathbf{v})$ as we did before for the one dimensional case. To do this we first expand $f(\mathbf{v})$ in harmonics of the angular part of $\mathbf{v}$. This means that that we write:

$$f(\mathbf{v}) = \sum_{n=0}^{\infty} f_n(v)\cos(n\theta) \tag{3.33}$$

where $\mathbf{v} = (v\cos(\theta), v\sin(\theta))$. We have taken the field to be in the x-direction, $\mathbf{E} = (E, 0)$, so that no term in $\sin(\theta)$ appears due to the symmetry of the system with respect to the x axis. Substituting in Eq. (3.32) we obtain:

$$\begin{aligned} &- E^2\nu v\partial_v f_n(v) - 2E^2\nu f_n(v) \\ &+ E\partial_v \frac{f_{n-1}(v) + f_{n+1}(v)}{2} - E\frac{(n-1)f_{n-1} - (n+1)f_{n+1}}{2v} \\ &= \left(\frac{(-1)^{n+1}}{4n^2-1} - 1\right)\lambda v f_n(v) \end{aligned} \tag{3.34}$$

where we have set $f_{-1} \equiv 0$. Since we expect that the distribution $f(\mathbf{v})$ will depend only on $|\mathbf{v}|$ when $E \to 0$ we can assume that $f_n(v) = Eg_n(v)$ for $n \neq 0$. When $n = 0$, the right hand side of Eq. (3.34) vanishes. The equation thus becomes, after simplifying a common E^2 factor,

$$\nu v\partial_v f_0(v) + 2\nu f_0(v) - \frac{1}{2}\partial_v g_1(v) - \frac{1}{2v}g_1(v) = 0 \tag{3.35}$$

while the leading term in E of the equation for $n = 1$ is

$$\partial_v f_0(v) = -\frac{4}{3}\lambda v g_1(v). \tag{3.36}$$

So that, at first order in E, we have:

$$f(\mathbf{v}) = C(1 + 2\nu\mathbf{E}\cdot\mathbf{v})\exp\left(-\frac{8}{9}\lambda\nu|\mathbf{v}|^3\right) \tag{3.37}$$

This is the same form as Eq. (3.27) with the limiting value for ν given by Eq. (3.28) with

$$K = \frac{1}{9}\frac{3^{\frac{3}{4}}2^{\frac{1}{2}}\pi^{\frac{3}{2}}}{\Gamma\left(\frac{2}{3}\right)^3} \tag{3.38}$$

3.6. Concluding Remarks

A similar analysis shows that the small $|\mathbf{E}|$ behavior of the NESS described by Eq. (3.4) will be of the same form as that given by Eq. (3.37) in all dimension d, in particular the zero order term will be an exponential in $-|\mathbf{v}|^3$. We do not however have any intuitive heuristic explanation for this behavior on the more microscopic many particles level.

Let us go back now to our original problem, the NESS for the N particles system in an electric field and Gaussian thermostat, described by Eqs. (3.1)-(3.3). As mentioned in the introduction, we expect that in the limit $N \to \infty$ this NESS will agree with our results for the stationary distribution of the Boltzmann equation (3.4) *i.e.* there will be a cut-off in the particle speed independent of N when N becomes very large. This indicates that the NESS for the original model will be, for large N, concentrated on those parts of the $dN - 1$ dimensional sphere in which the particle speeds are closer to each other than they would for a uniform distribution on the sphere. The latter is of course what gives rise to a Maxwellian distribution with the same mean energy. This is consistent with what we found in,[5] see in particular section III there.

Acknowledgments

The work of FB was supported in part by NSF grant 0604518. The work of JLL was supported in part by NSF grant DMR08021220 and by AFOSR grant AF-FA9550-07.

References

1. N.W. Ashcroft, N.D. Mermin, *Solid States Physics*, Holt, Rinehart and Winston, New York (1976)
2. D. Ruelle, *Smooth dynamics and new theoretical ideas in nonequilibrium statistical mechanics*, Journ. Stat. Phys. **95**, 393–468 (1999)
3. F. Bonetto, D. Daems, J.L. Lebowitz, *Properties of Stationary Nonequilibrium States in the Thermostatted Periodic Lorentz Gas I: the One Particle System*, Journ. Stat. Phys. **101**, 35-60 (2000)
4. D.J. Evans, G Morris, Statistical "Mechanics of Nonequilibrium Liquids", Cambridge University Press, 2006.
5. F. Bonetto, D. Daems, J.L. Lebowitz, V. Ricci *Properties of Stationary Nonequilibrium States in the Thermostatted Periodic Lorentz Gas I: the Many Particles System*, Journ. Stat. Phys. (2000)
6. F. Bonetto, G. Gallavotti, P.L. Garrido, *Chaotic principle: an experimental test*, Physica D **105**, 226–252 (1997).

Chapter 4

Propagation of Light in Complex Fluids: Embedded Solitons in Liquid Crystals

R.F. Rodríguez* and Jorge Fujioka

Instituto de Física, Universidad Nacional Autónoma de México
04510 México D.F.,México
zepeda@fisica.unam.mx

The soft nature of liquid crystals induces highly nonlinear couplings between the optical field and the director. This nonlinear dynamics makes it possible to have solitons of the optical fields. In this work the conditions under which a new type of solitons (embedded solitons, *ES*) may exist in liquid crystals waveguides is discussed. To describe this new type of solitary waves, we first review some basic concepts of the hydrodynamics of liquid crystals and then discuss how the *ES* were discovered and what is the mechanism that accounts for their existence. We use a model for a cylindric nematic waveguide and write the coupled basic equations for the electromagnetic modes in the guide and for the orientation field. Then we use a simplified multiple scale technique to derive the structure of the basic equations which describe the evolution of light pulses in liquid crystals at different scales. We study the *ES* of the complex modified Korteweg de Vries equation (*cmKdV*) and use a variational method to explain the stability of its solitons.

This work is dedicated to Professor Dr. Leopoldo García-Colín Scherer, pioneer in statistical physics in México, as a token of admiration for his work and accomplishments in his eightieth birthday.

Contents

*Also at FENOMEC, UNAM, México.

4.1. Introduction

The field of complex fluids extends over a broad range of materials, including liquid crystals, polymers, colloids and self-assembled surfactant systems. These fluids contain extended polyatomic structures of length scales larger than atomic or molecular scales, typically of the order of 10 angstroms to several microns.

Liquid crystal phases (mesophases) are complex fluids which exhibit a certain degree of order in the spatial arrangement of their molecules. The mesophase for which only the rotational invariance has been broken is called *nematic* and is characterized by an axis of preferred orientation, represented by a unit vector field (the director), which does not distinguishes head from tail. Uniaxial nematics are found for rod-shaped and disk-shaped molecules, and in thermotropic low molecular weight materials, their properties change predominantly as a function of temperature. As a result, there is anisotropy in their mechanical, electric, magnetic and optical properties.

Although liquid crystals combine some properties of a crystalline solid and an isotropic liquid, they exhibit very specific electro-optic phenomena which, as a rule, have no counterpart in solids or isotropic liquids. The basis for the physical mechanisms of these phenomena lies in the reorientation of the director under the influence of an externally applied field. Anisotropy of electric properties such as the electric conductivity or the dielectric susceptibility, is the origin of reorientation, whereas the dynamics of the process also depends on the viscoelastic properties of the medium and on the initial orientation of the director relative to the external fields.

The study of the propagation of electromagnetic fields in uniaxial nematics is a well established issue. It has been triggered by the many electro-optic effects existing in these liquids which have produced a variety of applications in display devices,[1–4] or for transport processes in these fluids,.[5,6] A complete set of electrohydrodynamic equations is derived by using rather general approaches based solely on the concepts of irreversible thermodynamics and broken symmetries.[7,8] As a consequence the qualitative features of most of the results also pertain to other anisotropic systems, such as smectics, cholesterics, crystals or two phases of supefluid ^3He, but the specific symmetries of a given system will permit to formulate specific constitutive equations.

All truly hydrodynamic variables are related to a hydrodynamic collective excitation whose frequency ω vanishes in the long wavelenght limit, i.e., $\lim_{k\to 0}\omega(k) = 0$. This condition defines the hydrodynamic regime, $\omega\tau_i << 1$, where τ_i includes any microscopic relaxation time, especially that of the polarization and magnetization. The hydrodynamic variables of a uniaxial, thermotropic nematic are conveniently classified in two groups[7,9,10] : i) quantities satisfying conservation laws, and ii) variables associated with spontaneously broken continuous symmetries satisfying balance equations. The first group contains the local mass density, $\rho(\vec{r},t)$, the entropy density $\sigma(\overrightarrow{r},t)$ and the the velocity field, $\vec{v}(\vec{r},t)$, whereas the director, $\overrightarrow{n}(\vec{r},t)$, a variable that characterizes the spontaneously broken continuous rotational symmetry, belongs to the second group. The entropy density σ is expressed in terms of the other variables like the temperature T and the pressure p through the corresponding Gibbs-Duhem relation.

The derivation of the hydrodynamic equations for these variables proceeds as follows: the first step is to establish their behavior under time reversal and spatial parity, Galilean covariance, and their invariance under rotations and translations. In a second step the Gibbs-Duhem relation is formulated to define the corresponding thermodynamic forces, which are then expanded in the hydrodynamic variables. Then the currents and quasi-currents appearing in the conservation and balance equations are expressed in terms of the thermodynamic forces. In the final step, the currents and quasi-currents are split into two contributions: a reversible part that leads to vanishing entropy production, and a dissipative (irreversible) component that is associated with positive entropy production.

External fields are often used to drive a nematic out of equilibrium into a new state through some kind of instability. As external fields one can have electric and magnetic fields, gravity, applied temperature and concentration gradients, or externally induced shear and vortex flow. Here we concentrate on dynamic electric fields $\overrightarrow{E}$. Light propagates in uniaxial nematics in two characteristic modes: the ordinary light ray, which behaves as in an isotropic fluid, and the extraordinary light mode, which possesses a direction-dependent index of refraction; their phase and group velocities are not equal to each other, and the electric field is not transverse. To discuss how an electromagnetic field can be combined with the hydrodynamic equations to get electro-hydrodynamic equations for a uniaxial nematic, it is essential to examine which part of the Maxwell equations must be combined with the hydrodynamic equations to get a consistent description in the hydrodynamic regime. For this purpose we follow Ref.,[7] although other schemes have been proposed.[8] Of the six Maxwell equations for the electric $\overrightarrow{E}$ and magnetic $\overrightarrow{H}$ fields, one dynamic degree of freedom is removed by gauge invariance, another one is the charge conservation law and the remaining four are given by the inhomogeneous equations

$$\left(curl\frac{\partial\overrightarrow{E}}{\partial t}\right)_i + c\nabla^2 H_i = -4\pi\left(curl\overrightarrow{j^e}\right)_i, \tag{4.1}$$

$$\left(curl\frac{\partial\overrightarrow{H}}{\partial t}\right)_i - c\nabla^2 H_i = 4\pi c\nabla_j\rho_e. \tag{4.2}$$

Since for many nematics the magnetic susceptibility is much smaller than the electric susceptibility, we assume a non-magnetic medium, i.e., $\overrightarrow{B} = \overrightarrow{H}$, and, for simplicity, it has been also assumed that $\overrightarrow{E} = \overrightarrow{D}$, being $\overrightarrow{D}$ the electric displacement vector. c is the speed of light, $\overrightarrow{j^e}$ is the charge current and $\rho_e = \frac{1}{4\pi}div\overrightarrow{D}$. In vacuum, where $\rho_e = \overrightarrow{j^e} = 0$, the two transverse electromagnetic waves that follow from these equations have frequencies far beyond any reasonable hydrodynamic description. Furthermore, in the presence of matter, the nonzero conductivity leads to relaxation, that is, to a non-hydrodynamic behavior. Thus, $curl\overrightarrow{E}$ and $curl\overrightarrow{H}$, which are neither related to conservation laws or spontaneous broken symmetries, are not hydrodynamic variables. Only $\overrightarrow{D}$ is to be considered as a hydrodynamic variable in group i) and there are no variables belonging to ii).

This procedure leads to the following complete set of electrohydrodynamic equations. The mass conservation,

$$\frac{\partial}{\partial t}\rho + div\overrightarrow{j} = 0, \tag{4.3}$$

where $\overrightarrow{j}(\overrightarrow{r},t) \equiv \rho\overrightarrow{v}(\overrightarrow{r},t)$ is the mass current which is identified with the momentum density $\overrightarrow{g}(\overrightarrow{r},t)$. The equation of motion for the velocity field $\overrightarrow{v}(\overrightarrow{r},t)$, is

$$\rho\left(\frac{\partial}{\partial t} + v_j\nabla_j\right)v_i + \nabla_j\sigma_{ij} = \rho_e E_i + P_j\nabla_j E_i; \tag{4.4}$$

the balance equation for the entropy density σ reads

$$\frac{\partial}{\partial t}\sigma + \nabla_j(\sigma v_j) + \nabla_k j_k^\sigma = \frac{R}{T} \tag{4.5}$$

and the balance equation for the director is given by

$$\left(\frac{\partial}{\partial t} + v_j\nabla_j\right)n_i + Y_i = 0. \tag{4.6}$$

In Eq. (4.4) the polarization, $P_i = \frac{1}{4\pi}(D_i - E_i)$, is defined in terms of $\overrightarrow{D}$, which obeys the conservation equation

$$\frac{\partial}{\partial t} D_i + v_i \nabla_j D_j + 4\pi j_i^e = 0. \tag{4.7}$$

The stress tensor σ_{ij}, and the irreversible part of the entropy current, j_i^σ, charge currents, j_i^e, and the quasi-current Y_i, are given in terms of the thermodynamic forces. The dissipation function R is positive for irreversible processes and can be interpreted as the energy per unit time and volume dissipated into the microscopic degrees of freedom. The explicit forms of Y_i and R will not be given explicitly, but are indeed requiered to gauge the validity of the approximations and of the final dynamical equation. Their explicit forms can be found in Ref.[7]

The coupling of the dynamics of the velocity and director fields to external optical fields such as $\overrightarrow{E}$, renders these equations highly nonlinear. The extent and consequences of these nonlinearities will be discussed and quantified in Section 4 (see coments following Eq. (4.32)). It is therefore possible, in principle, to have solutions of these equations in the form of solitons of the director or the optical fields, with or without involving the fluid motion.[11,12] Furthermore, the strong coupling of the director to light makes any director-wave more easily detectable by optical methods than it is for isotropic fluids, where only the flow field is observable. Lam and Proust[13] have described how theoretical studies on the existence of solitons in liquid crystals started in the late sixties and early seventies and how experimental confirmations were reported subsequently. In Ref. ([13]) it is also described how some nonlinear partial differential equations ($PDEs$) appearing in the liquid-crystal theory give rise to exact soliton solutions. These are the Korteweg-de Vries (KdV), nonlinear Schrödinger (NLS), and the sine-Gordon (sG) equations. The KdV equation describes a medium with weak nonlinearity and weak dispersion, whereas the NLS equation describes situations where weak nonlinearity and strong dispersion prevail, such as the propagation of signals in liquid-crystal optical fibers.

Solitons are solitary waves which are able to propagate in nonlinear media. Their discovery and subsequent history is well documented,[14] and at present we know that several categories of solitons exist: bright, dark, topological, non-topological, Bragg solitons, vector and vortex solitons, spatio-temporal solitons (optical bullets), lattice solitons, etc. Although most of these categories are known since the seventies, a brand-new category was discovered recently, namely, the *embedded solitons* (ES). The discovery of these solitary waves was a surprise because they exist under conditions in which, prior to 1997, it was considered that the propagation of solitons was impossible. In nonlinear systems where solitons can exist, the propagation of small-amplitude linear waves, which obey the linearized version of the nonlinear equations, is possible too. However, for a soliton to exist, it is absolutely necessary that no resonances occur between the soliton and these linear waves. Otherwise, the soliton would decay into radiation due to an energy transfer towards the linear waves. Based on this absence-of-resonance argument, it was frequently assumed that the solitons' internal frequencies could not be contained in the linear spectrum of the system, i.e., they could not lie within the band of frequencies allowed to linear waves. However, at the end of the nineties exceptions to this rule were found, and a special type of solitons were discovered, which do not resonate with linear waves, in spite of having frequencies immersed in the spectrum of them. In 1999 these peculiar solitary waves were named *embedded solitons* and in the following years a number of models supporting ES were identified. The distinctive feature of the ES is that their internal frequencies fall in a spectral region occupied by linear wave modes, where regular solitons cannot exist. In this sense ES are exceptional states that *do not* emit linear waves, despite being in resonance with them.

Most of the models which support ES describe continuous systems like nonlinear optical waveguides with quadratic nonlinearities,[15–17] hydrodynamic systems,[18] or as has been recently discovered, they may also exist in liquid crystals.[19,20] Furthermore, some examples of discrete ES have been also found in finite-difference versions of a higher-order nonlinear Schrödinger equation (NLS) and a complex modified Korteweg-de Vries equation ($cmKdV$).[21,22] A number of generic features of the ES found in continuous systems have been identified. Usually, ES are isolated solutions, although continuous families of ES may exist too, under special symmetry conditions.[19,23] The ES may be stable against small perturbations in the linear approximation, but general arguments and direct simulations demonstrate that they are usually unstable to second order in these perturbations, which is the reason why they are often called *semi-stable* solutions. Nevertheless, examples of completely stable ES have been also found,[19,23,24] and their stability has been demonstrated in Refs.[23,24]

The main purpose of the present work is to discuss what these embedded solitons are, and which are the different types of ES known to date for the case of nematic liquid crystals. We describe the main features of an approach which generates a hierarchy of nonlinear partial differential equations ($PDEs$) to describe the propagation of optical pulses in nematic liquid crystal waveguides, beyond the weakly nonlinear limit corresponding to the Kerr medium.[25] The different nonlinear regimes are conveniently described in terms of successive powers of a parameter, q, defined as the ratio between the electric energy density of the optical field and the elastic energy density of the liquid crystal, which measures the strength of the coupling between the optical field and the orientational configuration of the liquid crystal.

To this end the work is organized as follows. In Section II we define a model for a cylindric nematic waveguide and write down the basic dynamic equations for the electromagnetic modes and for the orientation field, assuming that initially the nematic is in the so called escape configuration. We use a multiple scale technique to derive the structure of the basic equations which describe the evolution of light pulses in liquid crystals at different scales. Then in Section III we review how the embedded solitons were discovered and what is the mechanism that accounts for their existence. Section IV is devoted to the study the ESs and $cmKdV$ equations and the a variational method is used in Section V to explain the stability of these solitons. We conclude the paper in Section VI by giving critical remarks of our approach and results.

4.2. Multiple-Scale Analysis

Consider a waveguide with a quiescent, uniaxial, nematic liquid crystal core and confined within a cylindrical region of radius d and length L. It is surrounded by an infinite homogeneous isotropic dielectric cladding with dielectric constant ϵ_c. Initially the nematic adopts the stable scape configuration, $\theta(x) = 2\arctan x$, where $x \equiv r/\rho$. An obliquely incident laser beam enters the guide with a linear polarization and forming an angle α with respect to the axis of the cylinder, as depicted in Fig. 1,

Since in general, the incident beam is neither plane nor Gaussian, and since under these conditions it can be shown that only the transverse magnetic (TM) components (E_r, E_z, H_ϕ) couple to the reorientation dynamics, we assume that these normal modes within the cavity are cylindrical plane waves propagating along the z axis with a propagation constant β,.[26] The transverse electric, TE modes, can be ignored in the discussion. For each specific value of β there is a specific field distribution described by $E_{r,z}(r, k_0)$ or $H_\phi(r, k_0)$ (modes of the waveguide), where $k_0 \equiv \frac{\omega}{c}$ is the free space wave number. For the present model, Maxwell's equations (without

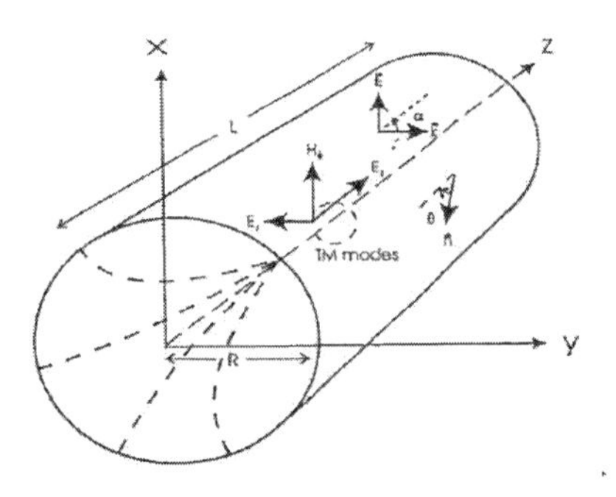

Fig. 4.1. Schematic of a linealrly polarized beam propagating through a nematic liquid.

sources) lead to the following dimensionless equation for the H_ϕ,

$$\begin{aligned}
&\epsilon_{rr}\frac{d^2H_\phi}{dx^2}+\left(\frac{\epsilon_{rr}}{x}+\frac{d\epsilon_{rr}}{dx}-2ipk_0d\epsilon_{rz}\right)\frac{dH_\phi}{dx}\\
&+\left\{(k_0d)^2\left(\epsilon_\parallel\epsilon_\perp-p^2\epsilon_{zz}\right)-\frac{\epsilon_{rr}}{x^2}-ipk_0d\left(\frac{d\epsilon_{rz}}{dx}+\frac{\epsilon_{rz}}{x}\right)\right\}H_\phi\\
=\,&0,
\end{aligned}\tag{4.8}$$

where we have identified $p \equiv \frac{\beta c}{\omega}$. The remaining TM modes E_r and E_z are expressed in terms of H_ϕ,[26,27] This equation contains the parameter k_0d, which to zero and first order defines the optical and WKB limits, respectively. Note that the components of the dielectric tensor $\epsilon_{ij}[\theta(x)]$ are $\epsilon_{rr} = \epsilon_\perp + \epsilon_a \sin^2\theta$, $\epsilon_{rz} = \epsilon_a \sin\theta\cos\theta$ and couple the above equation with the orientation configuration.

On the other hand, the spatially varying nematic's dielectric tensor $\epsilon_{ij}[\theta(x)]$ is determined by the director field, whose dynamics is determined by Eqs. (4.3) - (4.7). It can be shown that for the present model the stationary orientation configurations are specified by

$$\frac{d^2\theta}{dx^2}+q[(|E_r|^2-|E_z|^2)\sin 2\theta-2\left(E_r^*E_z+E_rE_z^*\right)\cos 2\theta]=0, \tag{4.9}$$

where the asterisk (*) denotes complex conjugate and $E_{r,z}$ stands for the amplitude of the corresponding fields. In the above equation we have used dimensionless field amplitudes dividing them by E_0, the amplitude of the incident field which is related to its intensity by $I \equiv \frac{cE_0^2}{8\pi}$. The above mentioned dimensionless parameter is $q \equiv \frac{E_0^2/8\pi}{K/R^2}$, being K the elastic constant in the equal elastic constants approximation. But in order to solve Eq.(4.9) the field distributions must be determined first from Eq. (4.8). The system of nonlinear, coupled equations (4.8) and (4.9) may then be solved iteratively to derive an equation for the envelope of a wave packet propagating through the guide.[25] An expansion in powers of the coupling parameter q leads to a specific hierarchy of $PDEs$ governing the evolution of the amplitude of propagating TM modes up to the order $O(q^4)$. To first order in q, this procedure actually generates the dispersion relation. Then, by rescaling the resulting equations to higher orders in q, in Ref. ([19]) it was shown that the standard NLS equation is obtained to order $O(q^3)$ and the $cmKdV$ equation up to $O(q^4)$.

However, a rather surprising result is that an equation for the structure of the basic equations that govern the evolution of light pulses at different scales in these fluids, can be derived in a simpler and more straight forward way, just from the knowledge of the dispersion relation. This relation links the wavenumbers (k), the frequencies (ω) and the intensities ($|A|^2$) of the pulses, and as a Taylor series it can be expressed in the form

$$k = k_0 + k_1 (\omega - \omega_0) + k_2 |A|^2 + \frac{1}{2!} \Big[k_{11} (\omega - \omega_0)^2 + 2 k_{12} |A|^2 (\omega - \omega_0) + k_{22} |A|^4 \Big] + \frac{1}{3!} \Big[k_{111} (\omega - \omega_0)^3 + ... + k_{222} |A|^6 \Big] + ..., \quad (4.10)$$

where the coefficients k_i, k_{ij} and k_{ijk} are the derivatives

$$k_1 = \frac{\partial k}{\partial \omega}, \quad k_2 = \frac{\partial k}{\partial |A|^2}, \quad k_{11} = \frac{\partial^2 k}{\partial \omega^2}, \quad k_{12} = \frac{\partial^2 k}{\partial \omega \, \partial |A|^2}, \quad k_{22} = \frac{\partial^2 k}{\partial (|A|^2)^2}, k_{111} = \frac{\partial^3 k}{\partial \omega^3}, \quad k_{222} = \frac{\partial^3 k}{\partial (|A|^2)^3}. \quad (4.11)$$

All these coefficients are evaluated at $|A|^2 = 0$ and $\omega = \omega_0$ (the frequency of the carrier wave). From this expansion we can obtain a hierarchy of $PDEs$ which govern the evolution of the pulse at different scales. To obtain such $PDEs$ we proceed in two steps: The first is based on the fact that the slowly varying amplitude $A(Z,T)$ of a quasi monochromatic pulse can be written as an inverse Fourier transform of the form

$$A(Z,T) = \mathcal{F}^{-1} \left[\widetilde{A} \right] = \frac{1}{(2\pi)^2} \iint_{-\infty}^{+\infty} \widetilde{A}(k,\omega) \, e^{i[(k-k_0)Z - (\omega - \omega_0)T]} dk \, d\omega, \quad (4.12)$$

thus implying that

$$\frac{\partial A}{\partial T} = \mathcal{F}^{-1} \left[-i (\omega - \omega_0) \, \widetilde{A}(k,\omega) \right] \quad \text{and} \quad \frac{\partial A}{\partial Z} = \mathcal{F}^{-1} \left[i (k - k_0) \, \widetilde{A}(k,\omega) \right]. \quad (4.13)$$

These equations show that multiplying a function $\widetilde{A}(k,\omega)$ by $-i(\omega-\omega_0)$ and $i(k-k_0)$ is related to taking the derivatives with respect to T and Z of the function $A(Z,T)$. In other words, the following correspondences exist between operors in Fourier and coordinate spaces,

$$(\omega-\omega_0) \leftrightarrow i\frac{\partial}{\partial T} \quad \text{and} \quad (k-k_0) \leftrightarrow -i\frac{\partial}{\partial Z}. \tag{4.14}$$

The second step introduces the scaled variables[28] $t=\epsilon T$ and $z_n=\epsilon^n Z$, which imply that

$$(\omega-\omega_0) \leftrightarrow i\frac{\partial}{\partial T} = i\epsilon\frac{\partial}{\partial t}, \tag{4.15}$$

$$(k-k_0) \leftrightarrow -i\frac{\partial}{\partial Z} = -i\sum \frac{\partial z_n}{\partial Z}\frac{\partial}{\partial z_n} = -i\sum \epsilon^n \frac{\partial}{\partial z_n}. \tag{4.16}$$

If we now replace the factors $k-k_0$ and $\omega-\omega_0$ in (4.10) by the differential operators shown in (4.15) and (4.16) we obtain the equation

$$\begin{aligned} -i\sum \epsilon^n \frac{\partial}{\partial z_n} &= i\epsilon k_1 \frac{\partial}{\partial t} + k_2 |A|^2 + \frac{1}{2!}\left[-\epsilon^2 \frac{\partial^2}{\partial t^2} + 2i\epsilon k_{12}|A|^2 \frac{\partial}{\partial t}\right. \\ &\left. + k_{22}|A|^4\right] + \frac{1}{3!}\left[-i\epsilon^3 k_{111}\frac{\partial^3}{\partial t^3} + \ldots + k_{222}|A|^6\right] + \ldots \end{aligned} \tag{4.17}$$

Finally, if we apply these operators to the field $A(z_1,z_2,...,t) = \epsilon\, u(z_1,z_2,...,t)$ and we group together the terms with the same powers of ϵ, we arrive at a hierarchy of $PDEs$, the first of which are

$$\epsilon^2 : u_{z_1} + k_1 u_t = 0, \tag{4.18}$$

$$\epsilon^3 : i\, u_{z_2} - \tfrac{1}{2!}k_{11}\, u_{tt} + k_2 |u|^2 u = 0, \tag{4.19}$$

$$\epsilon^4 : u_{z_3} - \tfrac{1}{3!}k_{111}\, u_{ttt} + k_{12}|u|^2 u_t = 0. \tag{4.20}$$

Obviously, the above multiple-scale technique does not yield the numerical values of the coefficients, nor it yields formulas which connect these coefficients to the properties of a particular liquid crystal. Such a connection cannot be done by mathematics alone, the *physics* of liquid crystals is needed here and is obtained by comparing $PDEs$ of the same order in q and in ϵ the iterative and in the multiple scales expansions.[19]

The equations obtained by the multiple-scale technique are particularly interesting. Note that to order ϵ^3 we obtain the NLS equation, which is known to describe the propagation of light pulses in liquid crystals in the picosecond range. On the other hand, to order ϵ^4, the method yields the $cmKdV$ equation, a feature that was overlooked until the year 2003,[19] when it was realized that the solitons of this equation are not regular, but *embedded*, and they are not isolated and unstable, as those of the eqs. (4.23) and (4.25),[15,30,31] but they form *a continuous family of stable ES*. Moreover, some of these solitons are not only embedded, but *double embedded*, as we shall explain in the following section.

4.3. Embedded Solitons

It is known that solitary light pulses (optical solitons) can propagate in liquid crystals, and depending on the spatial and time scales, and the intensity of the beam, the behavior of these solitary waves can be described by the different nonlinear $PDEs$ mentioned in the previous section. However, the optical solitons *par excellence* are the solutions of the NLS equation

$$i\frac{\partial u}{\partial z} + \frac{1}{2}\frac{\partial^2 u}{\partial t^2} + |u|^2 u = 0, \tag{4.21}$$

and these prototypical solitons have the form

$$u(z,t) = A\, sech\,(At)\, e^{ik_0 z}, \tag{4.22}$$

where A is an arbitrary real constant and the soliton's wavenumber k_0 is given by $k_0 = A^2/2$. The NLS wavenumbers are *always positive*, while the wavenumbers of the small-amplitude linear waves (the so-called radiation modes) are *always negative*, since they are given by the linear dispersion relation of the NLS equation, namely, $k(\omega) = -\omega^2/2$. Consequently, the NLS wavenumbers lie outside of the range of the dispersion relation. For many years it was believed that this was a *necessary* condition for the existence of an optical soliton: to possess a wavenumber which is not contained in the range of the linear dispersion relation. However, in 1997 it was found that the equation[15]

$$i\frac{\partial u}{\partial z} + \varepsilon_2\frac{\partial^2 u}{\partial t^2} + \varepsilon_4\frac{\partial^4 u}{\partial t^4} + \gamma_1 |u|^2 u - \gamma_2 |u|^4 u = 0, \tag{4.23}$$

which describes the propagation of ultrashort (sub-picosecond) pulses in optical fibers, has an exact soliton-like solution of the form

$$u\,(z,t) = \left(\frac{\gamma_1}{2\gamma_2}\right)^{1/2} sech\left(\frac{t}{\sqrt{14\varepsilon_4/\varepsilon_2}}\right)\, exp\left(i\,\frac{15}{196}\frac{\varepsilon_2^2}{\varepsilon_4}\,z\right), \tag{4.24}$$

and the wavenumber of this solution is immersed (or "embedded") in the range of the dispersion relation, which in this case is $k\,(\omega) = \varepsilon_4\omega^4 - \varepsilon_2\omega^2$. In the following years other systems with this unusual solitary pulses were found, and in 1999 the term "embedded soliton" was coined[29] to distinguish these peculiar light waves from the standard optical solitons.

For a number of years it was not understood why the embedded solitons do not resonate with the radiation modes, in spite of the fact of possessing wavenumbers which are immersed in the spectrum of these modes. The answer was found in 2003 while studying the equation[30]

$$i\frac{\partial u}{\partial z}+\varepsilon_2\frac{\partial^2 u}{\partial t^2}-i\varepsilon_3\frac{\partial^3 u}{\partial t^3}+\varepsilon_4\frac{\partial^4 u}{\partial t^4}+\gamma_1\left|u\right|^2 u-\gamma_2\left|u\right|^4 u=0. \quad (4.25)$$

This equation has an exact ES solution of the form:

$$u\left(z,t\right)=A\,sech\left(\frac{t-az}{w}\right)\,exp\left[i\left(qz+rt\right)\right], \quad (4.26)$$

where the soliton parameters are defined by

$$w=\left(\frac{24\,\varepsilon_4}{\gamma_2}\right)^{1/4}/A, \quad r=\varepsilon_3/4\,\varepsilon_4, \quad a=2\,\varepsilon_2 r+8\,\varepsilon_4 r^3, \quad (4.27)$$

$$A=\left(\frac{6}{5\,\gamma_2}\right)^{1/2}\left[\gamma_1-\left(2\,\varepsilon_2+\frac{3}{4}\frac{\varepsilon_3^2}{\varepsilon_4}\right)\left(\frac{\gamma_2}{24\,\varepsilon_4}\right)^{1/2}\right]^{1/2}, \quad (4.28)$$

$$q=-\varepsilon_2 r^2-3\,\varepsilon_4 r^4+\left(\varepsilon_2+6\,\varepsilon_4 r^2\right)\left(\frac{\gamma_2}{24\,\varepsilon_4}\right)^{1/2}A^2+\frac{\gamma_2}{24}A^4. \quad (4.29)$$

In Ref.,[30] it was shown that these solitons are embedded if ε_i and γ_i are positive, and satisfy the condition $A>0$. To understand why these solitons do not resonate with the linear waves we must pay attention to the double Fourier transform (in space and time) of eq. (4.25), which has the form

$$\widetilde{u}(k,\omega)=\frac{\pi wA\left[sech\left(\frac{\pi w(r+\omega)}{2}\right)\right]\sum_{n=0}^{4}\left(c_n\omega^n\right)\,\delta\left(k-q-a\left(r+\omega\right)\right)}{-\left(q+ar\right)-a\omega-\varepsilon_2\omega^2+\varepsilon_3\omega^3+\varepsilon_4\omega^4}, \quad (4.30)$$

where the coefficients c_n are functions of γ_1, γ_2, A, w and r. This implies that resonances between the soliton and the linear waves could occur at the values of the frequencies where the denominator vanishes and, as a consequence, the soliton would transfer energy to the corresponding radiation modes, and it would eventually disappear. However, note that both, numerator and denominator, contain fourth-order polynomials in ω, and it has been shown in Ref.[30] that they coincide precisely when the soliton parameters A, w, a, q, and r satisfy Eqs. (4.27) - (4.29), which are the necessary and sufficient conditions for the mutual cancellation of these polynomials.

Therefore, this striking cancellation is the fundamental reason for the existence of the embedded soliton. Note that the polynomial in the numerator depends *only* on the nonlinear coefficients (γ_1 and γ_2), while the denominator depends on the dispersive coefficients (ε_2, ε_3 and ε_4), but it does not depend explicitly on the nonlinear coefficients. Therefore, the mutual cancellation of these two polynomials is the result of a delicate balance between nonlinearity and dispersion, as is usually the case whenever solitons exist. It should be emphasized though, that this is only a mathematical explanation for the existence of embedded solitons, but it does not identify the underlying physical mechanism which accounts for this cancellation.

4.4. The Family of cmKdV Embedded Solitons

The multiple-scale analysis presented in the previous section shows that the behavior of a light pulse in dispersive nonlinear media like a liquid crystal obeys different equations at different scales. In particular, it also predicts that at smaller scales the correct equation will be the *cmKdV* equation. Although this fact has been known for several years, the role played by the *cmKdV* in liquid crystals had remained obscure until recently, when it was shown that this equation should be useful to describe sub-picosecond pulses with field intensities in the range of $10^4 - 10^5$ V/m.[32]

The *cmKdV* equation is an interesting equation since it is integrable by the inverse scattering method and it belongs to the hierarchy generated by the multiple-scale technique. However, the interest in this equation increased in 2003, Ref. ([19]), when it was realized that all the solitons of this equation are *embedded*, some of them are *double embedded*, and they are *stable* in spite of having wavenumbers immersed in the linear spectrum of the system. In the following we will clarify these issues.

In order to uniform our notation with that used in refs.[19] and,[32] in the following the coefficients of the *cmKdV* equation will be denoted as ε and γ (instead of $k_{111}/3!$ and $-k_{12}$) and the distance will be denoted just as z (instead of z_3). Therefore the *cmKdV* equation takes the form

$$u_z - \varepsilon\, u_{ttt} - \gamma \left|u\right|^2 u_t = 0. \tag{4.31}$$

In this equation $|u|$ is the amplitude of the electric field and is measured in volts/meter, i.e., $[u] = V/m$. Since $[z] = m$, $[t] = ps \equiv 10^{-12}s$, $[\varepsilon] = (ps)^3/m$, $[\gamma] = (ps)\,m/V^2$, Eq. (4.31) can be rewritten in dimensionless form by introducing the dimensionless variables $U \equiv u/u_0$, $\xi \equiv z/z_0$ and $\tau \equiv t/t_0$, where u_0, z_0 and t_0 are characteristic values of the electric field, distance and time. Thus,

$$U_\xi - \varepsilon' U_{\tau\tau\tau} - \gamma' |U|^2 U_\tau = 0, \tag{4.32}$$

where $\varepsilon' \equiv \varepsilon z_0/t_0^3$ and $\gamma' = \gamma u_0^2 z_0/t_0$. This equation shows that the nonlinear term is significant when the value of γ' is comparable (or greater than) the value of ε', and this occurs when $u_0 t_0 \gtrsim (\varepsilon/\gamma)^{1/2}$. A typical value of ε for a liquid crystal is $1.4 \times 10^{-8} (ps)^3/m$ and that of γ should be much higher than for fused silica, which is approximately $5 \times 10^{-17} (mps)/V^2$,.[31] Therefore, the nonlinearity in Eq. (4.32) will be important whenever $u_0 t_0 \gtrsim 10^4 psV/m$. This condition implies that the nonlinearity will be important in the propagation of optical pulses of $1ps$ when $|u_0| \gtrsim 10^4 V/m$, which roughly corresponds to a laser power of $10mW$.

It is worth noting that Eq. (4.32) reduces to the real modified Korteweg de Vries ($mKdV$) when $u(z,t)$ is restricted to be real, and all the real solutions of the $mKdV$ equation, including the N-soliton ones, are also solutions of the $cmKdV$ equation. On the other hand, the $cmKdV$ equation also has complex solutions of the form[19]

$$u(z,t) = A \operatorname{sech}\left(\frac{t - az}{w}\right) e^{i(qz+rt)}, \tag{4.33}$$

where the parameters A, a, w, q and r must satisfy the following conditions:

$$A^2 w^2 = \frac{6\varepsilon}{\gamma},\ a = 3\varepsilon r^2 - \frac{1}{6}\gamma A^2,\ q = \frac{1}{2}\gamma A^2 r - \varepsilon r^3. \tag{4.34}$$

Since (4.33) contains five free parameters, and we only have the three conditions (4.34), these equations define a two-parameter family of bright soliton solutions of the $cmKdV$ equation, which include the real soliton solutions of the $mKdV$ equation, which are obtained when $r = 0$. Equation (4.34) also implies that the bright soliton solutions only exists if $\epsilon\gamma > 0$.

In spite of their similarity to ordinary bright solitons, the bright soliton solutions of eq. (4.31) are special since they are *embedded solitons*. The *embedded* character of these solitons is obvious since the substitution of a plane wave $u = \exp[i(kz - \omega t)]$ into the linear part of eq. (4.31) yields the dispersion relation $k(\omega) = \varepsilon\omega^3$, whose range contains all the real axis $-\infty < \omega < \infty$. Consequently, the wavenumber of the soliton (4.33) (whatever its value might be) necessarily lies within the range of the function $k(\omega)$, thus implying that is embedded.

To understand why the embedded solitons of the *cmKdV* equation do not resonate with the radiation modes we can proceed as in Sec. III. We calculate the double Fourier transform (FT) of the *linear* equation

$$\frac{\partial u}{\partial z} - \varepsilon \frac{\partial^3 u}{\partial t^3} - \gamma \left|u_0\right|^2 \frac{\partial u_0}{\partial t} = 0, \tag{4.35}$$

where $u_0(z,t)$ is the function

$$u_0(z,t) = A \operatorname{sech}\left(\frac{t-az}{w}\right) e^{i(qz+rt)}, \tag{4.36}$$

with the result

$$\widetilde{u}(k,\omega) = \frac{\pi w A \operatorname{sech}\left[\frac{\pi}{2} w(r+\omega)\right]}{-(r+\omega)a - q + \epsilon\omega^3} \left\{ -\frac{A^2\gamma r}{3} - \frac{w^2 A^2 \gamma r^3}{3} + \frac{A^2\gamma\omega}{6} - \frac{w^2 A^2 \gamma r^2 \omega}{2} + \frac{w^2 A^2 \gamma \omega^3}{6} \right\} \delta\left\{\left[(r+\omega)a + q\right] - k\right\}. \tag{4.37}$$

This equation contains third order polynomials whose mutual cancellation explains the absence of resonances and leads to the existence of embedded solitons, in a similar way as argued in connection with Eq. (4.30).

4.5. Variational Analysis

In this section we will study the stability of the bright-soliton solutions of eq. (4.31) by using the variational technique introduced by Anderson long ago.[33] To this end we introduce a trial function of the following form

$$u(z,t) = A(z) \operatorname{sech}\left[\frac{t - V(z)}{W(z)}\right] \exp i\left[Q(z) + R(z)\,t + P(z)\,t^2\right], \tag{4.38}$$

in the Lagrangian density of the *cmKdV* equation, namely,

$$L = i\left(u_z u^* - u_z^* u\right) + i\varepsilon\left(u\, u_{ttt}^* - u^* u_{ttt}\right) + \frac{i\gamma}{2}\left[u^2 u^* u_t^* - (u^*)^2 u\, u_t\right]. \tag{4.39}$$

Then we integrate this function over the time, we obtain the averaged (effective) Lagrangian

$$\mathcal{L} = \int_{-\infty}^{\infty} L\, dt, \tag{4.40}$$

and from this functional we obtain a set of nonlinear Euler-Lagrange equations which control the evolution of the functions A, W, V, Q, R and P. We now look for the fixed points of these equations by setting

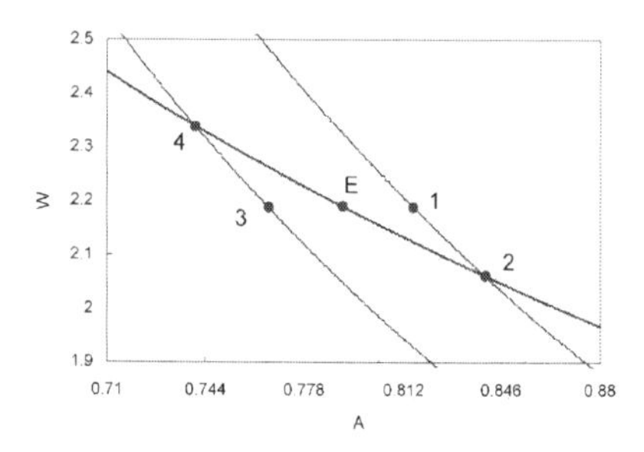

Fig. 4.2. The bold curve passing through the point E is the plot of eq. (4.41) with $\varepsilon = 1$ and $\gamma = 6$. The thin line passing through point 1 plots eq. (4.43) with $A(0) = 0.815 > A_E$ and $W(0) = W_E$. The thin line passing through point 3 is also a plot of eq. (4.43), with $A(0) = 0.765 < A_E$ and $W(0) = W_E$.

$A' = W' = R' = P' = P = 0$, $Q' = \text{const} \equiv q$, $V' = \text{const} \equiv a$. This process leads us to the following equations:

$$A^2 W^2 = \frac{18\,\varepsilon}{\gamma}, \tag{4.41}$$

$$a = 3\,\varepsilon R^2 - \frac{1}{6}\gamma A^2, \quad q = \frac{1}{2}\gamma A^2 R - \varepsilon R^3, \tag{4.42}$$

$$A^2 W = \text{const} = A^2(0)\, W(0). \tag{4.43}$$

Equation (4.41) is the variational counterpart of eq. (4.34), and the expressions for a and q coincide exactly with those in (4.34). On the other hand, eq. (4.43) applies not only to stationary solutions, but to solutions where $A(z)$ and $W(z)$ are changing with z, as it expresses the variational version of the energy conservation law.

In Fig. 2 we can see two thin curves which are two of the trajectories described by eq. (4.43), corresponding to two different initial conditions. This figure also shows the graph (the bold curve) of eq. (4.41), corresponding to $\varepsilon = 1$ and $\gamma = 6$. As we shall see in the following, this diagram permits us to understand how an initial pulse which is not an exact soliton of eq. (4.31) will evolve.

Let us consider eq. (4.31) in the particular case when $\varepsilon = 1$ and $\gamma = 6$, and let us fix our attention on a soliton solution of the form (4.33) with the following parameters:

$$A_E = \sqrt{\frac{5}{8}},\ w_E = \sqrt{\frac{8}{5}},\ r_E = 1/\sqrt{24},\ a_E = -1/2,\ q_E = \frac{11}{6\sqrt{24}}. \tag{4.44}$$

These values satisfy the conditions (4.34), as they correspond to an exact bright soliton of the *cmKdV* equation. Since $\varepsilon a_E < 0$, this soliton is a single-embedded one (*i.e.*, it is embedded solely according to its wavenumber). The amplitude A_E and width w_E of this soliton are the coordinates of the point **E** on the bold curve shown in Fig. 2. Now let us perturb this soliton by taking an initial amplitude A_1 which is slightly greater than the exact amplitude A_E, $A_1 = 0.815 > 0.790 = A_E$. This initial condition corresponds to the point **1** in Fig. 2. According to eq. (4.43), this perturbed pulse must evolve sliding along the thin curve which passes through point **1.** This thin curve intersects the equilibrium bold curve at point **2**, which is a fixed point. Therefore, we expect that the point which moves on the thin line will eventually end at the fixed point **2**. We can see that the amplitude corresponding to point **2** is greater than the amplitude which corresponds to point **1**, and therefore it seems that if we perturb an exact soliton by increasing its amplitude, the soliton will evolve to a new stationary configuration with an even greater amplitude.

In a similar way, if we start with the initial condition corresponding to point **3**, the soliton will slide along the thin line until it gets stuck at the stable fixed point **4.** Therefore, if we perturb the exact soliton by decreasing its amplitude, the soliton will evolve to a new stationary configuration with an even smaller amplitude.

These predictions, suggested by the variational approximation, were confirmed in ref.,[19] where the evolution of two perturbed solitons whose initial amplitudes were different from the exact amplitude $\mathbf{A}_E$ was calculated numerically. In Fig. 3 we can see the results of the numerical test.

The behavior shown in Fig. 3 implies that the soliton solutions of the *cmKdV* equation are indeed stable.

4.6. Conclusions

In summary, we collect the following conclusions. The main objective of this work has been to review the basic features of the electro-hydrodynamic description of nematic liquid crystals. Since this theory is described in detail in the quoted literature, only the basic principles were outlined, and only some results which have stem out of this development were described. In

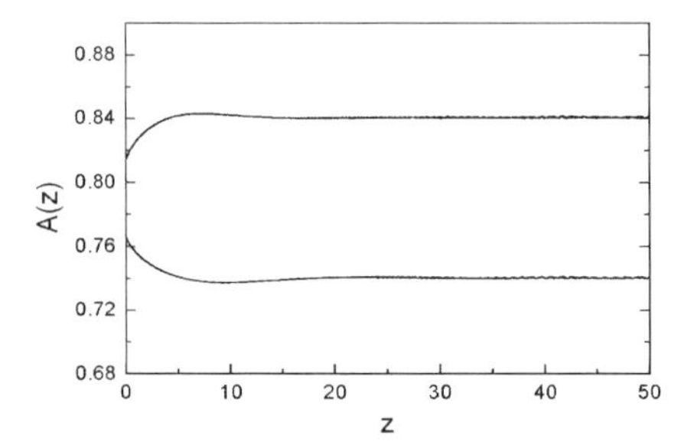

Fig. 4.3. Evolution of the amplitude of two perturbed single-embedded solitons of eq. (4.31) (with $\varepsilon = 1$ and $\gamma = 6$). The upper curve corresponds to an initial condition of the form (4.33) with $A_0 = 0.815 > A_s$, $w_0 = w_s$ and $r_0 = r_s$, where A_s, w_s and r_s are the values (4.44). The lower curve corresponds to a similar initial condition with $A_0 = 0.765 < A_s$.

particular, we have discussed why a new kind of nonlinear solitary waves (*embedded solitons, ES*) is expected to exist in nematics. The essential idea is that in these complex fluids, the nonlinear dependence of the refractive index on light intensity leads to partial differential equations with several nonlinear terms, which may balance the effects of the dispersive terms. Thus, the existence of ES is the result of the interplay between nonlinearity and dispersion. These solitary waves are sub-picosecond pulses, and are at least one order of magnitude shorter than the standard NLS solutions, which were known to exist in liquid crystals since the seventies.

The field is rather new, but it has grown faster than expected. The discovery of the ES has shown that the destructive resonances that frequently hinder the propagation of solitary waves in nonlinear systems, can be cancelled out (in some cases) by higher-order nonlinear terms. This idea suggests that new embedded solitons might be found in highly non-linear systems not yet studied. However, the experimental observation of these solitary waves, as well as their physical implications in liquid crystals and other complex systems, remains to be assessed.

It should be pointed out that the explicit consideration of hydrodynamic flows beyond the Kerr-medium approximation, was not dealt with here. The inclusion of the flow is unavoidable due to the fluid's nature of the system and will inevitably couple to the reorientation dynamics of the liquid crystal. Obviously, its inclusion will substantially complicate the problem. Also, the rigorous mathematical analysis of the systems with embedded solitons is an important part of this field,[34] but such mathematical-oriented aspects of the problem lie beyond the scope of this review.

Acknowledgements

We thank the *Departamento de Supercómputo* of DGSCA-UNAM (Dirección General de Servicios de Cómputo Académico de la Universidad Nacional Autónoma de México) for granting us access to their computers.

References

1. L. M. Blinov and V. G. Chigrinov, *Electro-optic Effects in Liquid Crystal Materials* (Springer-Verlag, New York, 1994)
2. W. A. Crossland and T. D. Wilkinson in *Handbook of Liquid Crystals*, D. Demus, J. Goodby, G. W. Gray, H.-.W. Spiess and V. Hill, editors (Wiley-VCH, Weinham, 1998) Vol. I, Chapter IX
3. I. C. Khoo, Prog. Optics, 26 (1988) 108
4. I. Jánnosy, *Optical Effects in Liquid Crystals* (Kulwer, Dordrecht, 1991)
5. R. F. Rodríguez and H. Híjar, Eur. Phys. J. B, 50 (2006) 105
6. R. F. Rodríguez and J. F. Camacho, in *Recent Developments in Mathematical and Experimental Physics, Vol. B Statistical Physics and Beyond*, A Macias, E. Díaz and F. Uribe, editors (Kluwer, New York, 2002) pp. 209-224
7. H. Pleiner and H. R. Brand in *Pattern Formation in Liquid Crystals*, A. Buka and L. Kramers, editors (Springer, Berlin, 1996)
8. M. Liu, Phys. Rev. E, 50 (1994) 2925
9. P.G. de Gennes and J. Prost, *The Physics of Liquid Crystals* (Oxford University Press, Oxford, 1993)
10. S. Chandrasekhar, *Liquid Crystals* (Cambridge University Press, Cambridge, 1992) 2nd. edition
11. J. F. Camacho, H. Híjar and R. F. Rodríguez, Physica A 348 (2005) 252
12. J. A. Reyes and R. F. Rodríguez in *Focus on Soliton Research*, L. V. Chen , editor, (Nova Science Publishers, New York, 2006) pp. 79-136
13. L. Lam and J. Prost, editors, *Solitons in Liquid Crystals* (Springer-Verlag, New York, 1992)
14. A.C. Newell, *Solitons in Mathematics and Physics* (SIAM, Philadelphia1, 985)
15. J. Fujioka and A. Espinosa-Cerón, J. Phys. Soc. Jpn. 66 (1997) 2601
16. Yang J., Malomed B. A., Kaup D. J., and Champneys A. R., Mathematics and Computers in Simulation 56 (2001) 585
17. A. R. Champneys, B. A. Malomed and M. J. Friedman, Phys. Rev. Lett. 80 (1998) 4169
18. J. Yang, Stud. Appl. Math. 106(2001) 337
19. R. F., Rodríguez, J.A. Reyes, A. Espinosa-Cerón, J. Fujioka and B. A. Malomed, Phys. Rev. E, 68 (2003) 036606

20. R. F., Rodríguez, J. Fujioka, A. Espinosa-Cerón and S. González-Pérez-Sandi, in *Recent Developments in Physics Research* (Transworld Research Network, Kerala, 2009) pp. 145-210, and references therein
21. S. González-Pérez-Sandi, J. Fujioka and B. A. Malomed, Physica D, 197(2004) 86
22. B. A. Malomed, J. Fujioka, A. Espinosa-Cerón, R. F.Rodríguez and S. González-Pérez-Sandi, Chaos, 16(2006) 013112
23. J. Yang, Phys. Rev. Lett. 91(2003) 143903
24. D. E. Pelinovsky and J. Yang, Chaos 15(2005) 037115
25. R. F. Rodríguez and J. A. Reyes, J. Mol. Liq. 71(1997) 115
26. R. F. Rodríguez and J. A. Reyes Opt. Commun. 197(2001) 103
27. J. A. Reyes and R. F. Rodríguez, J. Nolin. Opt. Phys. Mat. 6(1997) 81
28. A. Degasperis, S. V. Manakov and P. M. Santini, Physica D 100(1997) 187
29. J. Yang, B. A. Malomed and D. J. Kaup, Phys. Rev. Lett. 83(1999) 1958
30. A. Espinosa-Cerón, J. Fujioka and A. Gómez-Rodríguez, Physica Scripta 67(2003) 314
31. A. Espinosa-Cerón, J. Fujioka and A. Gómez-Rodríguez, Rev. Mex. Fis. 49(2003) 493
32. A. Espinosa-Cerón, J. Fujioka, B. A. Malomed and R.F. Rodríguez, Physica D 236(2007) 141
33. D. Anderson, Phys. Rev. A 27 (1983) 3135
34. D. E. Pelinovsky and J. Yang, Proc. R. Soc. Lond. A, 458 (2003) 1

Chapter 5

Hall Fluctuation Relations for a Charged Brownian Particle

J.I. Jiménez-Aquino, R.M. Velasco and F.J. Uribe

Department of Physics, Universidad Autónoma Metropolitana, Iztapalapa, Distrito Federal 09340, México

ines@xanum.uam.mx, rmvb@xanum.uam.mx, paco@xanum.uam.mx

The solution of the Fokker-Planck-Kramers equation for a charged Brownian particle in an electromagnetic field, is taken as a starting point to establish some fluctuation relations for the configuration and velocity space probability densities. These fluctuation relations are related with the work done by the external fields along the forward and backward trajectories. The electric field is in general time dependent and the magnetic field a constant. As a result we have found two kinds of fluctuation relations, both in configuration and velocity spaces. The first one is called the barotropic fluctuation relation and the second, and more interesting, measures how the magnetic and the electric fields couple to give place to the Hall fluctuation relations.

Contents

5.1. Introduction

It is well known that classical thermodynamics describes the macroscopic properties of a great variety of systems in their equilibrium state, the transfer of heat and the transformation of work into heat are only two particular examples among others. Linear irreversible thermodynamics is an extension of such concepts to systems that are close to equilibrium, its results and applications being in the frame of the local equilibrium hypothesis.[1–5] These traditional concepts are limited in their scope to large systems, in such a way that they are valid in the thermodynamic limit. It is important to emphasize that in usual thermodynamics as well as in statistical physics for equilibrium states, the temporal behavior does not play a role. Processes described by equilibrium physical schemes correspond to cuasi-static ones, taking an infinite time interval to be achieved. In contrast, modern technology has undergone the way to miniaturization, a fact which poses challenges going from the devices invention to the fundamental concepts in thermodynamics. However, as noted in the literature[7] the miniaturization has faced the fact that small machines are not rescaled versions of large ones. The work performed by a small machine in a cycle may be of the same order of magnitude of the thermal energy by degree of freedom consequently, the probability to convert thermal energy into useful work without any other change can not be neglected. Though such consideration may seems to be in contradiction with the contents of the second law of thermodynamics, it should be noted that very small machines operating in short time intervals are out of the thermodynamic limit.[31] These concepts have received a great interest in the study of nonequilibrium statistical mechanics of small systems performing transformations in a short time scale. Up to date, all these developments have lead to the discovery of several fluctuation theorems[6–43] that have given a new insight in our understanding and use of thermodynamics. Basically, the fluctuation theorems make explicit how macroscopic irreversibility appears in systems that in general obey time-reversible microscopic dynamics. Their main application touches the thermodynamic concepts for nanosized systems that are of interest to biologist, physicist, and engineers. The fluctuation theorems (FT) has been applied to a wide class of systems as well as several nonequilibrium quantities, including work,[20] heat,[21] entropy production,[6,31] the Jarzynski equality[9] and have generated some controversies about its universal validity.[14,16] They can also be applied to steady states[20,25] and transient situations,[20,23,24] and have been corroborated by both experiments[31–39] and computer simulations.[23,27,28] The work fluctuation theorems hold in a general way for Gaussian white noise or Gaussian colored noise.[11] However, it has been proved, for other type of noise such as the Lévy noise[12] and Poissonian shot noise,[13] that the stationary state work fluctuation theorems do not hold. Very recently Roy and Kumar,[17] established two fluctuation relations for a system of noninteracting electrons embedded in a thermal bath, under the influence of crossed electric and magnetic fields in the high friction limit.

One relation has been named as the longitudinal or barotropic-type fluctuation relation, while the other is named the transversal or Hall fluctuation relation. The Roy and Kumar's study[17] has been extended and generalized for a classical noninteracting harmonic oscillators in the presence of an electromagnetic field, also in a high friction limiting case.[43] Our aim in this work is to show that it is also possible to establish other fluctuation relations for a system of noninteracting charged Brownian particles in an electromagnetic field, through the solution of the Fokker-Planck-Kramers (FPK) equation associated with phase-space transition probability density, that is, beyond the high friction limit. The magnetic field is assumed to be constant and pointing along the z-axis and the electric field as uniform in space but in general a time-dependent vector. Under these conditions, the three-dimensional stochastic process described by the Langevin equation associated with a charged Brownian particle can be decomposed into two independent process: one takes place on the x-y plane perpendicular to the magnetic field and the other one, along the z-axis parallel to this field. The integration of the phase-space transition probability density over the velocity and configuration spaces, allows us to establish the respective fluctuation relations for the configuration-space probability density $f(\mathbf{x}, t)$, and for the velocity-space probability density $f(\mathbf{u}, t)$, where $\mathbf{x} \equiv (x, y)$ and $\mathbf{u} \equiv (u_x, u_y)$ are respectively the position and velocity vectors on the x-y plane. The fluctuation relations for the marginal probability densities $f(x, t)$, $f(y, t)$, $f(u_x, t)$, and $f(u_y, t)$ are also established through the respective marginal integrations of the functions $f(\mathbf{x}, t)$ and $f(\mathbf{u}, t)$. All the probability densities satisfy a fluctuation relation, which in a schematic way, can be written as $f(\Gamma, t)/f(-\Gamma, t) = \exp(\beta W(t))$, where $f(\Gamma, t)$ accounts for the probability density in the configuration-space or in the velocity-space during the forward process $(0, t)$ and, $f(-\Gamma, t)$ is the probability density for its backward counterpart $(t, 0)$. As stated, the ratio of these probabilities is related with the work done by the external field in the presence of the dissipative effects caused by the friction with the surrounding medium. As a result we have found that there are two kinds of contributions, one of them comes from the work done by the electric field and, the other has its origin in the presence of the magnetic field. This last effect can be seen as the work done by the torque associated with the electric field or alternatively as the work produced by the torque associated with the magnetic field. An effect which can be interpreted as a coupling between the electric and magnetic fields and, has the characteristics of the Hall effect. Several particular cases are analyzed to make explicit the corresponding fluctuation relations.

Our work is structured in the following way, in Sec. II, we show that the three dimensional Langevin equation written in the phase-space can be decomposed into two independent process. One process is described in the phase-space $(\mathbf{x}, \mathbf{u})$ perpendicular to the magnetic field and the other process occurs parallel to this field. We will concentrate on the process in x-y plane, described by its transition probability density (TPD) $P(\mathbf{x}, \mathbf{u}, t|\mathbf{x}_0, \mathbf{u}_0)$, which is the solution of the corresponding FPK equation. In Sec. III, we establish all the configuration-space fluctuation relations for the constant and time-dependent electric fields. The velocity-space fluctuation relations for constant and time-dependent electric field are established in Sec. IV. We finally end with some concluding remarks in Sec. V.

5.2. The Langevin and Fokker-Planck-Kramers Equations

The Langevin equation for a charged particle of charge q and mass m, embedded in a bath in the presence of an electromagnetic field can be written in the phase-space $(\mathbf{r}, \mathbf{v})$ as

$$\frac{d\mathbf{r}}{dt} = \mathbf{v}\,, \tag{5.1}$$

$$\frac{d\mathbf{v}}{dt} = -\gamma\mathbf{v} + \frac{q}{mc}\mathbf{v} \times \mathbf{B} + \frac{q}{m}\mathbf{E}(t) + \mathbf{A}(t)\,. \tag{5.2}$$

The interaction between the particle and the electromagnetic field is given through the Lorentz force, γ is the friction coefficient per unit mass, and $\mathbf{A}(t)$ is a Gaussian white noise with zero mean value and correlation function

$$\langle A_i(t)A_j(t')\rangle = 2\lambda\,\delta_{ij}\,\delta(t-t')\,, \tag{5.3}$$

λ being the noise intensity given by $\lambda = \gamma k_B T/m$, with k_B as the Boltzmann constant and T as the temperature of the bath. We consider a constant magnetic field pointing along the z-axis and the electric field as a homogeneous but in general a time-dependent vector. In this case Eqs. (5.1) and (5.2) can separately be written as two independent processes. First we have the usual Brownian motion along the z-axis, where the particle is affected by an electric field and it is completely decoupled from the dynamics in the x-y plane. Secondly, the motion in the x-y plane, perpendicular to the magnetic field is described as

$$\frac{d\mathbf{x}}{dt} = \mathbf{u}\,, \tag{5.4}$$

$$\frac{d\mathbf{u}}{dt} = -\gamma\mathbf{u} + \mathcal{W}\mathbf{u} + \mathbf{a}(t) + \mathcal{A}(t)\,, \tag{5.5}$$

where $\mathcal{W}$ is a real antisymmetric matrix given by

$$\mathcal{W} = \begin{pmatrix} 0 & \Omega \\ -\Omega & 0 \end{pmatrix}, \tag{5.6}$$

$\Omega = qB/mc$ being the Larmor's frequency. Here, $\mathbf{x} = (x, y)$ is the position vector, $\mathbf{u} = (v_x, v_y)$ the velocity vector, the vector $\mathbf{a}(t) = (a_x(t), a_y(t))$, is directly related to the electric field $a_i(t) = (q/m)E_i$, and $\mathcal{A}(t) = (A_x, A_y)$. Due to the fact that the motion in the x-y plane is independent of the dynamics in the z-axis, from now on we will consider only the motion in this plane.

The FPK equation for the transition probability $P(\mathbf{x}, \mathbf{u}, t|\mathbf{u}_0, \mathbf{x}_0)$ associated with the system of Eqs. (5.4) and (5.5) is given as follows[44–47]

$$\frac{\partial P}{\partial t} + \mathbf{u}\cdot\mathrm{grad}_{\mathbf{x}}P + \mathbf{a}\cdot\mathrm{grad}_{\mathbf{u}}P = \mathrm{div}_{\mathbf{u}}(\Lambda\mathbf{u}P) + \lambda\,\nabla^2_{\mathbf{u}}P\,. \tag{5.7}$$

The explicit solution $P(\mathbf{R}, \mathbf{S}) \equiv P(\mathbf{x}, \mathbf{u}, t|\mathbf{u}_0, \mathbf{x}_0)$ of Eq. (5.7) was obtained by Jiménez-Aquino and Romero-Bastida[44] and can be written as

$$P(\mathbf{R}, \mathbf{S}) = \frac{K_0}{4\pi^2(FG - H^2)}\exp\{-\big[F|\mathbf{S}|^2 - 2H\,\mathbf{R}\cdot\mathbf{S} \\ -2C\,H(\mathbf{S}\times\mathbf{R})_z + K_0\,G|\mathbf{R}|^2\big]/2(FG - H^2)\}\,, \tag{5.8}$$

for the initial condition $P(\mathbf{x}, \mathbf{u}, 0|\mathbf{u}_0, \mathbf{x}_0) = K_0\delta(\mathbf{x} - \mathbf{x}_0)\delta(\mathbf{u} - \mathbf{u}_0)$, $K_0 = (\gamma^2 + \Omega^2)/\gamma^2 = (1 + C^2)$ a constant and, $C = \Omega/\gamma = qB/mc\gamma$ is a dimensionless parameter. All functions appearing in Eq. (5.8) are defined as follows

$$F = \frac{\lambda}{\gamma^3}(2\gamma\,t - 3 + 4e^{-\gamma t} - e^{-2\gamma t})\,, \tag{5.9}$$

$$G = \frac{\lambda}{\gamma}(1 - e^{-2\gamma t})\,, \tag{5.10}$$

$$H = \frac{\lambda}{\gamma^2}(1 - e^{-\gamma t})^2\,, \tag{5.11}$$

$$\mathbf{R} = \mathbf{x} - \mathbf{x}_0 - \Lambda^{-1}\,(1 - e^{-\Lambda t})\mathbf{u}_0 - <\mathbf{x}>\,, \tag{5.12}$$

$$\mathbf{S} = \mathbf{u} - e^{-\Lambda t}\mathbf{u}_0 - <\mathbf{u}>\,, \tag{5.13}$$

$$< \mathbf{u} >= e^{-\Lambda t} \int_0^t e^{\Lambda\, s}\, \mathbf{a}(s)\, ds, \qquad < \mathbf{x} >= \int_0^t e^{-\Lambda\,(s-t)} < \mathbf{u}(s) > ds\,. \tag{5.14}$$

The matrix Λ reads as

$$\Lambda = \gamma \mathbf{I} - \mathcal{W} = \begin{pmatrix} \gamma & -\Omega \\ \Omega & \gamma \end{pmatrix}, \tag{5.15}$$

$\mathbf{I}$ being the unit matrix. The matrix $e^{-\Lambda t}$ can be factorized as $e^{-\Lambda t} = e^{-\gamma t}\mathcal{R}(t)$, where the rotation matrix is written as $\mathcal{R}(t) = e^{\mathcal{W}t}$. The rotation matrix, an orthogonal matrix, and the inverse of matrix Λ can both be expressed as

$$\mathcal{R}(t) = \begin{pmatrix} \cos\Omega t & \sin\Omega t \\ -\sin\Omega t & \cos\Omega t \end{pmatrix}, \qquad \Lambda^{-1} = \frac{1}{\gamma K_0}\begin{pmatrix} 1 & C \\ -C & 1 \end{pmatrix}. \tag{5.16}$$

Once the phase space probability density (TPD) is written, see Eq. (5.8), the marginal TPD in configuration space $P(\mathbf{R}) \equiv P(\mathbf{x}, t|\mathbf{x}_0, \mathbf{u}_0) = \int P(\mathbf{S}, \mathbf{R}) d\mathbf{S}$, is calculated by the integration, hence

$$P(\mathbf{x}, t|\mathbf{x}_0, \mathbf{u}_0) = \frac{K_0}{\pi F} \exp\Big\{ -\frac{K_0}{2F} |\mathbf{x} - \mathbf{x}_0 - \Lambda^{-1}(1 - e^{-\Lambda t})\, \mathbf{u}_0 - < \mathbf{x} > |^2 \Big\}. \tag{5.17}$$

On the other hand the TPD in the velocity space denoted as $P(\mathbf{S}) \equiv P(\mathbf{u}, t|\mathbf{u}_0)$, can also be obtained from Eq. (5.8), by the integration over the configuration space, giving as a result

$$P(\mathbf{u}, t|\mathbf{u}_0) = \frac{1}{2\pi G} \exp\Big\{ -\frac{|\mathbf{u} - e^{-\Lambda t}\mathbf{u}_0 - < \mathbf{u} >)|^2}{2G} \Big\}. \tag{5.18}$$

5.3. Configuration-Space Fluctuation Relations

In this section we will first formulate the fluctuation relations for the joint probability density $f(\mathbf{x}, t)$, which can be obtained by integration over the initial conditions. Such an integral is made with the TPD and the initial distribution function $f(\mathbf{x}_0, \mathbf{u}_0, 0)$, hence

$$f(\mathbf{x}, t) = \int d\mathbf{x}_0 \int d\mathbf{u}_0 f(\mathbf{x}_0, \mathbf{u}_0, 0) P(\mathbf{x}, t|\mathbf{x}_0, \mathbf{u}_0)\,. \tag{5.19}$$

The selection of the initial distribution function is a matter of choice and, in principle, we can propose the simplest expression by means of choosing $f(\mathbf{x}_0, \mathbf{u}_0, 0) = \delta(\mathbf{x}_0)\delta(\mathbf{u}_0)$. However, if we recall that the system is embedded in a thermal bath at temperature T, it will be convenient to suggest a more general initial distribution. Hence, we will assume that it can be written as a product of functions in the configuration and velocity spaces. In order to compare with results obtained in the high friction limit,[18] we will assume the position dependence as a gaussian function with a width determined by the temperature T and a delta function in the velocity $\mathbf{u}_0$, in such a way that

$$f(\mathbf{x}_0, \mathbf{u}_0, 0) = \frac{m\gamma^2 K_0}{2\pi k_B T} \exp\left(-\frac{m\gamma^2 K_0 \, |\mathbf{x}_0|^2}{2k_B T} \right) \delta(\mathbf{u}_0) \, . \tag{5.20}$$

The direct substitution of Eqs. (5.17) and (5.20) into Eq. (5.19), allows us to obtain

$$f(\mathbf{x}, t) = \frac{m\gamma^2 K_0}{2\pi k_B T(1 + F_0)} \exp\left(-\frac{m\gamma^2 K_0 \, |\mathbf{x} - \langle \mathbf{x} \rangle|^2}{2k_B T(1 + F_0)} \right), \tag{5.21}$$

where F_0 is a dimensionless time dependent function given by $F_0 = 2\gamma t - 3 + 4e^{-\gamma t} - e^{-2\gamma t}$.

The fluctuation relation for this joint probability density is constructed as a quotient of the distribution functions as given in Eq. (5.21) and the one corresponding to a position $-\mathbf{x}$, then

$$\frac{f(\mathbf{x}, t)}{f(-\mathbf{x}, t)} = \exp\left(\frac{2m\gamma^2 K_0 \, \mathbf{x} \cdot \langle \mathbf{x} \rangle}{k_B T(1 + F_0)} \right). \tag{5.22}$$

In order to understand the physical meaning contained in Eq. (5.22), let us study the behavior in the exponent in several cases. First of all, we consider a constant electric field $\mathbf{E} = (E_x, E_y)$, in this case the mean value of position is given through the deterministic solution coming from Eq. (5.14) in such a way that

$$\langle \mathbf{x} \rangle = \frac{q}{m} (\Lambda^{-1} \mathbf{E} t + e^{-\Lambda t} \Lambda^{-2} \mathbf{E} - \Lambda^{-2} \mathbf{E}) \, . \tag{5.23}$$

Now, after a direct substitution we obtain that the scalar product $\mathbb{W}(t) = \dfrac{2m\gamma^2 K_0 \mathbf{x} \cdot < \mathbf{x} >}{k_B T(1 + F_0)}$ can be expressed as follows

$$\mathbb{W}(t) = \frac{2q\mathbf{x} \cdot < \mathbf{x} >}{k_B T(1 + F_0)} = \Phi_1(t) \mathbf{x} \cdot \mathbf{E} + \Phi_2(t) |\mathbf{x} \times \mathbf{E}|, \tag{5.24}$$

where we have defined the functions $\Phi_1(t)$ and, $\Phi_2(t)$ as

$$\Phi_1(t) = \frac{2q}{k_B T(1 + F_0)} \left[\gamma t + \frac{1 - C^2}{1 + C^2} (e^{-\gamma t} \cos \Omega t - 1) - \frac{2C}{1 + C^2} e^{-\gamma t} \sin \Omega t \right], \tag{5.25}$$

$$\Phi_2(t) = \frac{2q}{k_BT(1+F_0)}\Big[C\gamma t + \frac{2C}{1+C^2}\big(e^{-\gamma t}\cos\Omega t - 1\big) + \frac{1-C^2}{1+C^2}e^{-\gamma t}\sin\Omega t\Big]. \tag{5.26}$$

These functions depend on time, the friction coefficient and the magnetic field. The time dependence has its origin in the rotation matrix which is affecting the electric field, an effect due to the presence of the magnetic field.

To understand the meaning of the fluctuation relation given in Eq. (5.22), let us notice that the exponent $\mathbb{W}(t)$ contains the scalar product $\mathbf{x}\cdot < \mathbf{x} >$. Now, we recall that the deterministic solution $< \mathbf{x} >$ defines a unique trajectory giving us the position at time t occupied by the brownian particle when it goes from $\mathbf{x}_0$ at time $t = 0$ to $< \mathbf{x} >$ at time t and, it can be said that in this case we follow the forward process through the deterministic trajectory. In the backward process the Brownian particle go from the position $(< \mathbf{x} >)$ up to $\mathbf{x}_0$ in a time interval given by $t \to 0$. In this case, the time t in the forward process becomes the initial time in the backward process, which is equivalent to take $- < \mathbf{x} >$ along the deterministic trajectory. However, in the scalar product $\mathbf{x}\cdot < \mathbf{x} >$ the backward process can be taken either with $(- < \mathbf{x} >, \ \mathbf{x})$ or $(< \mathbf{x} >, \ -\mathbf{x})$. Then, the fluctuation relation Eq. (5.22) can be interpreted as giving us the probability density to go over the backward trajectory in relation to the forward one. Besides, it is worth noticing that the probability density in the backward trajectory is related with the forward one through an exponential function. Also, the exponential function contains the contribution of the work done by the electric field and the torque of the magnetic field in the presence of the dissipative effects produced by the friction.

To complete the analysis we will study some particular limits which help to disentangle the terms in the general expression. In the small time regime, characterized by $\gamma t \ll 1$ and $\Omega t \ll 1$, the exponent $\mathbb{W}(\gamma t \ll 1, \Omega t \ll 1) \to 0$. If we recall that this regime corresponds to the ballistic region in the Brownian movement of the particle, we have that $\dfrac{f(\mathbf{x},t)}{f(-\mathbf{x},t)} \to 1$. Hence, both probability densities $f(\mathbf{x},t)$ and $f(-\mathbf{x},t)$ are equal. On the other hand, in the large time regime we have that

$$\mathbb{W}(\gamma t \gg 1) = \frac{q}{k_BT}\big(\mathbf{x}\cdot\mathbf{E} + C|\mathbf{x}\times\mathbf{E}|\big). \tag{5.27}$$

The first term in Eq. (5.27) is the dimensionless work done by a constant electric field and, the second term comes from the torque associated to it, such a torque is parallel to the magnetic field. In fact, this term can also be written as $qC|\mathbf{x} \times \mathbf{E}| = \frac{q}{mc\gamma}\mathbf{x} \cdot (\mathbf{E} \times \mathbf{B})$ which shows in an explicit way how the magnetic and electric fields couple themselves to give a contribution. In a similar way it is written as $qC|\mathbf{x} \times \mathbf{E}| = -\frac{q}{mc\gamma}\mathbf{E} \cdot (\mathbf{x} \times \mathbf{B})$, an expression which shows explicitly the torque caused by the magnetic field (this torque is the x-y plane). In this case the variance in the $\mathbf{x}$ has attained its equilibrium value.

When the magnetic field is not present $\Phi_2(t) \to 0$ and we obtain

$$\mathbb{W}(t)|_{(B=0)} \to \frac{2q\mathbf{x} \cdot \mathbf{E}}{k_B T}\left(\frac{\gamma t - 1 + e^{-\gamma t}}{1 + F_0}\right), \tag{5.28}$$

which shows the dimensionless work done by the constant electric field multiplied by a function of time. In the case of large times $\gamma t \gg 1$ we find that $\mathbb{W}(\gamma t \gg 1)|_{B=0} = \frac{q}{k_B T}\mathbf{x} \cdot \mathbf{E}$ which corresponds to the dimensionless work done by the electric field on the Brownian particle. Coming back to the joint probability density in configuration space, Eq. (5.21), we can calculate the variance of variable $\mathbf{x}$ defined as

$$\sigma_{\mathbf{x}} = \int_{-\infty}^{\infty} (\mathbf{x} - <\mathbf{x}>)^2 f(\mathbf{x}, t) d\mathbf{x} = \sigma_{\mathbf{x}} = \frac{k_B T(1 + F_0)}{m\gamma^2 K_0}, \tag{5.29}$$

it is affected by the magnetic field through the dependence of K_0 and, the time dependent function is caused by the friction presence. Now it is possible to write $\mathbb{W}(t) = \frac{2\mathbf{x} \cdot <\mathbf{x}>}{\sigma_{\mathbf{x}}}$, an expression which carries the effects of the deterministic trajectory induced by the electromagnetic field and the friction. Its average $<\mathbb{W}(t)>$ and variance $\Sigma_{\mathbb{W}}(t) = <\mathbb{W}^2(t)> - <\mathbb{W}(t)>^2$ can be defined in terms of the joint probability density for variable $\mathbf{x}$ and, after some algebraic steps we can show that

$$\Sigma_{\mathbb{W}}(t) = 2 <\mathbb{W}(t)> . \tag{5.30}$$

Also, the distribution function associated to $\mathbb{W}(t)$ is Gaussian due to the fact that $\mathbb{W}(t)$ is a linear function of variable $\mathbf{x}$, which is Gaussian.

Let us now establish the fluctuation relations for the marginal probability densities $f(x, t)$ and $f(y, t)$ to analyze some particularly interesting conditions. To obtain $f(x, t)$ we take an integration over the y coordinate of the joint probability density given in Eq. (5.21) and then calculate the quotient

$$\frac{f(x, t)}{f(-x, t)} = \exp\left(\frac{2q\gamma^2 K_0 x <x>}{k_B T(1 + F_0)}\right). \tag{5.31}$$

In a similar way we can calculate $f(y,t)$ with the integration in Eq. (5.21) over the x coordinate and we obtain

$$\frac{f(y,t)}{f(-y,t)} = \exp\left(\frac{2q\gamma^2 K_0 y < y >}{k_B T(1+F_0)}\right). \tag{5.32}$$

Both expressions become interesting when we consider the particle as an electron $q = -e$, the electric field as a constant pointing along the x-axis, $E_x = E$, $E_y = 0$ and the large time limit $\gamma t \gg 1$. In this case $C = -eB/mc\gamma$ and Eq. (5.31) reduces to

$$\frac{f(x,t)}{f(-x,t)} = \exp\left(-\frac{eEx}{k_B T}\right), \tag{5.33}$$

which is exactly the longitudinal or barotropic-type fluctuation relation established by Roy and Kumar,[17] for a two-dimensional system of noninteracting electrons under the action of crossed electromagnetic fields in a high friction limit. On the other hand, Eq. (5.32) allows us to write

$$\frac{f(y,t)}{f(-y,t)} = \exp\left(-\frac{e^2 EBy}{k_B T mc\gamma}\right), \tag{5.34}$$

corresponding to the transversal or Hall-fluctuation relation obtained in the same reference.[17] Both fluctuation relations, Eqs. (5.33,5.34) were also obtained for a two-dimensional system of noninteracting harmonic oscillators in an electromagnetic field in a high friction limit.[43] It is clear that in the small time regime, both marginal fluctuation relations satisfy $f(x,t) = f(-x,t)$ and $f(y,t) = f(-y,t)$ as expected.

As a second case we will take an electric field which depends explicitly on time $\mathbf{E}(t) = (\mathcal{E}_x, \mathcal{E}_y) = (\mathcal{E}, \mathcal{E})\phi(t)$, where $\mathcal{E}$ is a constant and $\phi(t)$ a dimensionless function of time. Now the exponent in Eq. (5.22) can be written as

$$\mathbb{W}(t) = \frac{2\gamma^2 K_0\, \mathbf{x}\cdot\langle\mathbf{x}\rangle}{k_B T(1+F_0)} = \frac{2q\mathcal{E}K_0}{k_B T(1+F_0)}\left[xI_x(\gamma t) + yI_y(\gamma t)\right], \tag{5.35}$$

where $I_x(\gamma t)$ and $I_y(\gamma t)$ are the x and y components of the vectorial function

$$\mathbf{I}(\gamma t) = \frac{1}{\gamma^2}\mathbb{I}\int_0^{\gamma t} e^{-(\Lambda/\gamma)\xi'}\, d\xi' \int_0^{\xi'} e^{(\Lambda/\gamma)\xi}\phi(\xi)\, d\xi\,, \tag{5.36}$$

and $\mathbb{I}$ is a vector with components $(1,1)$. Notice that the integrals depend on the specific functionality contained in the function $\phi(t)$. Also, they have a dependence with the magnetic field which appears the in rotation matrix contained in Λ. The explicit expression can be calculated once the function $\phi(t)$ is specified. Obviously they reduce to the given results when the electric field becomes constant.

5.4. Velocity Space Fluctuation Relations

Let us concentrate on the behavior in the velocity space, then we need the joint probability density $f(\mathbf{u}, t)$, which can be calculated from the TPD given in Eq. (5.18) and the initial conditions $f(\mathbf{u}_0, 0)$,

$$f(\mathbf{u}, t) = \int d\mathbf{u}_0 f(\mathbf{u}_0, 0) P(\mathbf{u}, t|\mathbf{x}_0). \tag{5.37}$$

We will assume that the initial conditions are described by a Maxwellian distribution function with a width proportional to the bath temperature, so

$$f(\mathbf{u}_0, 0) = \frac{m}{2\pi k_B T} \exp\left(-\frac{m|\mathbf{u}_0|^2}{2k_B T}\right). \tag{5.38}$$

The direct susbtitution of Eqs. (5.18), (5.37), (5.38) drives directly to

$$f(\mathbf{u}, t) = \frac{m}{2\pi k_B T} \exp\left(-\frac{m|\mathbf{u} - <\mathbf{u}>|^2}{2k_B T}\right), \tag{5.39}$$

which tells us that the probability density to find the Brownian particle with velocity $\mathbf{u}$ at time t, follows the deterministic solution $<\mathbf{u}>$ as it should be. In this case the fluctuation relation is given by

$$\frac{f(\mathbf{u}, t)}{f(-\mathbf{u}, t)} = \exp\left(\frac{2m\mathbf{u}\cdot <\mathbf{u}>}{k_B T}\right). \tag{5.40}$$

In order to begin the analysis in Eq. (5.40) we will consider a constant electric field for which $\mathbf{a} = (q/m)\mathbf{E}$. The deterministic solution for the velocity according to Eq. (5.14) is

$$<\mathbf{u}> = \frac{q}{m}(\Lambda^{-1}\mathbf{E} - e^{-\Lambda t}\Lambda^{-1}\mathbf{E}). \tag{5.41}$$

Now the exponent in Eq. (5.40) is defined as

$$\mathbb{W}_{\mathbf{u}}(t) = \frac{2m\mathbf{u}\cdot <\mathbf{u}>}{k_B T} = \frac{2q}{k_B T}\mathbf{u}\cdot(\Lambda^{-1}\mathbf{E} - e^{-\Lambda t}\Lambda^{-1}\mathbf{E}), \tag{5.42}$$

which after direct substitution of matrix Λ, can be written as

$$\mathbb{W}_{\mathbf{u}}(t) = \frac{2q}{k_B T\gamma K_0}\left\{\Psi_1(t)(\mathbf{u}\cdot\mathbf{E}) + \Psi_2(t)|\mathbf{u}\times\mathbf{E}|\right\}, \tag{5.43}$$

where we have defined two functions of time which depend on the friction coefficient and the magnetic field, namely

$$\Psi_1(t) = 1 - e^{-\gamma t}(\cos \Omega t - C \sin \Omega t), \tag{5.44}$$

$$\Psi_2(t) = C - e^{-\gamma t}(\sin \Omega t + C \cos \Omega t). \tag{5.45}$$

Again, the interpretation of the fluctuation relation, Eq. (5.40), is given in terms of the forward and backward processes. The difference being that now we are working in the velocity space instead of the configuration space. The exponent in the fluctuation relation as written in Eq. (5.43) contains the power generated by the electric field $\mathbf{u}\cdot\mathbf{E}$ multiplied by the characteristic time associated to the friction γ^{-1}. Also, there appears the contribution to the power produced by the torque caused by the electric field, multiplied by the same characteristic time. Both terms have in their structure the contributions coming from the magnetic field through the Larmor frequency and, are modulated by the friction.

In the short time regime $\gamma t \ll 1$, $\Omega t \ll 1$, functions $\Psi_1(t) \to \gamma t$ and $\Psi_2(t) \to 0$, then $\mathbb{W}_{\mathbf{u}} \to \dfrac{2q\mathbf{u}\cdot\mathbf{E}t}{k_B T K_0}$. If we recall that the Brownian particle moves in the ballistic regime when we consider short times after the beginning of its motion, we can interpret this result taking into account that in such a case the displacement is essentially given as $\mathbf{x} = \mathbf{u}t$. Now, according to Eq. (5.40) we can write that

$$f(-\mathbf{u},t)|_{ballistic} = f(\mathbf{u},t)\mathrm{exp}\Big(-\frac{2q\mathbf{u}\cdot\mathbf{E}t}{k_B T K_0}\Big) = f(\mathbf{u},t)\mathrm{exp}\Big(-\frac{2q\mathbf{x}\cdot\mathbf{E}}{k_B T K_0}\Big). \tag{5.46}$$

In this case, it is the work done by the electric field the quantity in the exponent of the fluctuation relation. The friction does not appear because we are in the ballistic region. On the other hand, when we consider the large time limit $\gamma t \gg 1$, the functions $\Psi_1(\gamma t \gg 1) \to 1$, $\Psi_2(\gamma t \gg 1) \to C$ and the fluctuation relation (5.40) reduces to

$$\frac{f(\mathbf{u},t)}{f(-\mathbf{u},t)} = \exp\left(\frac{2q}{k_B T K_0 \gamma}\Big[\mathbf{E}\cdot\mathbf{u} - C|\mathbf{E}\times\mathbf{u}|\Big]\right), \tag{5.47}$$

where we can see in a direct way how the magnetic field cause an effect through the dimensionless quantity K_0. Also, there is the presence of a torque coming from the coupling between the electric and the magnetic field. A fact which can be seen by means of the rewriting of $\dfrac{2q}{k_B T K_0} C|\mathbf{E}\times\mathbf{u}|$

as $\dfrac{-2q^2}{mc\gamma K_0}\mathbf{E}\cdot(\mathbf{x}\times\mathbf{B}) = \dfrac{2q^2}{mc\gamma K_0}\mathbf{x}\cdot(\mathbf{E}\times\mathbf{B})$, both manners give us the contribution of the torque produced by the electric or the magnetic fields and the direct effect of the coupling among themselves to give place to the Hall effect.

Also, we can establish the fluctuation relations for the marginal densities $f(u_x, t)$ and $f(u_y, t)$, which are calculated by means of an integration of Eq. (5.39) over the components u_y or u_x respectively. This procedure yields to

$$\frac{f(u_x,t)}{f(-u_x,t)} = \exp\left(\frac{2mu_x < u_x >}{k_B T}\right), \tag{5.48}$$

$$\frac{f(u_y,t)}{f(-u_y,t)} = \exp\left(\frac{2mu_y < u_y >}{k_B T}\right). \tag{5.49}$$

If we consider the Brownian particle as an electron $q = -e$, a constant electric field pointing in the x direction $\mathbf{E} = (E_x = E, E_y = 0)$, $C = -eB/mc\gamma$, the fluctuation relations written in Eqs. (5.48) reduce to the following

$$\frac{f(u_x,t)}{f(-u_x,t)} = \exp\left(-\frac{2eEu_x}{k_B T\gamma(1+C^2)}\right), \tag{5.50}$$

$$\frac{f(u_y,t)}{f(-u_y,t)} = \exp\left(-\frac{2e^2EBu_y}{k_B T mc\gamma(1+C^2)}\right). \tag{5.51}$$

The comparison between Eqs. (5.33), (5.50) allows us to conclude that the fluctuation relation in the configuration space, named as the barotropic-type work fluctuation relation, is associated with the work done by the electric field, whereas in velocity space is related to the power dissipated by the electric field in a characteristic time γ^{-1}. Also, if we compare Eqs. (5.34), (5.51), which are called as the Hall fluctuation relations, we can see that they are related with a power done in a characteristic time γ^{-1}. They correspond to the effect caused by both the electric and the magnetic fields. When we realize that in this case the magnetic (z-axis) and the electric field (x-axis) are perpendicular to each other, we see that the fluctuation relation represents a crossed effect of both fields. A characteristic typical of the Hall effect, that is why such equations are called as Hall fluctuation relations. On the other hand, in the configuration space we have the work and in the space velocity we have the power in the characteristic time γ^{-1}.

Lastly, it must be said that it is also possible to write the corresponding fluctuation relations in the case where the electric field is time dependent. However, its explicit time dependence is needed to extract the explicit expressions.

5.5. Concluding Remarks

In this work, we have shown that all the fluctuation relations constructed for a charged Brownian particle in an electromagnetic field, can be written as $f(\Gamma, t)/f(-\Gamma, t) = \exp(W(t)/k_BT)$.[15] The quantity Γ carries the trajectories induced by the external fields in the configuration or in the velocity space. This relation quantifies the irreversible behavior of the system, by comparing the probabilities during the forward and backward processes. The important quantity to measure in this scheme is related to the work done by the external fields, when immersed in the thermal bath. It means that the friction represented by the Stokes expression in the Langevin dynamics affects the work done by the external fields, giving place to the dissipative effects. Also, it can be said that the fluctuation relations measure how difficult is for the system to go backwards in the presence of irreversible phenomena. It is a remarkable fact the way in which the effect caused by the magnetic field manifests. Such an effect is given through the torque produced by the magnetic field, the torque is in the x-y plane, perpendicular to the magnetic field. In all cases the work done by the external field appears modified by the friction contributions so, it seems to be adequate to call such combinations as the irreversible work. Interesting particular cases are obtained when we consider the short and large time regimes. In the first case we have the ballistic behavior in which the friction does not play a role and, we have obtained that the probability densities in the forward and backward processes become equal. A characteristic interpreted either as a kind of reversibility or as a negligible friction effect. In the large time regime the fluctuation relations become stationary and the work done by the electric field plays a fundamental role.

Finally, we have been able to obtain the same fluctuation relations recently established by Roy and Kumar[17] and by us[43] by other alternative methods. Also, in the velocity-space, we have established what we have named as the velocity-space barotropic-type and velocity-space Hall-type fluctuation relations, similar to the ones in configuration space.

Acknowledgments

There are men who struggle for a day and they are good.
There are men who struggle for a year and they are better.
There are men who struggle many years, and they are better still.
But there are those who struggle all their lives:

These are the indispensable ones.

Bertolt Brecht

This work is dedicated to Leopoldo García-Colín Scherer, our professor, colleague, mostly friend and an indispensable scientist for Mexico.

References

1. R. Balescu, *E*quilibrium and Nonequilibrium Statistical Mechanics. (John Wiley and Sons, 1975).
2. H. B. Callen, and T. A. Welton, Irreversibility and generalized noise, *P*hys. Rev. **83**(1), 34–40, (1951). ISSN 0031-899X.
3. L. Onsager, Reciprocal relations in irreversible processes. I., *P*hys. Rev. **37**(4), 405–426, (1931); Reciprocal relations in irreversible processes. II. *P*hys. Rev. **38**(12), 2265–2279, (1931). ISSN: 0031-899X
4. L. Onsager and S. Machlup, Fluctuations and irreversible processes, *P*hys. Rev. **91**(6), 1505–1512, (1953). ISSN: 0031-899X.
5. S. de Groot, and P. Mazur, *N*onequilibrium Thermodynamics. (Dover, New York, 1984).
6. A. Saha, S. Lahiri, and A. M. Jayannavar, Entropy production theorems and some consequences, *P*hys. Rev. E **80**(1), 011117 (2009). ISSN: 1539-3755.
7. E.M. Sevick, R. Prabhakar, S.R. Williams, and D.J. Searles, Fluctuation theorems, *A*nnu. Rev. Phys. Chem. **59**, 603–633 (2008). ISSN: 0066-426X.
8. S. Joubaud, N.B. Garnier, and S. Ciliberto, Fluctuation theorems for harmonic oscillators, *J*. Stat. Mech. P09018 (2007). ISSN: 1742-5468.
9. C. Jarzynski, Nonequilibrium Fluctuatons of a Single Biomolecule, *L*ect. Notes Phys. **711**, 201 (2007); Nonequilibrium equality for free energy differences, *P*hys. Rev. Lett. **78**(14), 2690–2693 (1997). ISSN: 0031-9007.; Equilibrium free-energy differences from nonequilibrium measurements: A master-equation approach, *P*hys. Rev. E **56**(5), 5018–5035 (1997). ISSN: 1063-651X.
10. J. Horowitz and C. Jarzynski, Comparison of work fluctuation relations, *J*. Stat. Mech., P11002, (2007). ISSN: 1742-5468.
11. T. Mai and A. Dhar, Nonequilibrium work fluctuations for oscillators in non-Markovian baths, *P*hys. Rev. E **75**(6), 061101, (2007). ISSN: 1539-3755.
12. H. Touchette and E.G.D. Cohen, Fluctuation relation for a Levy particle, *P*hys. Rev. E **76**(2), 020101(R), (2007). ISSN: 1539-3755.
13. A. Baule and E.G.D. Cohen, Fluctuation properties of an effective nonlinear system subject to Poisson noise, *P*hys. Rev. E **79**(13), 030103(R), (2009). ISSN: 1539-3755.
14. J.M.G. Vilar and J.M. Rubi, Failure of the work-Hamiltonian connection for free-energy calculations, *P*hys. Rev. Lett. **100**(2), 020601 (2008); Comment on "Failure of the Work-Hamiltonian Connection for Free-Energy Calculations" - Reply, *P*hys. Rev. Lett. **101**(9), 098902 (2008).; Comment on

"Failure of the Work-Hamiltonian Connection for Free-Energy Calculations" - Reply, *Phys.* Rev. Lett. **101**(9), 098904, (2008). ISSN: 0031-9007.
15. J. Horowitz and C. Jarzynski, Comment on "Failure of the Work-Hamiltonian Connection for Free-Energy Calculations", *Phys.* Rev. Lett. **101**(9), 098901 (2008). ISSN: 0031-9007.
16. L. Peliti, Comment on "Failure of the Work-Hamiltonian Connection for Free-Energy Calculations", *Phys.* Rev. Lett. **101**(9), 098903 (2008). ISSN: 0031-9007.
17. D. Roy and N. Kumar, Langevin dynamics in crossed magnetic and electric fields: Hall and diamagnetic fluctuations, *Phys.* Rev. E **78**(5), 052102 (2008). ISSN: 1539-3755.
18. J. I. Jiménez-Aquino, R. M. Velasco, and F. J. Uribe, submitted
19. C. Bustamante, J. Liphardt, and F. Ritort, The nonequilibrium thermodynamics of small systems, *Phys.* Today **58** (7), 43–48, (2005). ISSN: 0031-9228.
20. R. van Zon and E.G.D. Cohen, Stationary and transient work-fluctuation theorems for a dragged Brownian particle, *Phys.* Rev. E **67**(4), 046102, (2003). ISSN: 1063-651X.
21. R. van Zon and E.G.D. Cohen, Extended heat-fluctuation theorems for a system with deterministic and stochastic forces, *Phys.* Rev. E **69**(5), 056121 (2004). ISSN: 1063-651X.
22. U. Seifert, Entropy production along a stochastic trajectory and an integral fluctuation theorem, *Phys.* Rev. Lett. **95**(4), 040602 (2005). ISSN: 0031-9007.
23. D.J. Evans and D.J. Searles, Equilibrium microstates which generate 2nd law violating steady states, *Phys.* Rev. E **50**(2), 1645–1648 (1994). ISSN: 1063-651X.
24. G.E. Crooks, Entropy production fluctuation theorem and the nonequilibrium work relation for free energy differences, *Phys.* Rev. E **60**(3), 2721–2726, (1999); Path-ensemble averages in systems driven far from equilibrium, *Phys.* Rev. E **61**(3), 2361–2366 (2000). ISSN: 1063-651X. Nonequilibrium measurements of free energy differences for microscopically reversible Markovian systems, *J.* Sthat. Phys. **90**(5–6), 1481–1487 (1998). ISSN: 0022-4715.
25. G. Gallavoti and E.G.D. Cohen, Dynamical ensembles in nonequilibrium statistical mechanics, *Phys.* Rev. Lett. **74**(14), 2694–2697 (1995). ISSN: 0031-9007.; Dynamical ensembles in stationary states, *J.* Stat. Phys. **80**(5–6), 931–970 (1995). ISSN: 0022-4715.
26. S. Lepri, L. Rondoni, and G. Benetitin, The Gallavotti-Cohen fluctuation theorem for a nonchaotic model, *J.* Stat. Phys. **99**(3–4), 857–872 (2000). ISSN: 0022-4715.
27. D.J. Searles and D.J. Evans, Ensemble dependence of the transient fluctuation theorem, *J.* Chem. Phys. **113**(9), 3503–3509 (2000). ISSN: 0021-9606.
28. D.J. Evans, E.G.D. Cohen, and G.P. Morris, Probability of 2nd law violations in shearing steady–states, *Phys.* Rev. Lett. **71**(15), 2401–2404 (1993). ISSN: 0031-9007.
29. T. Hatano, Jarzynski equality for the transitions between nonequilibrium states, *Phys.* Rev. E **60**, R5017–R5020 (1999). ISSN: 1063-651X.

30. T. Hatano and S. I. Sasa, Steady-state thermodynamics of Langevin systems, *Phys. Rev. Lett.* **86**, 3463-3466 (2001). ISSN: 0031-9007.
31. G.M. Wang, E.M. Sevick, E. Mittag, D.J. Searles, and D.J. Evans, Experimental demonstration of violations of the second law of thermodynamics for small systems and short time scales, *Phys. Rev. Lett.* **89**(5), 050601 (2002). ISSN: 0031-9007.
32. V. Blickle, T. Speck, L. Helden, U. Seifert, C. Bechinger, Thermodynamics of a colloidal particle in a time-dependent nonharmonic potential, *Phys. Rev. Lett.* **96**(7), 070603 (2006). ISSN: 0031-9007.
33. W. Lechner, H. Oberhofer, C. Dellago, and P.L. Geissier, Equilibrium free energies from fast-switching trajectories with large time steps, *J. Chem. Phys.* **124**(4), 044113 (2006). ISSN: 0021-9606.
34. J. Liphardt, S. Dumont, S. B. Smith, I. Tinoco, and C. Bustamente, Equilibrium information from nonequilibrium measurements in an experimental test of Jarzynski's equality, *Science* **296**(5574), 1832–1835 (2002). ISSN: 0036-8075.
35. O. Narayan and A. Dhar, Reexamination of experimental tests of the fluctuation theorem, *J. Phys. A* **37**(1), 63–76 (2004). ISSN: 0305-4470.
36. F. Douarche, S. Ciliberto, A. Patrosyan, and I. Rabbiosi, An experimental test of the Jarzynski equality in a mechanical experiment, *Europhys. Lett.* **70**(5), 593–599 (2005). ISSN: 0295-5075.
37. D. Collin, F. Ritort, C. Jarzynski, S. B. Smith, I. Tinoco, Jr., and C. Bustamante, *V*erification of the Crooks fluctuation theorem and recovery of RNA folding free energies, Nature (London) **437**(7056), 231–234 (2005). ISSN: 0028-0836.
38. C. Tietz, S. Schuler, T. Speck, U. Seifert, and J. Wrachtrup, Measurement of stochastic entropy production, *Phys. Rev. Lett.* **97**(5), 050602 (2006). ISSN: 0031-9007.
39. E. H. Trepagnier, C. Jarzynski, F. Ritort, G.E. Crooks, C.J. Bustamante, and J. Liphardt, Experimental test of Hatano and Sasa's nonequilibrium steady-state equality, *Proc. Natl. Acad. Sciences* **101**(42), 15038–15041, (2004). ISSN: 0027-8424.
40. A.M. Jaynnavar and M. Sahoo, Charged particle in a magnetic field: Jarzynski equality, *Phys. Rev. E* **75**(3), 032102 (2007). ISSN: 1539-3755.
41. A. Saha and A. M. Jaynnavar, Nonequilibrium work distributions for a trapped Brownian particle in a time-dependent magnetic field, *Phys. Rev. E* **77**(2), 022105 (2008). ISSN: 1539-3755.
42. J. I. Jiménez-Aquino, R. M. Velasco, and F. J. Uribe, Dragging of an electrically charged particle in a magnetic field, *Phys. Rev. E* **78**(3), 032102 (2008). ISSN: 1539-3755.
43. J.I. Jiménez-Aquino, R. M. Velasco, and F. J. Uribe, Fluctuation relations for a classical harmonic oscillator in an electromagnetic field, *Phys. Rev. E* **79**(6), 061109 (2009). ISSN: 1539-3755.
44. J.I. Jiménez-Aquino and M. Romero-Bastida, Fokker-Planck-Kramers equations of a heavy ion in presence of external fields, *Phys. Rev. E* **76**(2), 021106 (2007). ISSN: 1539-3755.

45. S. Chandrasekhar, Stochastic problems in physics and astronomy, *R*ev. Mod. Phys. **15**91), 0001–0089, (1943). ISSN: 0034-6861.
46. N. van Kampen, *S*tochastic Processes in Physics and Chemistry. (North-Holland, Amsterdam 1992), revised and enlarged edition.
47. H. Risken, *The Fokker-Planck equation: Methods of solution and Applications* (Springer-Verlag, Berlin, 1984).

Chapter 6

Thermodynamic Properties and Model for Vapor-Liquid Azeotropic Binary Mixtures

S.M.T. de la Selva, Pablo Lonngi and Eduardo Piña

Departamento de Física
Universidad Autónoma Metropolitana - Iztapalapa,
P. O. Box 55 534
Mexico, D. F., 09340 Mexico
tere@xanum.uam.mx

A study of the binary mixture liquid-vapor azeotrope within the frame of basic equilibrium thermodynamics is presented. In the introduction we review the defining properties of the azeotrope. In a second section we demonstrate a new general property that links, in the limit of the azeotrope, the pressure as a function of the liquid composition and the pressure as a function of the vapor composition, both computed at a given value of temperature, to the composition of the vapor as a function of the liquid composition at that same temperature and, analogously for the temperature as a function of the liquid composition and of the vapor composition for a fixed value of the pressure. We also give a demonstration of the fact that whenever the azeotrope occurs as a maximum in the pressure vs composition coordinates it will be a minimum in the temperature vs composition coordinates and viceversa. In a third section we construct a polynomial model for the above mentioned coexistence pressure functions of composition of liquid, and of vapor, based on the requisites that are to be met by the pressure at the extremes of the pure components and at the azeotrope.

Dedicated to Prof. Leopoldo García-Colín Scherer.

Contents

6.1. Introduction

The liquid-vapor coexistence of a binary mixture whith an azeotrope is either treated rather briefly in most texts of general thermodynamics or physical chemistry[1–3] or, it is extensively studied in more specialised texts and treatises[4–6] or in chemical engineering applications books[7,8] were the methods presented are rather involved for the nonspecialist. At the same time the experimental data on binary azeotropic mixtures[9] and[10,11,13] can be considered abundant but by no means complete. In what follows we present a study of the azeotrope within the bounds of basic thermodynamics with new demonstrations and new adjustable parameters depending correlations accesible to the nonspecialist. To make this paper self contained we review first the essential facts that define the azeotrope and the derivation of the so called Gibbs-Konowalow laws.[4,6]

In the experimental study of the equilibrium coexistence of liquid and vapor of binary mixtures, the most commonly measured variables are the absolute temperature T, the pressure p, the molar fraction of component 1 in the liquid x, and the molar fraction of the same component in the vapor y. Both molar fractions are defined within the closed interval $[0, 1]$. The molar fractions of component 2 are then given by $1 - x$ and $1 - y$ in liquid and vapor respectively.

Gibbs' phase rule tells us that for such a system there are only two degrees of freedom. Then, out of the four intensive variables T, p, x, y that are readily measured, only two are independent, so that only two of them can be chosen to be given preassigned values.

When the results of experiments are reported at a given temperature, the other independent variable is usually chosen as either one of the molar fractions. If one chooses let us say the molar fraction in the liquid x, the independent variables are T, x; the pressure p, being a dependent variable, is obtained from measurements and reported as the function $p = p_l(x, T)$ shown in a graph of coordinates (x, p_l) at constant T. For every value of the pressure, there coexists a value of the dependent variable y in the vapor phase (in general $y \neq x$) in such a way that the same pressure value p is given by a different function $p_g(y, T)$.

In Fig. 6.1 this fact is exhibited by the horizontal tie connecting two equal pressure points, a and b and different compositions x and y. The experimental results are thus represented graphically in diagrams for a given

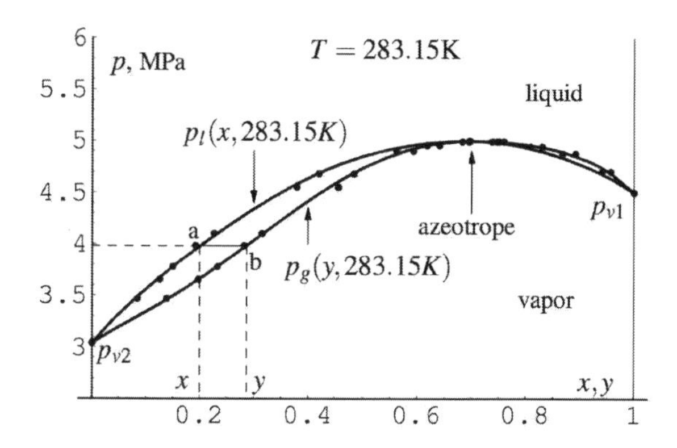

Fig. 6.1. Carbon dioxide (1) and ethane (2) at 283.15K. Experimental coexistence data from Ref.[9] and functions $p_l(x)$ and $p_g(y)$ as calculated in section 3.

value of absolute temperature T where both functions $p_l(x,T)$ for the liquid and $p_g(y,T)$ for the vapor are shown in the same diagram. An example of such a diagram is Fig. 6.1.

The two coexistence curves $p_l(x,T)$ and $p_g(y,T)$ at constant T meet at the value $x = y = 0$, the system is constituted in this extreme case only by the component 2 with the vapor pressure value P_{v2} of the pure component 2 at the stated temperature. The two curves meet also at $x = y = 1$ where the system becomes the pure component 1 with the vapor pressure P_{v1} of this component at the stated temperature. At those two extremes only one degree of freedom is left.

In many mixtures the coexistence curves $p_l(T,x)$ and $p_g(T,y)$ meet at an intermediate composition point, $x = y \equiv y_a$, called the azeotrope, with zero slope:

$$\left(\frac{\partial p}{\partial x}\right)_T = 0 \quad \text{and} \quad \left(\frac{\partial p}{\partial y}\right)_T = 0\,, \qquad (x = y \equiv y_a)\,, \tag{6.1}$$

corresponding to a maximum as illustrated in Fig. 6.1 and less frequently, in some other mixtures, to a minimum see for example Fig. 6.2.[11] The experimental results are shown as in figures Fig. 6.1 and Fig. 6.2 because then the graph represents not only how the pressure, function of x, T varies for the liquid and how it varies as a function of y, T for the vapor, but because it permits the use of the thermodynamic equilibrium requisite that

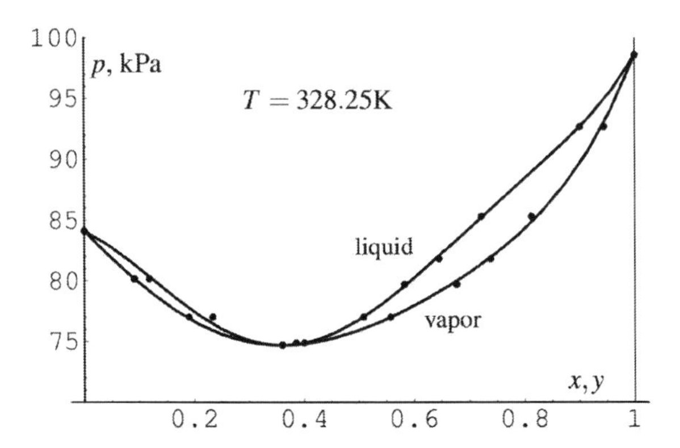

Fig. 6.2. Acetone (1) and chloroform (2) at 328.25K. Experimental data from[11] and correlations $p_l(x)$ and $p_g(y)$.

the pressure has the same value for both phases,

$$p = p_l(T, x) = p_g(T, y) \tag{6.2}$$

to obtain the values of the mole fractions x and y in coexistence at the given p and T or to establish the relation between the variables T, x, y.

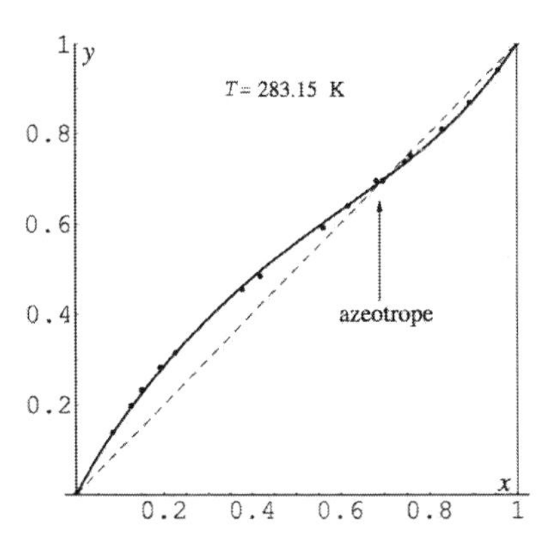

Fig. 6.3. Carbon dioxide (1) and ethane (2) at 283.15K. Experimental coexistence data from Ref.[9] and fitted function $y = y(T, x)$.

In fact the second equation in (6.2) expresses the relation among the three variables x, y, T that is generally represented graphically as $y = y(x, T)$ at $T =$ constant, see Fig. 6.3.

It is usual to find that the experimental results are given as a graph of $T_l(x,p)$ and as a graph of $T_g(y,p)$ at a given p in the same diagram. Then the roles of the variables T and p in the above discussion become interchanged, and the thermodynamic equilibrium is satisfied with the requisite,

$$T = T_l(x,p) = T_g(y,p)\,. \tag{6.3}$$

From the second equation in (6.3) we find the coexistence relation among the variables x, y, p, usually expressed and represented as $y = y(p,x)$ at constant p. Again, at the composition values $x = 0$ and $x = 1$ the curves $T_l(x,p)$ and $T_g(y,p)$ meet at a point of pure component 2 with boiling temperature T_{b2}, or pure component 1 with boiling temperature T_{b1} respectively, with only one degree of freedom left.

The occurence of an intermediate point, the azeotrope, presents itself also in the case of the curves of temperature at a given pressure, where at the same molar fraction for liquid and for vapor, $x = y \equiv y_a$, the extreme occurs with,

$$\left(\frac{\partial T}{\partial x}\right)_p = 0 \quad \text{and} \quad \left(\frac{\partial T}{\partial y}\right)_p = 0\,, \qquad (x = y \equiv y_a), \tag{6.4}$$

corresponding to a maximum or to a minimum.

The fact that the azeotrope, namely a coexistence point in which $x = y$ should occur with properties (6.1) and (6.4) is contained within the consequences of the laws of thermodynamics. For the sake of completeness we review next the steps that lead to the equations from where those properties of the azeotrope become apparent[4-6] and from where some others will be demonstrated.

Apart from the equilibrium thermodynamic coexistence requisites on presure and temperature, given by equations (6.2) and (6.3), there are two additional conditions that require the chemical potential μ_j of each component, $j = 1, 2$ to have the same value for both phases. These requisites are formalized by equating the chemical potentials of each phase, which depend on three independent intensive variables as,

$$\mu_{1,l}(T,p,x) = \mu_{1,g}(T,p,y)\,, \quad \text{and} \quad \mu_{2,l}(T,p,x) = \mu_{2,g}(T,p,y)\,. \tag{6.5}$$

It is worthwhile to emphasize that even though p and T take the same value on both sides of these equations the functions $\mu_{j,l}(T,p,x)$, $\mu_{j,g}(T,p,y)$, $(j = 1, 2)$ are different, because one side refers to the liquid and the other to the vapor at coexistence and in general $x \neq y$. The two equations in (6.5) among the four intensive variables T, p, x, y imply that at coexistence, there are only two independent intensive variables in compliance with Gibbs' phase rule. Notice that if p is solved from both equations and the resulting functions are equated one obtains again the relation among T, x, y. If instead T is solved and the resulting functions are equated one finds the relation among p, x, y.

In differential form equations in (6.5) are,

$$\left(\frac{\partial \mu_{j,l}}{\partial T}\right)_{p,x} dT + \left(\frac{\partial \mu_{j,l}}{\partial p}\right)_{T,x} dp + \left(\frac{\partial \mu_{j,l}}{\partial x}\right)_{T,p} dx$$

$$= \left(\frac{\partial \mu_{j,g}}{\partial T}\right)_{p,y} dT + \left(\frac{\partial \mu_{j,g}}{\partial p}\right)_{T,y} dp + \left(\frac{\partial \mu_{j,g}}{\partial y}\right)_{T,p} dy \quad (j = 1, 2)\,. \tag{6.6}$$

Let us now consider the thermodynamics of a single phase. From

$$dG = -SdT + Vdp + \mu_1 dn_1 + \mu_2 dn_2,$$

the following Maxwell relations are obtained,

$$\left(\frac{\partial \mu_{j,l}}{\partial T}\right)_{p,n_1,n_2} = -\left(\frac{\partial S_l}{\partial n_j}\right)_{p,T,n_{i\neq j}} \equiv -S_{j,l}\,, \quad (j = 1, 2) \tag{6.7}$$

$$\left(\frac{\partial \mu_{j,g}}{\partial T}\right)_{p,n_1,n_2} = -\left(\frac{\partial S_g}{\partial n_j}\right)_{p,T,n_{i\neq j}} \equiv -S_{j,g}\,, \quad (j = 1, 2) \tag{6.8}$$

$$\left(\frac{\partial \mu_{j,l}}{\partial p}\right)_{T,n_1,n_2} = \left(\frac{\partial V_l}{\partial n_j}\right)_{p,T,n_{i\neq j}} \equiv V_{j,l}\,, \quad (j = 1, 2) \tag{6.9}$$

$$\left(\frac{\partial \mu_{j,g}}{\partial p}\right)_{T,n_1,n_2} = \left(\frac{\partial V_g}{\partial n_j}\right)_{p,T,n_{i\neq j}} \equiv V_{j,g}\,, \quad (j = 1, 2)\,, \tag{6.10}$$

that involve the partial molar entropies $S_{j,l}$ and $S_{j,g}$ and the partial molar volumes $V_{j,l}$ and $V_{j,g}$, for liquid and for gas. Notice that in the left hand side of equations (6.7) and (6.8) there appears the same kind of derivative of the two different functions that are assumed to have equal value at coexistence in (6.5), nevertheless the two derivatives are different, *i.e.*, at coexistence the two partial molar entropies are generally different and an analogous assertion is to be made concerning the partial molar volumes thus,

$$S_{j,l} \neq S_{j,g}\,, \qquad V_{j,l} \neq V_{j,g}\,. \tag{6.11}$$

Since that the derivatives of the chemical potentials in equations (6.7)-(6.10), either with respect to the temperature or with respect to the pressure, at constant mole numbers, are derivatives of intensive variables with respect to intensive variables, they can only depend on the number of moles through their quotient. For this reason, the two extensive variables n_1, n_2 maybe replaced by the intensive variable x for the liquid and the intensive variable y for the vapor. This allows us to write equations (6.7)-(6.10) for $(j = 1, 2)$ simply as,

$$\left(\frac{\partial \mu_{j,l}}{\partial T}\right)_{p,x} = -S_{j,l}\,, \tag{6.12}$$

$$\left(\frac{\partial \mu_{j,g}}{\partial T}\right)_{p,y} = -S_{j,g}\,, \tag{6.13}$$

$$\left(\frac{\partial \mu_{j,l}}{\partial p}\right)_{T,x} = V_{j,l}\,, \tag{6.14}$$

$$\left(\frac{\partial \mu_{j,g}}{\partial p}\right)_{T,y} = V_{j,g}\,. \tag{6.15}$$

Notice that in these equations both terms represent the same function of only three independent intensive variables (T, p, x) or (T, p, y).

The substitution of these last four equations into (6.6) lead us after rearrangement to the following two equations, one for each component,

$$\left(\frac{\partial \mu_{j,l}}{\partial x}\right)_{T,p} dx - \left(\frac{\partial \mu_{j,g}}{\partial y}\right)_{T,p} dy$$
$$= -\left(S_{j,g} - S_{j,l}\right) dT + \left(V_{j,g} - V_{j,l}\right) dp \qquad j = 1, 2\,. \tag{6.16}$$

To reduce the number of intensive variables we proceed as follows. We recall that each phase satisfies a Gibbs-Duhem equation,

$$S_i dT - V_i dp + n_{1,i} d\mu_1 + n_{2,i} d\mu_2 = 0\,, \quad i = l, g\,.$$

After dividing by the total number of moles in the phase, $n_1 + n_2 = n$ and differentiating with respect to the molar fraction of the phase, at constant temperature and pressure, the following two equations that relate the derivatives of the chemical potentials of the two components in the same phase are obtained,

$$x\left(\frac{\partial \mu_{1,l}}{\partial x}\right)_{T,p} + (1 - x)\left(\frac{\partial \mu_{2,l}}{\partial x}\right)_{T,p} = 0 \tag{6.17}$$

for the liquid, and

$$y\left(\frac{\partial \mu_{1,g}}{\partial y}\right)_{T,p} + (1 - y)\left(\frac{\partial \mu_{2,g}}{\partial y}\right)_{T,p} = 0 \tag{6.18}$$

for the vapor. Let us now multiply the equation (6.16) with $j = 1$ by y and and with $j = 2$ by $(1 - y)$, and add them. The substitution of equations (6.17) and (6.18) on the resulting expression leads to,

$$\left(\frac{y-x}{1-x}\right)\left(\frac{\partial \mu_{1,l}}{\partial x}\right)_{T,p} dx = -\left[y(S_{1,g} - S_{1,l}) + (1-y)(S_{2,g} - S_{2,l})\right] dT$$

$$+\left[y(V_{1,g} - V_{1,l}) + (1-y)(V_{2,g} - V_{2,l})\right] dp\,. \tag{6.19}$$

From equation (6.19) after taking the partial derivative with respect to x, once at constant T and once at constant p, we find the following two equations that we are looking for, valid at coexistence,

$$\left(\frac{y-x}{1-x}\right)\left(\frac{\partial \mu_{1,l}}{\partial x}\right)_{T,p} = \left[y(V_{1,g} - V_{1,l}) + (1-y)(V_{2,g} - V_{2,l})\right]\left(\frac{\partial p}{\partial x}\right)_T \tag{6.20}$$

and

$$\left(\frac{y-x}{1-x}\right)\left(\frac{\partial \mu_{1,l}}{\partial x}\right)_{T,p} = -\left[y(S_{1,g} - S_{1,l}) + (1-y)(S_{2,g} - S_{2,l})\right]\left(\frac{\partial T}{\partial x}\right)_p \tag{6.21}$$

Since the factors multiplying the derivative on the right hand side of these last equations are different from zero (6.11), we find that at the azeotrope when $x = y \equiv y_a$, they become the first property in (6.1) and the first property in (6.4)

$$\left(\frac{\partial p}{\partial x}\right)_T = 0\,, \quad \text{and} \quad \left(\frac{\partial T}{\partial x}\right)_p = 0\,.$$

Of course we may as well multiply equation (6.16) with $j = 1$ by x and with $j = 2$ by $(1 - x)$, add them, apply (6.17) and (6.18) to find,

$$\left(\frac{y-x}{1-y}\right)\left(\frac{\partial \mu_{1,g}}{\partial y}\right)_{T,p} dy = -\left[x(S_{1,g} - S_{1,l}) + (1-x)(S_{2,g} - S_{2,l})\right] dT$$

$$+\left[x(V_{1,g} - V_{1,l}) + (1-x)(V_{2,g} - V_{2,l})\right] dp \tag{6.22}$$

From equation (6.22) after partial differentiation with respect to y once at constant T and once at constant p, we obtain the following two additional equations valid at coexistence,

$$\left(\frac{y-x}{1-y}\right)\left(\frac{\partial \mu_{1,g}}{\partial y}\right)_{T,p} = \left[x(V_{1,g} - V_{1,l}) + (1-x)(V_{2,g} - V_{2,l})\right]\left(\frac{\partial p}{\partial y}\right)_T \tag{6.23}$$

and

$$\left(\frac{y-x}{1-y}\right)\left(\frac{\partial \mu_{1,g}}{\partial y}\right)_{T,p} = -\left[x(S_{1,g}-S_{1,l})+(1-x)(S_{2,g}-S_{2,l})\right]\left(\frac{\partial T}{\partial y}\right)_p \tag{6.24}$$

Taking into acount again inequalities (6.11) we find at the azeotrope, when $x = y \equiv y_a$, the second property in (6.1) and the second property in (6.4),

$$\left(\frac{\partial p}{\partial y}\right)_T = 0\,, \quad \left(\frac{\partial T}{\partial y}\right)_p = 0\,.$$

Expresions (6.20), (6.21), (6.23) and (6.24) known as Gibbs-Konowalow equations are the starting basis of this work.

6.2. Additional Properties of the Azeotrope

In this section we show further properties of the azeotrope.

For the set of variables (p, T, x, y) from the rules of calculus we have that

$$\left(\frac{\partial y}{\partial x}\right)_T = \frac{\left(\frac{\partial y}{\partial p}\right)_T}{\left(\frac{\partial x}{\partial p}\right)_T} \qquad x, y \neq y_a, \tag{6.25}$$

but in the limit in which x and y both tend to the value y_a of the azeotrope, the right-hand side of expression (6.25) becomes indeterminate because both terms in the quotient become infinite. In order to calculate it we must resort to L'Hôpital's rule,

$$\lim_{x\to y_a, y\to y_a}\left(\frac{\partial y}{\partial x}\right)_T = \lim_{x\to y_a, y\to y_a}\frac{\left(\frac{\partial y}{\partial p}\right)_T}{\left(\frac{\partial x}{\partial p}\right)_T} = \lim_{x\to y_a, y\to y_a}\frac{\left(\frac{\partial^2 y}{\partial p^2}\right)_T}{\left(\frac{\partial^2 x}{\partial p^2}\right)_T}\,. \tag{6.26}$$

We transform now the derivatives in the right-hand side of this equation, whose independet variables are p, T, to the sets x, T and y, T in order to make it more useful for the correlations we propose in the next section for the functions $p_l(x, T)$ and $p_g(y, T)$:

$$\left(\frac{\partial^2 y}{\partial p^2}\right)_T = \left(\frac{\partial}{\partial p}\frac{1}{\left(\frac{\partial p}{\partial y}\right)_T}\right)_T = \left(\frac{\partial y}{\partial p}\right)_T\left(\frac{\partial}{\partial y}\frac{1}{\left(\frac{\partial p}{\partial y}\right)_T}\right)_T = -\left(\frac{\partial y}{\partial p}\right)_T^3\left(\frac{\partial^2 p}{\partial y^2}\right)_T\,. \tag{6.27}$$

Similarly, we have that,

$$\left(\frac{\partial^2 x}{\partial p^2}\right)_T = -\left(\frac{\partial x}{\partial p}\right)_T^3\left(\frac{\partial^2 p}{\partial x^2}\right)_T\,. \tag{6.28}$$

From the last expression in 6.27 and from 6.28 it follows, when they are applied to 6.26 that,

$$\lim_{x,y\to y_a} \frac{\left(\frac{\partial y}{\partial p}\right)_T}{\left(\frac{\partial x}{\partial p}\right)_T} = \lim_{x,y\to y_a} \frac{\left(\frac{\partial y}{\partial p}\right)_T^3 \left(\frac{\partial^2 p}{\partial y^2}\right)_T}{\left(\frac{\partial x}{\partial p}\right)_T^3 \left(\frac{\partial^2 p}{\partial x^2}\right)_T}. \tag{6.29}$$

From this last expression we obtain

$$\lim_{x,y\to y_a} \frac{\left(\frac{\partial y}{\partial p}\right)_T}{\left(\frac{\partial x}{\partial p}\right)_T} = \lim_{x,y\to y_a} \sqrt{\frac{\left(\frac{\partial^2 p}{\partial x^2}\right)_T}{\left(\frac{\partial^2 p}{\partial y^2}\right)_T}}, \tag{6.30}$$

and then at the azeotrope

$$\lim_{x,y\to y_a} \left(\frac{\partial y}{\partial x}\right)_T = \lim_{x,y\to y_a} \sqrt{\frac{\left(\frac{\partial^2 p}{\partial x^2}\right)_T}{\left(\frac{\partial^2 p}{\partial y^2}\right)_T}}. \tag{6.31}$$

In an analogous way, one finds that

$$\lim_{x,y\to y_a} \left(\frac{\partial y}{\partial x}\right)_p = \lim_{x,y\to y_a} \sqrt{\frac{\left(\frac{\partial^2 T}{\partial x^2}\right)_p}{\left(\frac{\partial^2 T}{\partial y^2}\right)_p}}. \tag{6.32}$$

Up to our knowledge equations (6.31) and (6.32) have not been published, and we believe them to be new general results. They show that, at the azeotrope composition, the slope of the function $y(x,T)$ at a constant T, see an example of it in Fig. 6.3, is related to the curvatures of the functions $p = p_l(x)$ and $p = p_g(y)$, at that same temperature, and similarly that the slope of the function $y(x,p)$ at a constant p is related to the curvatures of the functions $T = T_l(x)$ and $T = T_g(y)$ at that same pressure.

Our aim is now to rewrite equations (6.31) and (6.32) which involve the data of coexistence experiments, such as those shown in Fig. 6.1, Fig. 6.2, Fig. 6.3 and Fig. 6.4 in a way involving other thermodynamic quantities. To do so for the case of (6.31), let us start by considering first the second derivative with respect to x at constant T of equation (6.20),

$$\left(\frac{y-x}{1-x}\right)\left(\frac{\partial}{\partial x}\left(\frac{\partial \mu_{1,l}}{\partial x}\right)_{T,p}\right)_T + \left(\frac{\partial \mu_{1,l}}{\partial x}\right)_{T,p}\left(\frac{\partial \frac{y-x}{1-x}}{\partial x}\right)_T$$

$$= [y(V_{1,g} - V_{1,l}) + (1-y)(V_{2,g} - V_{2,l})]\left(\frac{\partial^2 p}{\partial x^2}\right)_T$$

$$+\left(\frac{\partial\left[y(V_{1,g}-V_{1,l})+(1-y)(V_{2,g}-V_{2,l})\right]}{\partial x}\right)_T\left(\frac{\partial p}{\partial x}\right)_T$$

In the limit of the azeotrope,

$$x=y=y_a \quad \text{and} \quad \left(\frac{\partial p}{\partial x}\right)_T=0\,,$$

the only terms left are,

$$\lim_{x\to y_a, y\to y_a}\left\{\left(\left(\frac{\partial y}{\partial x}\right)_T-1\right)\frac{\left(\frac{\partial \mu_{1,l}}{\partial x}\right)_{T,p}}{(1-x)}\right\}$$

$$=\lim_{x\to y_a, y\to y_a}\left\{\left[y(V_{1,g}-V_{1,l})+(1-y)(V_{2,g}-V_{2,l})\right]\left(\frac{\partial^2 p}{\partial x^2}\right)_T\right\}. \quad (6.33)$$

In an analogous way we find, in the limit of the azeotrope,

$$x=y=y_a \quad \text{and} \quad \left(\frac{\partial p}{\partial y}\right)_T=0\,,$$

from equation (6.23) that,

$$\lim_{x\to y_a, y\to y_a}\left\{\left(\left(\frac{\partial y}{\partial x}\right)_T-1\right)\frac{\left(\frac{\partial x}{\partial y}\right)_T\left(\frac{\partial \mu_{1,g}}{\partial y}\right)_{T,p}}{(1-y)}\right\}$$

$$=\lim_{x\to y_a, y\to y_a}\left\{\left[x(V_{1,g}-V_{1,l})+(1-x)(V_{2,g}-V_{2,l})\right]\left(\frac{\partial^2 p}{\partial y^2}\right)_T\right\}. \quad (6.34)$$

The quotient of equations (6.33) and (6.34) gives,

$$\lim_{x\to y_a, y\to y_a}\frac{\left(\frac{\partial^2 p}{\partial x^2}\right)_T}{\left(\frac{\partial^2 p}{\partial y^2}\right)_T}=\lim_{x\to y_a, y\to y_a}\frac{\left(\frac{\partial \mu_{1,l}}{\partial x}\right)_{T,p}}{\left(\frac{\partial \mu_{1,g}}{\partial y}\right)_{T,p}\left(\frac{\partial x}{\partial y}\right)_T}$$

which upon substitution in (6.31) yields,[6]

$$\lim_{x\to y_a, y\to y_a}\left(\frac{\partial y}{\partial x}\right)_T=\lim_{x\to y_a, y\to y_a}\frac{\left(\frac{\partial \mu_{1,l}}{\partial x}\right)_{T,p}}{\left(\frac{\partial \mu_{1,g}}{\partial y}\right)_{T,p}}. \quad (6.35)$$

Equation 6.35 together with 6.31 gives us

$$\lim_{x\to y_a, y\to y_a}\frac{\left(\frac{\partial \mu_{1,l}}{\partial x}\right)_{T,p}}{\left(\frac{\partial \mu_{1,g}}{\partial y}\right)_{T,p}}=\lim_{x,y\to y_a}\sqrt{\frac{\left(\frac{\partial^2 p}{\partial x^2}\right)_T}{\left(\frac{\partial^2 p}{\partial y^2}\right)_T}} \quad (6.36)$$

This last equation embodies a new general thermodynamic result. It establishes a connection between the experimental functions $p = p_l(x, T)$ and $p = p_g(y, T)$ at a given T, which may be obtained from some fitting procedure applied to the experimental points, see Fig. 6.1, with the chemical potentials of both phases. In other words, for a model of the chemical potential to be valid, it must comply with equation (6.36). In an analogous way we find that in the limit of the azeotrope,

$$x = y = y_a, \quad \left(\frac{\partial T}{\partial x}\right)_p = 0 \quad \text{and} \quad \left(\frac{\partial T}{\partial y}\right)_p = 0\,,$$

the second derivatives of equations (6.21) and (6.24) with respect to x and y at constant p are respectively,

$$\lim_{x\to y_a, y\to y_a}\left\{\left(1-\left(\frac{\partial y}{\partial x}\right)_p\right)\frac{\left(\frac{\partial \mu_{1,l}}{\partial x}\right)_{T,p}}{(1-x)}\right\}$$
$$= \lim_{x\to y_a, y\to y_a}\left\{[y(S_{1,g}-S_{1,l})+(1-y)(S_{2,g}-S_{2,l})]\left(\frac{\partial^2 T}{\partial x^2}\right)_p\right\} \tag{6.37}$$

and

$$\lim_{x\to y_a, y\to y_a}\left\{\left(1-\left(\frac{\partial y}{\partial x}\right)_p\right)\frac{\left(\frac{\partial x}{\partial y}\right)_p\left(\frac{\partial \mu_{1,g}}{\partial y}\right)_{T,p}}{(1-y)}\right\}$$
$$= \lim_{x\to y_a, y\to y_a}\left\{[x(S_{1,g}-S_{1,l})+(1-x)(S_{2,g}-S_{2,l})]\left(\frac{\partial^2 T}{\partial y^2}\right)_p\right\} \tag{6.38}$$

which upon substitution in (6.32) give

$$\lim_{x\to y_a, y\to y_a}\left(\frac{\partial y}{\partial x}\right)_p = \lim_{x\to y_a, y\to y_a}\frac{\left(\frac{\partial \mu_{1,l}}{\partial x}\right)_{T,p}}{\left(\frac{\partial \mu_{1,g}}{\partial y}\right)_{T,p}}\,. \tag{6.39}$$

Comparison of expressions (6.32) and (6.39) leads to,

$$\lim_{x,y\to y_a}\frac{\left(\frac{\partial \mu_{1,l}}{\partial x}\right)_{T,p}}{\left(\frac{\partial \mu_{1,g}}{\partial y}\right)_{T,p}} = \lim_{x,y\to y_a}\sqrt{\frac{\left(\frac{\partial^2 T}{\partial x^2}\right)_p}{\left(\frac{\partial^2 T}{\partial y^2}\right)_p}}\,, \tag{6.40}$$

which is a new general thermodynamic result analogous to (6.36), establishing a connection between the experimentally fitted functions $T = T_l(x, p)$ and $T = T_g(y, p)$ with the chemical potential of the mixture.

Comparison of equations (6.35) and (6.39) leads us to another previously unnoticed, up to our knowledge result,

$$\lim_{x \to y_a, y \to y_a} \left(\frac{\partial y}{\partial x} \right)_T = \lim_{x \to y_a, y \to y_a} \left(\frac{\partial y}{\partial x} \right)_p . \tag{6.41}$$

A consequence from equation (6.41) is the following. Consider first the equations (6.33) and (6.37), in them

$$\lim_{x \to y_a} (1 - x) > 0 \, , \quad \left(\frac{\partial \mu_{1,l}}{\partial x} \right)_{T,p} > 0 \tag{6.42}$$

and, in the case of the liquid-vapor coexistence,

$$[y_a(V_{1,g} - V_{1,l}) + (1 - y_a)(V_{2,g} - V_{2,l})] > 0 \tag{6.43}$$

and

$$[y_a(S_{1,g} - S_{1,l}) + (1 - y_a)(S_{2,g} - S_{2,l}) > 0] \, . \tag{6.44}$$

Then, thanks to (6.41) for the same azeotropic mixture, the second derivative of the pressure in (6.33) and the second derivative of the temperature in (6.37) are of opposite sign. Thus, if in (6.33) the second derivative of the pressure happens to be negative, the azeotrope is a maximum of the functions $p = p_l(x, T)$ and $p = p_g(y, T)$, see for example Fig. 6.1, and the slope

$$\left(\frac{\partial y}{\partial x} \right)_T < 1 \, ,$$

as shown in Fig. 6.3, but because of (6.41)

$$\left(\frac{\partial y}{\partial x} \right)_p < 1$$

too, and from (6.37) the azeotrope is a minimum of the functions $T = T_l(x, p)$ and $T = T_g(y, p)$. And viceversa for

$$\left(\frac{\partial y}{\partial x} \right)_T > 1 \, ,$$

as shown in Fig. 6.2 and Fig. 6.4. The same conclusions are drawn from equations (6.34) and (6.38), with the same arguments. The fact that the sign of the extreme of curves p vs x and of p vs y is opposite to the sign of the extreme of the curves of T vs x and of T vs y has been obtained in a different way in reference.[6]

6.3. A Model for the Azeotropic Binary Mixture

A recurrent topic in the study of binary mixtures with an azeotrope is the calculation of correlations that claim to accurately represent the experimental data and which are the functions $p_l(x,T), p_g(y,T)$ and of $y(x,T)$ presented in section 1. Some correlations available in the literature are based in theoretical equations of state or semiempirical equations of state with or without considerations from statistical mechanics models,[14] while others may be the result of a simple fitting procedure. Some authors,[3,4] assume that for many mixtures it is possible to write the function $p_l(x,T)$ as a sum of two functions called partial pressures. The simplest correlation, the so-called Raoult's law, presupposes that these partial-pressure representations are two straight lines that tie the origin of the graph (0 pressure, 0 mol fraction) with the vapor pressure value corresponding to the mol fraction value 1. A more elaborated model accepts the existence of partial pressures for the liquid phase, but instead of accepting a linear behavior with slope equal to the vapor pressure, it accepts this behavior only asymptotically near the mol fraction value 1.[3,4]

In this section we build correlations for $p_l(x,T)$, $p_g(y,T)$ and $y(x,T)$ with a fixed T from the data of coexistence pressure and molar fraction in the liquid and molar fraction in the vapor at a fixed T. These correlations are based on those properties that the azeotropic binary mixture always comply with. The relation betwen the coefficients of $y(x,T)$ and those of the pressure correlations is worked out.

The requisites upon which the correlations are built are on the one hand the conditions at the azeotrope, namely that the values of the molar fraction y_a and of the pressure p_a are the same for both liquid and vapor and that the isothermal slopes $\frac{\partial p}{\partial x} = \frac{\partial p}{\partial y} = 0$. Because of these facts, we start by expressing the pressure as a function of either the liquid or the vapor composition in a form that fullfils them,

$$p_l(x) = p_a + (x - y_a)^2 F_l(x) \quad \text{and} \quad p_g(x) = p_a + (y - y_a)^2 F_g(y) \tag{6.45}$$

where we postulate $F_l(x)$ and $F_g(y)$ are well behaved functions at the azeotropic point $x = y = y_a$, in general different for each phase. On the other hand at $x = y = 0$, pure component 2, and at $x = y = 1$, pure component 1, the presure must be the vapor pressure p_{v2} and p_{v1} respectively. Notice that the form (6.45) already guarantees that the conditions required at the extremes of composition are fullfilled since at $x = y = 0$,

$$p_l(0) = p_{v2} = p_a + y_a^2 F_l(0) \quad \text{and} \cdot \quad p_g(0) = p_{v2} = p_a + y_a^2 F_g(0) \tag{6.46}$$

implying that the values of the functions $F_l(x)$ and $F_g(y)$ are known and the same at each extreme. In fact, at $x = y = 0$,

$$F_l(0) = F_g(0) = \frac{p_{v2} - p_a}{y_a^2}, \tag{6.47}$$

and at $x = y = 1$,

$$p_l(1) = p_{v1} = p_a + (1 - y_a)^2 F_l(1) \quad \text{and} \quad p_g(1) = p_{v1} = p_a + (1 - y_a)^2 F_g(1) \tag{6.48}$$

implying now that $F_l(1)$ and $F_g(1)$ are the same value for both phases:

$$F_l(1) = F_g(1) = \frac{p_{v1} - p_a}{(1 - y_a)^2}. \tag{6.49}$$

Next, we put forward that the functions $F_l(x)$ and $F_g(y)$ may be conveniently represented by fourth degree polynomials of the form,

$$F_l(x) = E + D(x^4 + x^3 + x^2 + x) + x(1 - x)(C_l x^2 + B_l x + A_l),$$

$$F_g(y) = E + D(y^4 + y^3 + y^2 + y) + y(1 - y)(C_g y^2 + B_g y + A_g), \tag{6.50}$$

where only the coefficients A_l, B_l, C_l of $F_l(x)$ are in general different from the coefficients A_g, B_g, C_g of $F_g(y)$ because the coefficients E and D are common to both phases. This is readily seen,

$$F_l(0) = F_g(0) = E, \tag{6.51}$$

and

$$F_l(1) = F_g(1) = E + 4D. \tag{6.52}$$

so that from of the two previous relations (6.47) and (6.49),

$$E = \frac{P_{v2} - P_a}{y_a^2} \tag{6.53}$$

and

$$D = \frac{P_{v1} - P_a}{4(1 - y_a)^2} - \frac{E}{4}. \tag{6.54}$$

The correlation for the pressure as a function of the mol fraction x in the liquid is therefore given after subtitution of the first equation in (6.50) into the first one in (6.45) by

$$p_l(x) = p_a + (x - y_a)^2[E + D(x^4 + x^3 + x^2 + x) + x(1 - x)(C_l x^2 + B_l x + A_l)], \tag{6.55}$$

and similarly, the pressure as a function of the mol fraction y in the vapor, after substitution of the second equation in (6.50) into the second one in (6.45) is given by

$$p_g(y) = p_a + (y - y_a)^2[E + D(y^4 + y^3 + y^2 + y) + y(1 - y)(C_g y^2 + B_g y + A_g)]\,. \tag{6.56}$$

Then it is only necessary to perform two linear least-squares fits of the experimental pressure data, one as a function of the mol fraction in the liquid to obtain the values of the three coefficients A_l, B_l, C_l; the other as a function of the mol fraction in the vapor to find the values of the three other coefficients A_g, B_g, C_g. In both cases the same values p_a, y_a, D, and E are used. Two examples are given in the next section.

In order to determine the mol fraction y as a function of the mol fraction x, at a given temperature namely,

$$y = G(x)\,, \tag{6.57}$$

we introduce a new correlation for it. This new correlation must take into account first that the three points where the composition of both phases is the same, namely that the extremes of the interval $(0, 1)$ and the azeotrope y_a, are fixed points of the function $G(x)$:

$$G(0) = 0\,, \quad G(y_a) = y_a\,, \quad G(1) = 1\,, \tag{6.58}$$

and second, that the value of the slope at the three fixed points is found from the coefficients of the correlations (6.55) and (6.56). This last fact is shown next.

The values of the slope at the extremes of composition are easily obtained from (6.25) with (6.55) and (6.56) because of (6.57). These values of $G'(x)$ are therefore

$$G'(0) = \frac{y_a(D + A_l) - 2E}{y_a(D + A_g) - 2E} \tag{6.59}$$

and

$$G'(1) = \frac{2E + 8D + (1 - y_a)(10D - A_l - B_l - C_l)}{2E + 8D + (1 - y_a)(10D - A_g - B_g - C_g)}\,. \tag{6.60}$$

To find the slope at the azeotrope we use (6.31) with the help of (6.45) to obtain,

$$G'(y_a) = \sqrt{\frac{F_l(y_a)}{F_g(y_a)}}$$

or in terms of our correlations (6.55) and (6.56),

$$G'(y_a) = \sqrt{\frac{E + Dy_a(y_a^3 + y_a^2 + y_a + 1) + y_a(1 - y_a)(C_l y_a^2 + B_l y_a + A_l)}{E + Dy_a(y_a^3 + y_a^2 + y_a + 1) + y_a(1 - y_a)(C_g y_a^2 + B_g y_a + A_g)}}. \tag{6.61}$$

In order to satisfy the requisites in (6.58) we put forward for $G(x)$ the following form

$$G(x) = x + x(1 - x)(y_a - x)H(x). \tag{6.62}$$

where $H(x)$ may be a third order polynomial with three out of its four coefficients such that they may be determined from the three conditions (6.59-6.61) leaving just one coefficient to be found by with a least squares fit. Therefore we put forward,

$$G(x) = x + x(1 - x)(y_a - x)[\alpha + \beta(x + x^2 + x^3) + x(1 - x)(\gamma + \delta x)]. \tag{6.63}$$

The coefficients α and β are found from the slopes at the extremes of composition upon equating the derivatives at $x = 0$ and $x = 1$ obtained from the form (6.63) with the ones expressed in (6.59) and (6.60) respectively. The results are,

$$\alpha = \frac{A_l - A_g}{y_a(D + A_g) - 2E}, \tag{6.64}$$

and

$$\beta = \frac{A_g + B_g + C_g - A_l - B_l - B_l}{3[2E + 8D + (1 - y_a)(10D - A_g - B_g - C_g)]} - \frac{\alpha}{3}. \tag{6.65}$$

Finally upon equating the derivative at the azeotrope given by (6.63) with the value in (6.61), a relation between the constants γ and δ is found:

$$G'(y_a) = 1 - y_a(1 - y_a)[\alpha + \beta(y_a + y_a^2 + y_a^3) + y_a(1 - y_a)(\gamma + \delta y_a)]. \tag{6.66}$$

The polynomials here proposed for the functions $p_l(x, T)$, $p_g(y, T$ and $G(x)$ are the simplest possible that satisfy the imposed thermodynamic conditions.

6.4. Particular Examples

As a first example we have studied the carbon dioxide (1), and ethane (2), mixture at 283.15 K from[9] where the pressure is given in atmospheres. After the transformation to units of MPa, we have that the values for the vapor pressure of the pure components P_{v1} and P_{v2} and the pressure P_a and mol fraction y_a at the azeotropic point are the following ones

$$P_{v1} = 4.49984\,, \quad P_{v2} = 3.03874\,, \tag{6.67}$$

$$P_a = 4.99532\,, \quad y_a = 0.6971 \tag{6.68}$$

Table 6.1. Carbon dioxide (1) and Ethane (2) at 283.15 K. Mol fractions of component (1), in the liquid x_i, and in the vapor y_i and coexistence pressure in MPa p_i from.[9] Calculated values $p_l(x_i)$, $p_g(y_i)$, and $G(x_i)$.

i	x_i	y_i	p_i	$p_l(x_i)$	$p_g(y_i)$	$G(x_i)$
1	0.0000	0.0000	3.03874	3.03874	3.03874	0.0000
2	0.0847	0.1389	3.46633	3.48798	3.45554	0.1376
3	0.1266	0.1979	3.65479	3.67526	3.65070	0.1954
4	0.1504	0.2333	3.77638	3.77570	3.77824	0.22588
5	0.1924	0.2824	3.98106	3.94509	3.96485	0.2763
6	0.22680	0.315	4.09961	4.07730	4.09197	0.31468
7	0.3791	0.4557	4.54848	4.58118	4.60476	0.46142
8	0.4199	0.4843	4.67919	4.68645	4.69041	0.49557
9	0.5621	0.5930	4.89805	4.92974	4.92352	0.60270
10	0.6185	0.6410	4.95479	4.97425	4.97493	0.6421
11	0.6828	0.6959	4.99026	4.99466	4.99531	0.68692
12	0.6971	0.6960	4.99634	4.99532	4.99532	0.6971
13	0.7466	0.7376	4.99026	4.98758	4.98538	0.73358
14	0.7598	0.7506	4.99026	4.98288	4.97815	0.74375
15	0.8304	0.8092	4.94061	4.93610	4.92340	0.80283
16	0.8923	0.8680	4.87373	4.85212	4.83507	0.86371
17	0.9568	0.9406	4.70047	4.68649	4.67908	0.93987
18	1.0000	1.0000	4.49984	4.49984	4.49984	1.0000

The parameters D and E, which are common to both phases, take the values

$$D = -0.34352, \quad E = -4.02632\,. \tag{6.69}$$

A linear least-squares fit applied to the data gave the following values of the parameters A_i, B_i, C_i $i = l, g$, in the polynomials for the pressure, Eqs.(6.55) and (6.56),

$$A_l = 1.4798\,, \qquad B_l = -14.1886\,, \quad C_l = 31.1619 \tag{6.70}$$

$$A_g = -4.2692\,, \quad B_g = -25.0489\,, \quad C_g = 28.9174\,. \tag{6.71}$$

The parameters for the composition correlation $G(x)$ equation (6.63) were found to be, from (6.64) and (6.65),

$$\alpha = 1.18851\,, \quad \beta = 0.14006 \tag{6.72}$$

while γ and δ were fitted to the values

$$\gamma = -1.2487\,, \quad \delta = 1.39256\,. \tag{6.73}$$

The experimental pressure *vs* composition data from[9] and the corresponding correlation curves $p_l(x)$ and $p_g(y)$ are shown in Fig. 6.1 while the Fig. 6.3 shows the experimental data for (x, y) from[9] and the corresponding correlation $G(x)$.

Table 6.1 shows the experimental data from[9] and the calculated values from Eqs.(6.55) and (6.56). The mean square relative error in the calculated values of the pressure is less than 0.45% for the liquid, and less than 0.46% for the vapor. The maximum error is less than, 0.9% in absolute value, for the liquid and less than 1.24% in absolute value for the vapor. The mean square relative error in the calculation of the mol fraction of vapor from $G(x)$ is 0.59% and the maximum error is 1.13% in absolute value.

As a second example we consider the Acetone (1), and Chloroform (2), mixture at 328.25 K from[13] where the pressure is given in mm-Hg. In this case the azeotrope is a minimum. The values for the molar fraction y_a and the pressure p_a at the azeotrope, and the vapor pressure of the pure components after conversion to kPa are

$$y_a = 0.36\,, \quad P_a = 74.687 \tag{6.74}$$

$$P_{v1} = 98.565495\,, \quad P_{v2} = 84.0862\,, \tag{6.75}$$

The parameters D and E, common to both phases, take the values

$$D = -3.5577, \quad E = 72.5247\,. \tag{6.76}$$

The parameters A_i, B_i, C_i $i = l, g$, in the polynomials for the pressure, (6.55) and (6.56), were found by the least-squares fit to be

$$A_l = 244.487\,, \quad B_l = -102.431\,, \quad C_l = -187.287\,. \tag{6.77}$$

$$A_g = 49.1167\,, \quad B_g = -127.4399\,, \quad C_g = -60.5324\,. \tag{6.78}$$

The experimental pressure vs composition data from[13] with the corresponding correlation curves $p_l(x)$, and $p_g(y)$ are shown in Fig. 6.2 while Fig. 6.4 shows the experimental data for (x, y) from[13] and the corresponding results from the correlation $G(x)$.

The parameters for the composition correlation $G(x)$, (6.63), were found to be

$$\alpha = -1.51864\,, \quad \beta = 0.33539\,, \tag{6.79}$$

while γ and δ were fitted to the values

$$\gamma = 3.2889\,, \quad \delta = -7.33323\,. \tag{6.80}$$

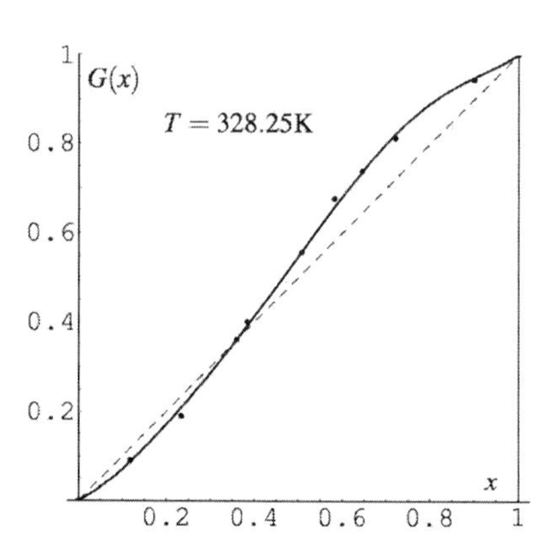

Fig. 6.4. Cetone (1) and Chloroform (2) at 328.25K. Experimental data from[13] and correlation curve $G(x)$.

Table 6.2. Acetone (1) and Chloroform (2) at 328.25K. Mol fractions of component (1), in the liquid x_i, and in the vapor y_i and coexistence pressure in kPa, p_i from.[13] Calculated values $p_l(x_i)$, $p_g(y_i)$, and $G(x_i)$.

i	x_i	y_i	p_i	$p_l(x_i)$	$p_g(y_i)$	$G(x_i)$
1	0.000	0.000	84.0862	84.0862	84.0862	0.0000
2	0.118	0.091	80.1265	80.3071	80.1308	0.0872
3	0.234	0.190	76.9935	76.4195	76.8599	0.2084
4	0.36	0.360	74.6870	74.6870	74.6870	0.36
5	0.385	0.400	74.8736	74.7572	74.7949	0.3922
6	0.508	0.557	76.9801	76.9895	76.9555	0.5571
7	0.582	0.677	79.6599	79.5011	79.9569	0.6566
8	0.645	0.738	81.7531	82.0129	81.8904	0.7365
9	0.721	0.812	85.2728	85.1851	84.7894	0.8211
10	0.900	0.944	92.6455	92.6404	92.9555	0.9485
11	1.000	1.000	98.56495	98.56495	98.56495	1.0000

Table 6.2 shows the calculated values from Eqs.(55) and (56) and the experimental data from.[13] The mean square relative error in the calculated values of the pressure is less than 0.30% for the liquid, and less than 0.27% for the vapor. The maximum error is less than, 0.75% in absolute value, for the liquid and less than 0.57% in absolute value for the vapor. The mean square relative error in the calculation of the mol fraction of vapor from $G(x)$ is 1.02% and the maximum error is 2.04% in absolute value.

The main purpose of the present paper has been to contribute to the understanding of the azeotropic mixtures and its main result is the new general thermodynamic condition (36) that must be satisfied by all existing methods or treatments of any azeotrope. In particular notice that the evaluation of its right-hand side from experimental data, independent of any model, affords the possibility of testing the compliance of any proposed model for the chemical potential, which is the function at the root of the several well known correlation methods, with the thermodynamics of the azeotrope. An advantage of the polynomials presented here, in addition to their simple implementation is the evaluation of the right-hand side of equation (36) directly from experimental data. We believe that the test for the compliance of the many methods and treatments there are in the literature with equation (36) as well as further improvements should be considered for future work.

References

1. A. Münster 1970 *Classical Thermodynamics* (London: Wiley-Interscience) p 124-125
2. D. Kondepudi and I. Prigogine 1998 *Modern Thermodynamics* (New York: John Wiley and Sons) pp 184-185, 221
3. P.W. Atkins 1986 *Physical Chemistry* (New York: W. H. Freeman and Co.) p 180
4. K. Denbigh 1966 *The Principles of Chemical Equilibium* (London: Cambridge University Press) pp 215-225
5. J.S. Rowlinson 1969 *Liquids and Liquid Mixtures* (London: Butterworth and Co.) pp 178-187
6. I. Prigogine and R. Defay 1967 *Chemical Thermodynamics* translated by Everett D H (Longmans, Green and Co LTD) pp 278-284 and p 465
7. R.C. Reid, J.M. Prausnitz and B.E. Poling 1988 *The Properties of Gases and Liquids* (New York: McGraw-Hill Book Co.) pp 307-309, p 83
8. J.M. Prausnitz, R.N. Lichtenthaler and E. Gomes de Azevedo *Termodinámica Molecular de los Equilibrios de Fases*(Madrid: Prentice Hall) pp 189-217
9. K. Ohgaki and T. Katayama, 1977 *Fluid Phase Equilibria* **1**: 27, p 29

10. E.W. Washburn (Ed.) 1928 *International Critical Tables of Numerical Data, Physics, Chemistry and Technology, 3 Partial Pressures* (New York: Mc Graw Hill Book Co) p 286
11. L.H. Horsley *et al* (Eds) 1973 *Advances in Chemistry Series* (Washington D C: American Chemical Society)
12. J. Weishaupt (Ed.) 1975 *Landolt-Börnstein, Numerical Data and Funtional Relationships in Science and Technology, 3 Thermodynamic equilibria of boiling Mixtures* (Springer- Verlag, Berlin)
13. J. Gmehling, U. Onken and W. Arlt (Eds) 1979 *Chemistry Data Series, 1-(3,4), Aldehydes and Ketones, Ethers* (Frankfurt am Main: DECHEMA)
14. P. Benedek and F. Olti 1985 *Computer Aided Chemical Thermodynamics of Gases and Liquids: theory, models and programs* (New York: John Wiley)

Chapter 7

Entropy Generation in Oscillatory Flow between Parallel Plates

Mariano López de Haro*

Departamento de Física, Universidad de Extremadura, Badajoz 06071 Spain

malopez@servidor.unam.mx

Federico Vázquez

Facultad de Ciencias, UAEM, Cuernavaca, Morelos, 62209, México

Miguel Ángel Olivares-Robles

Sección de Posgrado e Investigación, ESIME-Culhuacán, IPN, 04430, México

Sergio Cuevas

Centro de Investigación en Energía, Universidad Nacional Autónoma de México, Temixco, Morelos 62580, México

The problem of a zero-mean oscillatory flow of a Newtonian fluid between infinite parallel plates with thermal boundary conditions of the third kind is considered. With the analytic solutions for the velocity and temperature fields at hand, the local and global time-averaged entropy production are computed. The consequences of having different convective heat transfer coefficients in each plate are assessed for this problem and some conditions that lead to entropy generation minimization are determined.

Contents

*On leave from Centro de Investigación en Energía, Universidad Nacional Autónoma de México, Temixco, Morelos 62580, México.

7.1. Introduction

Most processes occurring in nature are irreversible and their thermodynamic description is by no means trivial. While the phenomenological aspects of the thermodynamics of irreversible processes started to be studied systematically already in the nineteenth century, the theoretical developments took much longer to reach a solid status, notably through the contributions in the first half of the twentieth century of (amongst others) Onsager, Eckart, Meixner, Prigogine and de Groot. These authors assumed that even in systems far from equilibrium the Gibbs relation and concepts like temperature and entropy are meaningful and that all equilibrium thermodynamic relations remain locally valid. The entropy change for each macroscopic region in the system is then decomposed into an external and an internal part where, in accordance with the second law of thermodynamics, the internal part is postulated to be positive definite. The entropy production may thus be expressed as a product of 'fluxes' and the corresponding 'generalized thermodynamic forces'. The phenomenological laws relating these fluxes and forces must satisfy the restriction of the second law. The so called linear irreversible thermodynamics implies a linear relationship between the forces and the fluxes with the proviso that for isotropic systems only fluxes and forces of the same tensorial character may be coupled. The scheme is completed with the reciprocity relations linking the phenomenological coefficients of cross effects due to the presence of two or more independent forces of the same tensorial character.

It is interesting to point out that both Onsager and Prigogine were awarded the Nobel Prize for Chemistry in 1968 and 1977, respectively, for their contributions to irreversible thermodynamics. Therefore, by the mid 1960's their work was already well known by the physics and chemistry communities and two excellent monographs were produced around that period.[1,2] Its permeation to the engineering community did not happen until much later and owes much to the work and efforts of A. Bejan (*c.f.* for instance his two books on the subject[3,4]). In his many writings, Bejan has explained how linear irreversible thermodynamics can be used to systematically improve engineering processes through minimizing the entropy generation. He has also advocated the use of this systematic methodological approach in the areas of conceptual process design and equipment design. In the case of Mexico, undoubtedly the introducer and key figure of the Mexican School of Irreversible Thermodynamics is Prof. Leopoldo García-Colín Scherer[5] and in the occasion of his eightieth birthday we find it appropriate to honor him with a small piece dealing with this line of research.

The role of oscillatory flows for the enhancement of transport processes has been known for a while. For instance, the effective thermal diffusivity of a Newtonian fluid in a duct subjected to a zero-mean oscillatory flow may reach a maximum for a specific oscillation frequency.[6,7] Since the design of many traditional engineering devices (such as heat exchangers and cooling modules) and the performance of processes involving heat removal from components such as electronic chips and from other similar high energy devices rely on heat transfer enhancement, one may reasonably wonder whether one can profit from a sensible combination of oscillatory flows, thermal boundary conditions and characteristics of the working fluid. Recently, attention has been directed to oscillatory flows at high frequencies under conditions where inertial effects are negligible with the aim at using them in microfuidic applications.[8]

This work is geared towards providing an analysis of the problem of a zero-mean oscillatory flow of a Newtonian fluid between infinite parallel plates with thermal boundary conditions of the third kind. Such an analysis will hopefully shed some more light on the the irreversible behavior of oscillating fluid flow systems used in some engineering devices. The emphasis will be placed on the computation of the entropy production for the system and its subsequent analysis using the method of the entropy generation minimization.[3,4] In view of the above, the paper represents yet another step in our previous efforts[9–13] to understand and determine optimum operating conditions, in the sense of minimum irreversible energy losses, of processes and devices.

The chapter is organized as follows. In the next section we present the transport problem to be analyzed, the governing equations and their solution. Section 7.3 deals with the computation of the entropy production and with illustrative results for particular choices of the values of the parameters of the system. The chapter is closed in Sect. 7.4 with further discussion and some concluding remarks.

7.2. Hydrodynamic Equations

We consider the flow of a Newtonian fluid between two infinite parallel plates separated by a distance $2a$. We assume that a zero-average time-periodic pressure gradient is established in the system producing an oscillatory flow in the axial x-direction. With the former approximations, the flow becomes fully developed with all quantities depending on the transversal coordinate y and the time t, except for the pressure, which varies with x and t. We also assume that the fluid is incompressible and monocomponent, so that mass diffusion phenomena are disregarded. In addition, all physical properties of the fluid are assumed to be constant.

The continuity and momentum equations for this system are

$$\nabla \cdot \mathbf{u} = 0, \tag{7.1}$$

$$\rho \frac{\partial \mathbf{u}}{\partial t} + (\mathbf{u} \cdot \nabla) \mathbf{u} = -\nabla p + \eta \nabla^2 \mathbf{u}, \tag{7.2}$$

where $\mathbf{u}$ and p are the velocity and pressure fields while ρ and η are the mass density and shear viscosity of the fluid, respectively. In turn, the energy balance equation reads

$$\rho C_p \left(\frac{\partial T}{\partial t} + (\mathbf{u} \cdot \nabla) T \right) = k \nabla^2 T + \eta \nabla \mathbf{u} : \nabla \mathbf{u}, \tag{7.3}$$

where T is the temperature, and C_p and k are the specific heat at constant pressure and the thermal conductivity of the fluid, respectively.

Assuming a unidirectional flow that depends only on the transversal coordinate and time, from the momentum balance equation [Eq.(7.2)] one gets

$$\rho \frac{\partial u}{\partial t} = -\frac{\partial p}{\partial x} + \eta \frac{\partial^2 u}{\partial y^2}. \tag{7.4}$$

The oscillatory pressure gradient that produces the motion can be expressed as the real part of $(\partial p / \partial x) = G e^{i\omega t}$, where G is a (real) constant. Therefore, we assume that the velocity is also a harmonic function of time so that it can be expressed as the real part of $u = u_o(y) e^{i\omega t}$. Under these circumstances the equation satisfied by u_o reads

$$\frac{\partial^2 u_o}{\partial y^{*2}} = i R_\omega u_o + K \tag{7.5}$$

where $y^* = y/a$ is the dimensionless coordinate, $K = Ga^2/\eta$, and $R_\omega = \omega a^2 \rho/\eta$ is the oscillation Reynolds number which compares the characteristic length with the viscous penetration depth. The corresponding nonslip boundary conditions for Eq. (7.5) are $u_o(\pm 1) = 0$. Using these conditions, the analytic solution of Eq.(7.5) is

$$u_o(y^*) = \frac{iK}{R_\omega}\left\{1 - \frac{\cosh\left[(1+i)\sqrt{R_\omega/2}y^*\right]}{\cosh\left[(1+i)\sqrt{R_\omega/2}\right]}\right\}, \tag{7.6}$$

which coincides with the usual result given in textbooks, for instance in the one by Currie[14]. For later developments, it is convenient to consider the dimensionless variable $u_o^*(y^*) = u_o(y^*)/U_o$, where U_0 is the average velocity in the cross-section given by

$$U_o = \frac{1}{4}\int_{-1}^{+1}(u_o + \overline{u_o})dy^*. \tag{7.7}$$

where the bar denotes complex conjugation. Using solution (7.6), we find that

$$U_o = -\left(\frac{i^{1/2}K\left\{\tan\left[\sqrt{iR_\omega}\right] - \tanh\left[\sqrt{iR_\omega}\right]\right\}}{{R_\omega}^{3/2}}\right). \tag{7.8}$$

Therefore, the dimensionless velocity component can be expressed as the real part of $u^*(y^*) = u_o^*(y^*)e^{it^*}$, where the dimensionless time t^* has been normalized by $1/\omega$. Notice that u_o^* does not depend on K.

We have checked that the real part of $u^*(y^*)$ correctly tends to $-(3/2)(-1+y^{*2})$ when $R_\omega \to 0$, which is the dimensionless expression for the ordinary Poiseuille flow velocity between the infinite parallel plates.

Once the velocity field is obtained, one may proceed to solve the energy balance equation. We note that in general different thermal boundary conditions lead to a different heat transfer process between the fluid and its surroundings and hence the solution of the energy equation depends on which boundary conditions are considered. The simplest thermal boundary condition, referred to as a boundary condition of the first kind since it only involves the value of the temperature at the boundaries, would be to assume that the temperature of the fluid at the plates coincides with the (constant) external ambient temperature. In turn, a boundary condition of the second kind assumes a prescribed value of the spatial derivative at the boundaries, that is, a constant wall heat flux. A more general condition, referred to as a boundary condition of the third kind, that is a boundary condition involving both the value of the temperature and of its spatial derivative at the boundaries, is to assume that at the boundaries there is convective heat transfer between the fluid and the surroundings. Here we solve the energy equation using boundary conditions of the third kind that indicate that the normal temperature gradient at any point in the boundary is assumed to be proportional to the difference between the temperature at the surface and the external ambient temperature. Hence, the amount of heat entering or leaving the system depends on the external temperature as well as on the convective heat transfer coefficient. We consider conditions where the flow is fully thermally developed so that convective heat transfer within the fluid is disregarded. Then, the energy balance equation [Eq. (7.3)] may be written, in dimensionless form, as

$$\mathrm{Pr}\, R_\omega \frac{\partial \Theta}{\partial t^*} = \frac{\partial^2 \Theta}{\partial y^{*2}} + \left(\frac{\partial u^*}{\partial y^*} \right)^2, \tag{7.9}$$

where $\mathrm{Pr} = \eta C_p / k$, is the Prandtl number and the dimensionless temperature is given by $\Theta = kT/\eta U_o^2$.

Note that the viscous dissipation contribution involves squared terms of harmonic functions of time. Consequently, the heat source term contains time harmonic terms with twice the frequency of oscillation as well as a steady contribution. Therefore, except for the case $R_\omega = 0$ that was already analyzed in the paper by Ibáñez *et al*[10] and that follows directly from Eq.(7.9) setting the left hand side equal to zero, one may assume that the dimensionless temperature will be the real part of

$$\Theta(y^*, t^*) = \Theta_u(y^*) e^{2it^*} + \Theta_s(y^*), \tag{7.10}$$

where the subindexes u and s refer to the unsteady and steady contributions, respectively. Introducing Eq.(7.10) into Eq.(7.9), the equations satisfied by Θ_u and Θ_s are found to be

$$\frac{d^2 \Theta_u}{dy^{*2}} - 2i\, \mathrm{Pr}\, R_\omega \Theta_u = -\frac{1}{4} \left(\frac{\partial u_0^*}{\partial y^*} \right)^2, \tag{7.11}$$

and

$$\frac{d^2 \Theta_s}{dy^{*2}} = -\frac{1}{2} \left(\frac{\partial u_0^*}{\partial y^*} \right) \left(\frac{\partial \overline{u_0^*}}{\partial y^*} \right). \tag{7.12}$$

The solution to Eqs.(7.11) and (7.12) must satisfy the boundary conditions of the third kind, namely

$$\frac{d\Theta_s}{dy^*} + Bi_1(\Theta_s - \Theta_A) = 0, \quad \text{at } y^* = 1, \tag{7.13}$$

$$\frac{d\Theta_s}{dy^*} - Bi_2(\Theta_s - \Theta_A) = 0, \quad \text{at } y^* = -1, \tag{7.14}$$

$$\frac{d\Theta_u}{dy^*} + Bi_1\Theta_u = 0, \quad \text{at } y^* = 1, \tag{7.15}$$

$$\frac{d\Theta_u}{dy^*} - Bi_2\Theta_u = 0, \quad \text{at } y^* = -1, \tag{7.16}$$

where Θ_A is the dimensionless ambient temperature and the Biot numbers $Bi_1 = h_1 a/k$ and $Bi_2 = h_2 a/k$ are the dimensionless expressions of the convective heat transfer coefficients of the upper and lower surfaces, h_1 and h_2, respectively, which in general are different. Note that in the limits $Bi_1 \to \infty$, $Bi_2 \to \infty$ conditions (7.13)-(7.16) reduce to boundary conditions of the first kind, that is, constant wall temperature conditions. In turn, if $Bi_1 \to 0$, $Bi_2 \to 0$, these conditions reduce to insulating wall conditions.

Apart from the spatial or the spatial and temporal dependence, since u_0^* does not depend on K, such solutions (derived with the aid of Mathematica), are functions of Θ_A, Pr , R_ω, Bi_1, and Bi_2. On the one hand, the stationary solution is given by

$$\begin{aligned}\Theta_s &= A + By^* \\ &-(R_\omega(\cos[by^*](4 + 2\cos[2c] + 4\cos[2ic] + \cos[2b] + \cosh[2b]) \\ &+4\cosh^2[c](\cos[2c]\cos[by^*] + 4\cos^2[c]\cosh[by^*])))/ \\ &(8(\cos[b] + \cosh[b])(\sin[b] - \sinh^2[b])),\end{aligned} \tag{7.17}$$

with $b = \sqrt{2R_\omega}$ and $c = \sqrt{iR_\omega}$. The constants A and B, that are complicated, lengthy and not very illuminating functions of Θ_A, R_ω, Bi_1, and Bi_2 and hence will not be written down explicitly,[15] are determined by the boundary conditions, Eqs. (7.13) and (7.14). On the other hand, the nonstationary solution reads

$$\Theta_u = Ce^{(-1-i)\sqrt{\Pr R_\omega}y^*} + De^{(1+i)\sqrt{\Pr R_\omega}y^*}$$
$$+(8iR_\omega(2 - \Pr(1 - \cosh[(1+i)by^*]))\cos^2[c]\cosh^2[c])/$$
$$((-2+\Pr)\Pr(2 - 4\cos[2c] - (2-2i)\cos[2b] + 4\cosh[2c]$$
$$+i\cosh[(1+3i)b] - (2+2i)\cosh[2b] + 2\cosh[(2+2i)b]$$
$$-i\cosh[(3+i)b])), \quad (7.18)$$

where the values of the constants C and D are obtained from conditions (7.15) and (7.16) leading to complicated and lengthy functions this time of Θ_A, Pr, R_ω, Bi_1, and Bi_2 whose explicit form will once more be ommitted[15].

It must be mentioned, moreover, that the limit of the stationary temperature $\Theta_s(y^*, R_\omega, Bi_1, Bi_2, \Theta_A)$ when $R_\omega \to 0$ is given by

$$\lim_{R_\omega \to 0} \Theta_s = \left\{24 + Bi_2\left(15 + 8\Theta_A + 12y^* - 3y^{*4}\right)\right.$$
$$+Bi_1\left[8\Theta_A + 2Bi_2\left(3 + 8\Theta_A - 3y^{*4}\right)\right.$$
$$\left.\left.-3\left(-5 + 4y^* + y^{*4}\right)\right]\right\} / \left\{8\left(Bi_1 + Bi_2 + 2Bi_1Bi_2\right)\right\}. \quad (7.19)$$

This limit corresponds to the temperature profile between the plates for the non-oscillatory (time-independent) case with the same boundary conditions.[10] Note that the complete stationary limit $R_\omega \to 0$ for the real part of the total temperature $\Theta(y^*, t^*)$, as given by Eq. (7.10), does include the limit of Θ_s given above and additional contributions coming from the solution of Eq. (7.11) when the corresponding limit is taken. This reflects the fact that, as stated earlier, in such limit the assumption of the form for the total temperature involved in Eq. (7.10) does not hold and should be avoided.

7.3. Entropy Production

The velocity and temperature fields already obtained will be used for the determination of the entropy generation rate for this problem. In dimensionless terms and assuming the validity of the local equilibrium assumption, the local entropy generation rate, $\dot{S}^*$, that takes into account irreversibilities due to heat conduction and viscous losses is given by[2]

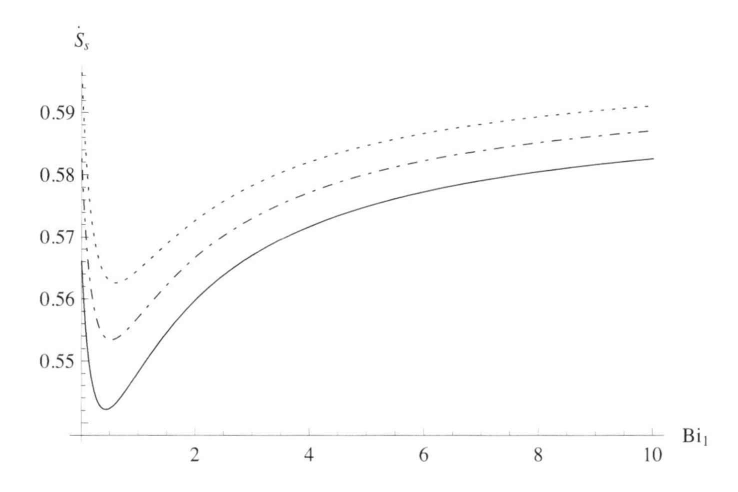

Fig. 7.1. The global entropy production $\dot{S}_s$ as a function of Bi_1 in the stationary case for different values of Bi_2. Continuous line, $Bi_2 = 10$; dot-dashed line, $Bi_2 = 20$; dotted line, $Bi_2 = 100$. In all cases $\Theta_A = 10$.

$$\dot{S}^* = \frac{1}{\Theta^2}\left(\frac{\partial \Theta}{\partial y^*}\right)^2 + \frac{1}{\Theta}\left(\frac{\partial u^*}{\partial y^*}\right)^2, \tag{7.20}$$

where $\dot{S}^*$ is normalized by k/a^2. Evidently, $\dot{S}^*$ is a function of space and time and, in fact, contains both time harmonic and steady parts. The global entropy generation rate per unit length in the axial direction, $\dot{S}$, is obtained by integrating $\dot{S}^*$ from $y^* = -1$ to $y^* = 1$ and over one period of time. In this particular problem, it is interesting to explore the entropy generation in the system by considering different heat transfer conditions between the fluid and the external ambient as well as different oscillation frequencies. While one could attempt to derive the explicit (complicated) form of (7.20) as a function of Θ_A, Pr, R_ω, Bi_1, and Bi_2, it seems much more practical to compute it numerically and to analyze the behavior of the entropy production of a Newtonian fluid subjected to a time periodic pressure gradient. A particular choice of the values of the parameters defining the system was made for the sake of illustration. So from here onwards we will take $\Theta_A = 10$ and $\mathrm{Pr} = 10$. First, we show in Fig. 7.1 the global dimensionless entropy production $\dot{S}_s$ as a function of the upper wall Biot number Bi_1 (spanning a range from 0 to 10) for different values of Bi_2 in the stationary case ($R_\omega = 0$). We remark that in such a case, as follows

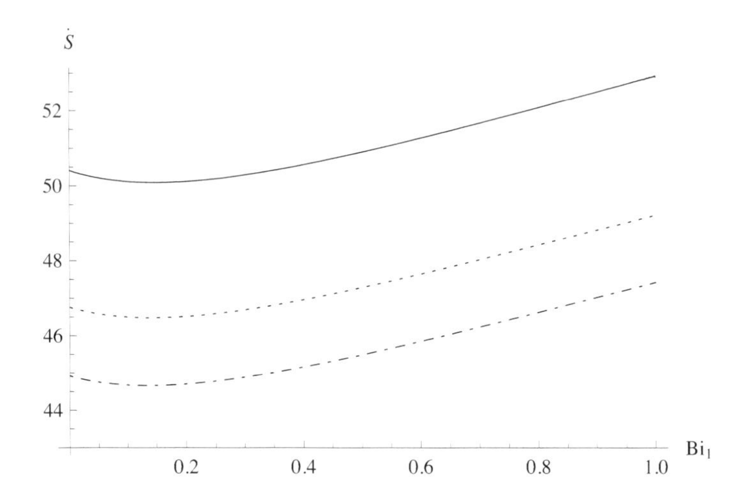

Fig. 7.2. The global entropy production $\dot{S}$ as a function of Bi_1 in the non stationary case for different values of R_ω. Continuous line, $R_\omega = 26$; dot-dashed line, $R_\omega = 27$; dotted line, $R_\omega = 29$. In all cases $\Theta_A = 10$, Pr $= 10$ and $Bi_2 = 100$.

from Eqs. (7.6), (7.8), (7.9) and (7.20), $\dot{S}_s$ does not depend on the Prandtl number. Note the minima at an approximate value of $Bi_1 \sim 0.6$ in the three cases displayed. These results are consistent with the ones previously reported in the work of Ibáñez *et al*[10] for other values of Θ_A, Bi_1 and Bi_2.

The time-dependent case for different values of the oscillation Reynolds number and $Bi_2 = 100$ can be seen in Fig. 7.2, where clearly the entropy production also displays well defined minima which may indicate an optimal device performance in the sense of minimum irreversible losses. The minimum of $\dot{S}$ as a function of Bi_1 that is displayed in Fig. 7.2 disappears for an oscillation Reynolds number of about 0.1 and the same values of Θ_A, Pr and Bi_2; similar behavior is observed for smaller values of R_ω. On the other hand, the global entropy production has an asymptotic value when $Bi_1 \to \infty$ as can be seen in Fig. 7.3, where $R_\omega = 0.1$ and again $Bi_2 = 100$.

As exemplified in Fig. 7.4, we have furthermore found that, for fixed values of Θ_A, Pr, Bi_1 and Bi_2, the global entropy production $\dot{S}$ may also exhibit a minimum with respect to the oscillation Reynolds number R_ω. In particular, if $Bi_1 = 0.1$ and $Bi_2 = 100$ such minimum occurs at $R_\omega \simeq 0.075$. This means that under these conditions, an optimal oscillation frequency exists that minimizes the flow irreversibilities.

The features of the optimal state of the system's performance, under-

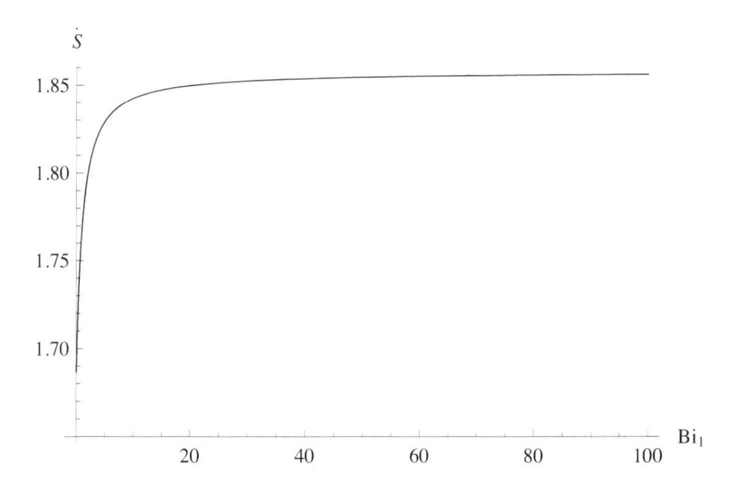

Fig. 7.3. The global entropy production $\dot{S}$ as a function of Bi_1 as $Bi_1 \to \infty$. The oscillation Reynolds number R_ω equals 0.1 while $\Theta_A = 10$, Pr $= 10$ and $Bi_2 = 100$ in this case.

stood as the one with minimum irreversible losses, may be further assessed by comparing the thermal and viscous phenomena responsible for the entropy production, represented in Eq. (7.20) by the first and the second terms on the right hand side, respectively. Consequently, in Fig. 7.5, the ratio of the first to the second term on the right hand side of Eq. (7.20), namely

$$R = \frac{1}{\Theta}\left(\frac{\partial \Theta}{\partial y^*} / \frac{\partial u^*}{\partial y^*}\right)^2, \tag{7.21}$$

was plotted against Bi_1 (taking the fixed value of $Bi_2 = 100$), for both the stationary and time-dependent cases. Observe how the oscillatory regime makes the thermal term dominate over the viscous one as Bi_1 increases. The stationary case shows exactly the opposite behavior.

Figure 7.6 displays the variation of the parameter R with respect to the oscillation Reynolds number for $Bi_1 = 10$ and $Bi_2 = 100$. A noteworthy feature in this case is the maximum of R at $R_\omega \simeq 0.8$.

To close this section, it must be mentioned that, although the corresponding plots will not be shown, the same previous analysis was performed for a Newtonian fluid under a time-periodic pressure gradient with boundary conditions of the first kind, that is $Bi_1 \to \infty$ and $Bi_2 \to \infty$. The main outcome of such analysis indicates that in this instance the global entropy

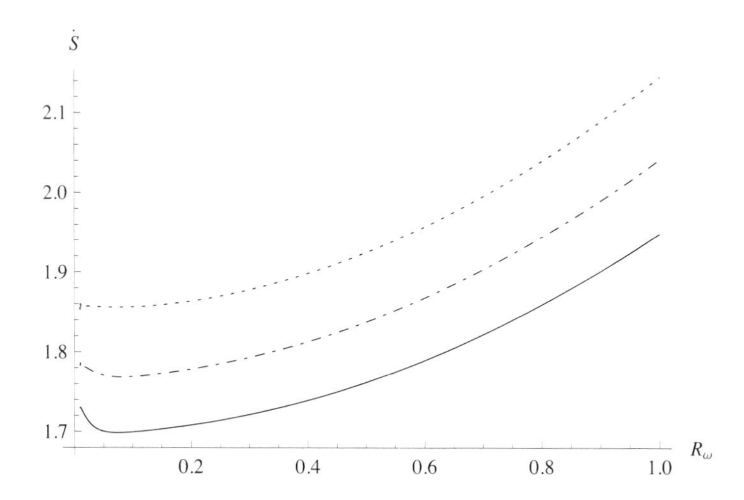

Fig. 7.4. The global entropy production $\dot{S}$ as a function of R_ω for different values of the upper plate Biot number Bi_1. Continuous line, $Bi_1 = 0.1$; dot-dashed line, $Bi_1 = 1$; dotted line, $Bi_1 = 100$. In all cases $\Theta_A = 10$, $\Pr = 10$ and $Bi_2 = 100$.

production does not show minima with respect to the oscillation frequency (or equivalently with respect to R_ω).

7.4. Concluding Remarks

In this chapter we have analyzed the problem of an oscillatory flow of a Newtonian fluid between infinite parallel plates and considered the fully thermally developed heat transfer problem with boundary conditions of the third kind. After deriving the velocity and temperature fields, we computed the global entropy production in the system assuming local equilibrium and found that, for fixed values of Pr, Θ_A and Bi_2, there exist minima with respect to the Biot number Bi_1 beyond a certain threshold value for the oscillation Reynolds number R_ω. In particular, if $\Theta_A = 10$, $\Pr = 10$ and $Bi_2 = 100$, these minima exist for $R_\omega \geq 0.1$. On the other hand and in contrast to previous cases[10,11] in which asymmetric convective cooling was required for a minimum global entropy production, for fixed values of Θ_A and Pr, in oscillatory flows with heat transfer boundary conditions of the third kind $\dot{S}$ may also display minima for particular values of the oscillating Reynolds number even in the case where $Bi_1 = Bi_2$. This difference in behavior between the stationary and oscillating cases, along with the one referring to the relative importance of thermal dissipation as compared

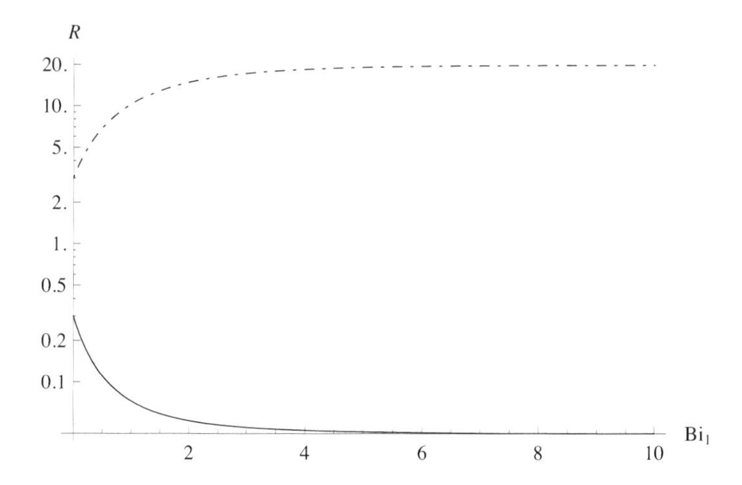

Fig. 7.5. The ratio R, as defined in Eq. (7.21), as a function of Bi_1. The continuous line corresponds to the stationary case and the dot-dashed line to the nonstationary one. The oscillation Reynolds number is 10 in the nonstationary case. The values of the other parameters are $\Theta_A = 10$, $\Pr = 10$ and $Bi_2 = 100$.

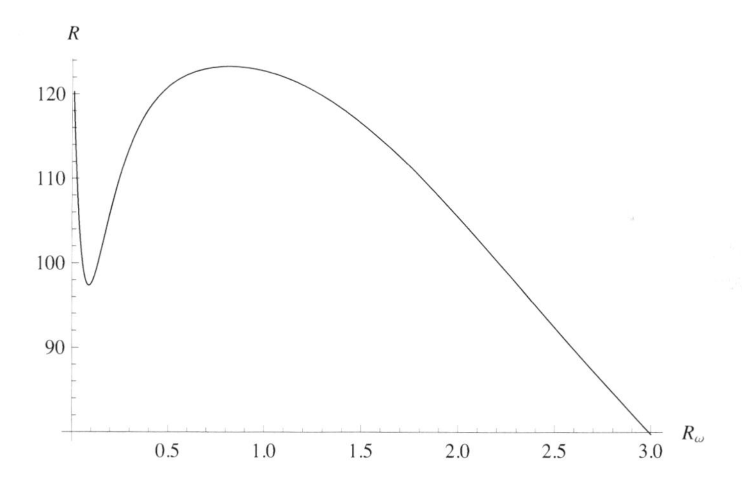

Fig. 7.6. The ratio R as a function of the oscillation Reynolds number R_ω. The plot was obtained with $Bi_1 = 10$, $Bi_2 = 100$, $\Pr = 10$ and $\Theta_A = 10$.

to viscous dissipation mentioned above, lie beyond the scope of this chapter but certainly deserve to be studied more thoroughly. Also worth of further study is the full heat transfer problem in the oscillating system. For instance, in order to include the convective term, one could attempt a different form to the one in Eq. (7.10) for the solution of the energy balance equation [Eq. (7.9], say $\Theta = \gamma(x^* + f(y^*)e^{it^*}$ where $\gamma \equiv \frac{\partial\Theta}{\partial x^*}$ is the (constant) dimensionless longitudinal temperature gradient and $x^* = x/a$; on the other hand, even with the present solution, one could also compute the Nusselt number to determine the heat transfer at the plates. This would then allow one to investigate the conditions where maximum heat transfer may be achieved and elucidate whether it is compatible with minimum entropy generation. Finally, the imposed boundary conditions and the oscillatory nature of the flow seem to enhance the thermal performance of the system around the minima of the global entropy production state where thermal effects dominate over viscous dissipative effects. In fact, one may find out the optimal working conditions (taking as the criterion a minimum global entropy production) which involve a careful selection of the frequency of oscillation and the Biot numbers for fixed values of Pr and Θ_A. Observe from Figs. 7.2, 7.3 and 7.6 that, if $\Theta_A = 10$, Pr $= 10$ and $Bi_2 = 100$, these conditions would be $R_\omega \sim 0.8$ and $Bi_1 \leq 1.0$.

Table 7.1. Greek symbols.

Symbol	Purpose
γ	dimensionless longitudinal temperature gradient, $\frac{\partial\Theta}{\partial x^*}$
η	shear viscosity, kg m^{-1} s^{-1}
ρ	mass density, kg m^{-3}
ω	angular frequency, radians s^{-1}
Θ	dimensionless temperature, $kT/\eta U_o^2$
Θ_A	dimensionless ambient temperature
Θ_s	steady contribution to the dimensionless temperature
Θ_u	unsteady contribution to the dimensionless temperature

Table 7.2. Subscripts.

Subscripts	Purpose
1	upper plate
2	lower plate
A	ambient
p	constant pressure
s	steady contribution
u	unsteady contribution
ω	oscillatory

Table 7.3. Nomenclature.

Symbol	Name
a	half the distance of separation between the parallel plates, m
A	constant of integration
b	auxiliary constant, $\sqrt{2R_\omega}$
B	constant of integration
Bi_1	upper plate Biot number, $h_1 a/k$
Bi_2	lower plate Biot number, $h_2 a/k$
c	auxiliary constant, $\sqrt{iR_\omega}$
C	constant of integration
C_p	specific heat at constant pressure, J kg^{-1} K^{-1}
D	constant of integration
G	magnitude of oscillatory pressure gradient, N m^{-3}
h_1	upper plate convective heat transfer coefficient, W m^{-2} K^{-1}
h_2	lower plate convective heat transfer coefficient, W m^{-2} K^{-1}
k	fluid thermal conductivity, W m^{-1} K^{-1}
K	auxiliary quantity, Ga^2/η, m s^{-1}
p	pressure, N m^{-2}
Pr	Prandtl number, $\eta C_p/k$
R	ratio of thermal to viscous dissipation, $\frac{1}{\Theta}\left(\frac{\partial\Theta}{\partial y^*}/\frac{\partial u^*}{\partial y^*}\right)^2$
R_ω	Oscillation Reynolds number, $\omega a^2\rho/\eta$
$\dot{S}^*$	dimensionless local entropy generation rate
$\dot{S}$	dimensionless global entropy generation per unit length in the x-direction
t	time, s
t^*	dimensionless time, ωt
T	fluid temperature, K
$\mathbf{u}$	fluid velocity vector, m s^{-1}
u_o	axial fluid velocity, m s^{-1}
u^*	axial dimensionless velocity, $u_0^*(y^*)e^{it^*}$
u_o^*	auxiliary dimensionless velocity, $u_0(y^*)/U_0$
U_o	average fluid velocity in the cross section, m s^{-1}
x	axial coordinate, m
x^*	dimensionless axial coordinate, x/a
y	transversal coordinate, m
y^*	dimensionless transversal coordinate, y/a

Acknowledgments

The work of M.L.H. has been supported by the Ministerio de Educación y Ciencia (Spain) through Grant No. FIS2007-60977 (partially financed by Feder funds) and by the Junta de Extremadura through Grant No. GRU09038. F. V. akcnowledges financial support of CONACYT and PROMEP (Mexico) and M. A. O.-R. that of Project PIFI CGPI No. 20090832 from IPN (Mexico).

References

1. I. Prigogine, Introduction to Thermodynamics of Irreversible Processes, Interscience, New York, 1961.
2. S. R. de Groot and P. Mazur, Non-Equilibrium Thermodynamics, North Holand, Amsterdam, 1962; reprinted by Dover, New York, 1984.
3. A. Bejan, Entropy Generation through Heat and Fluid Flow, Wiley, New York, 1994.
4. A. Bejan, Minimization of Entropy Generation, CRC Press, Boca Raton, 1996.
5. It would be too lengthy to quote all the work Prof. García-Colín has done on this subject. Suffice it to mention here his relatively recent book with P. Goldstein (P. Goldstein and L. S. García-Colín, La Física de los Procesos Irreversibles, Vols. I and II, El Colegio Nacional, México, 2003) and the three volumes of his collected works edited by El Colegio Nacional (L. García-Colín Scherer, Obras 3: Fundamentación microscópica de los procesos irreversibles I, El Colegio Nacional, México, 2003; Obras 4: Fundamentación microscópica de los procesos irreversibles II, El Colegio Nacional, México, 2005; and Obras 5: Termodinámica irreversible extendida, El Colegio Nacional, México, 2005).
6. U.H. Kurzweg, Enhanced heat conduction in fluids subjected to sinusoidal oscillations, J. Heat Transfer 107 (1985) 459-462.
7. U.H. Kurzweg, Enhanced heat conduction in oscillating viscous flows within parallel-plate channels, J. Fluid Mech. 156 (1985) 291-300.
8. V. Yakhot, C. Colosqui, Stokes' second flow problem in a high-frequency limit: application to nanomechanical resonators, J. Fluid Mech. 586 (2007) 249-258.
9. G. Ibáñez, S. Cuevas, M. López de Haro, Optimization analysis of an alternate magnetohydrodynamic generator, Energy Conversion and Management 43 (2002) 1757-771.
10. G. Ibáñez, S. Cuevas, M. López de Haro, Minimization of entropy generation by asymmetric convective cooling, Int. J. Heat Mass Transfer 46 (2003) 1321-1328.
11. G. Ibáñez, S. Cuevas, M. López de Haro, Heat transfer in asymmetric convective cooling and optimized entropy generation rate, Rev. Mex. Fís. 49(2) (2003) 338-343.
12. G. Ibáñez, S. Cuevas, Optimum wall conductance ratio in magnetoconvective flow in a long vertical rectangular duct, Int. J. Thermal Sci. 47 (2008) 1012-1019.
13. F. Vázquez, M. A. Olivares-Robles, S. Cuevas, Viscoelastic effects on the entropy production in oscillatory flow between parallel plates with convective cooling, Entropy 11 (2009) 4 - 16.
14. I. G. Currie, Fundamental Mechanics of Fluids, Second Edition, McGraw-Hill, New York, 1993.
15. Should one require them, the actual lengthy expressions for these constants, as well as the Mathematica notebook that was used to derive them, are available upon request.

Chapter 8

Wien's Law with Zero-Point Energy Implies Planck's Law Unequivocally

A. Valdés, L. de la Peña and A. M. Cetto

Instituto de Física, Universidad Nacional Autónoma de México, Apartado Postal 20-364, 01000 México, D.F., México

andreavh, luis, ana@fisica.unam.mx

The assumption of a zero-point energy to derive Planck's law within a continuous (non-quantum) perspective and a thermodynamic context has been studied previously. However, some of the arguments involved in such derivations, even though physically sustained, lack of a demonstrative character. In this paper we address these deficiencies. In doing so, we complete a sound demonstration showing that Planck's law —and hence the ensuing discrete description— is a necessary consequence of the existence of the zero-point zero-point energy. We also disclose the relevance of Wien's law for reaching this conclusion.

It is with great pleasure that we pay homage to our highly estimated friend and colleague Professor Leopoldo García Colín.

Contents

8.1. Introduction

In a recent paper[1] we have shown that Planck's distribution can be derived within an approach in which no explicit discontinuity is introduced, neither in the structure of the radiation field nor in the mechanism of its interaction with matter, by considering instead an *extended* form of classical thermodynamics along with a statistical description of a system of harmonic oscillators.

The starting point for extending the classical thermodynamic description consists in writing the mean energy of the oscillators of frequency ω in equilibrium at temperature T, $U(\omega, T)$, in the form of Wien's law,

$$U(\omega, T) = \omega f(\omega/T), \tag{8.1}$$

a result that follows directly from the thermodynamic relations of the system of harmonic oscillators[2]. The observation that constitutes the definitive point of deviation from the usual classical description consists simply in recognizing, as is done in Ref. 2, that at $T = 0$ Wien's law (8.1) accepts a solution of the form

$$\mathcal{E}_0 \equiv U(\omega, 0) = A\omega, \tag{8.2}$$

with A a non-zero constant. In the customary form of the theory the constant A is taken as zero, thereby excluding *arbitrarily* the existence of a non-thermal energy and allowing *only* for a thermal energy of the oscillators. In contrast, the extended treatment admits a zero-point energy for the oscillators that is different from zero.

Clearly this simple and natural introduction of the zero-point energy, being contrary to classical equipartition, falls outside the domain of classical physics and thus opens the door to interesting physical consequences. These have been the subject of an initial exploration in Refs. 1 and 2 from a thermodynamic approach. In Ref. 2 Boyer derived Planck's law using an interpolation procedure between energy equipartition at high temperatures and zero-point energy at low temperatures. However, the relevance of the result obtained makes the interpolation procedure somewhat unsatisfactory and so, in order to avoid it, a statistical analysis was developed in Ref. 1 to show that the mere existence of the zero-point energy is sufficient to recover Planck's distribution.

However, the derivation carried out in Ref. 1 can still not be considered entirely flawless. As is made clear below, arguments in the cited reference were given in due place to sustain that a function that appears along the calculations is identically null, even though this was not formally demonstrated. In the present paper we pay attention to this flaw and provide a demonstration of the legitimacy of our previous choice. In doing so, we complete the proof needed to give sound support to the idea of the reality of the zero-point energy, while also exhibiting the richness of Wien's law, as is discussed in the concluding section.

The paper is organized as follows. In section 8.2 we briefly present the basic contents of Ref. 1, limiting ourselves essentially to explain how the mean energy $U(\beta)$ —and hence Planck's distribution— can be obtained within the present approach. Then in section 8.3 we show the point at which the unknown function, arguably taken to be identically zero, appears. Thereafter, in section 8.4, we offer two independent demonstrations of the validity of this assumption. Finally, in the last section we add some concluding remarks.

8.2. Role of the Energy Variance in Determining the Statistical Properties of the System

To determine the statistical properties of the field oscillators at equilibrium inside a blackbody cavity we look for a probability distribution $W(E)$ such that the entropy S,

$$S = -k_B \int W \ln W dE, \tag{8.3}$$

(k_B is the Boltzmann constant) is consistent with the formalism of maximum entropy. This requirement leads to write $W(E)$ in the general form[3,4]

$$W_g(E)dE = \frac{1}{Z_g(\beta)} g(E) e^{-\beta E} dE, \tag{8.4}$$

$$Z_g(\beta) = \int g(E) e^{-\beta E} dE, \tag{8.5}$$

where $\beta = 1/(k_B T)$, $Z_g(\beta)$ is the partition function and $g(E)$ is a weight function representing a possible intrinsic probability for the states with energy E. The distribution (8.4) is a generalization of the (classical) Boltzmann distribution, which corresponds to the selection of equal intrinsic probability for all energy states, $g(E) = 1$,

$$W_1(E) \equiv W_{g=1}(E) = \frac{1}{Z_1(\beta)} e^{-\beta E}, \tag{8.6a}$$

$$Z_1(\beta) = \int_0^\infty e^{-\beta E} dE = \frac{1}{\beta} = k_B T, \tag{8.6b}$$

$$\langle E \rangle = U = -\frac{1}{Z_1} \frac{dZ_1}{d\beta} = \frac{1}{\beta} = k_B T. \tag{8.6c}$$

This last equation shows that the selection $g(E) = 1$ reproduces the classical description, since it implies $U(T = 0) = \mathcal{E}_0 = 0$ and hence the absence of a zero-point energy. By contrast, the presence of a non-thermal energy demands that $g(E)$ be a nontrivial function. Its form for the present case is derived in Ref. 1 and given in Eq. (8.22).

From the general distribution $W_g(E)$ it follows that (r is a positive integer)

$$\langle E^r \rangle' = -\frac{Z_g'}{Z_g} \langle E^r \rangle - \frac{1}{Z_g} \int_0^\infty E^{r+1} g(E) e^{-\beta E} dE = -\frac{Z_g'}{Z_g} \langle E^r \rangle - \left\langle E^{r+1} \right\rangle, \tag{8.7}$$

where the prime denotes derivation with respect to β. Noting that

$$\langle E \rangle = U = \frac{1}{Z_g} \int_0^\infty E g(E) e^{-\beta E} dE = -\frac{Z_g'}{Z_g}, \tag{8.8}$$

Eq. (8.7) gives the following recurrence relation

$$\langle E^{r+1} \rangle = U \langle E^r \rangle - \langle E^r \rangle'. \tag{8.9}$$

In particular, for $r = 1$ this expression gives for the variance σ_E^2 of the energy

$$\sigma_E^2 \equiv \left\langle (E - U)^2 \right\rangle = \left\langle E^2 \right\rangle - U^2 = -\frac{dU}{d\beta}, \tag{8.10}$$

which allows us to cast the recurrence relation in the form

$$\left\langle E^{r+1} \right\rangle = U \left\langle E^r \right\rangle + \sigma_E^2 \frac{d \left\langle E^r \right\rangle}{dU}. \tag{8.11}$$

From this result it follows that all moments $\langle E^r \rangle$ for $r > 2$ are determined by the first two, U and $\left\langle E^2 \right\rangle$, which means that the distribution W_g is completely determined once we know these two quantities, or equivalently once we know U and σ_E^2.

Equation (8.10) allows us to rewrite σ_E^2 in the customary form[4]

$$\sigma_E^2 = -\left(\frac{\partial U}{\partial T}\right)_\omega \frac{dT}{d\beta} = k_B T^2 \left(\frac{\partial U}{\partial T}\right)_\omega = k_B T^2 C_\omega, \tag{8.12}$$

where $C_\omega = \left(\frac{\partial U}{\partial T}\right)_\omega$ is the heat capacity at constant volume (and constant ω, since the frequency is being treated as a fixed parameter). Because C_ω is a finite quantity at any temperature, including $T = 0$, Eq. (8.12) gives

$$\sigma_E^2(T = 0) = 0, \tag{8.13}$$

hence the fluctuations of the energy described by the probability distribution W_g are purely thermal. Hence, according to this thermodynamic description, even if the energy $E(T)$ is a fluctuating quantity, $E(0)$ is a fixed, nonfluctuating one. This substantial point, and the possibility and implications of recovering the zero-point (non-thermal) fluctuations, are discussed at length in Ref. 1.

Inverting Eq. (8.10) we observe that once the variance of the energy σ_E^2 is determined as a function of the mean energy U, $\sigma_E^2(U)$ (subject to condition (8.13)), we may write

$$d\beta = -\frac{dU}{\sigma_E^2(U)}, \tag{8.14}$$

so that an integration allows us to obtain $\beta = \beta(U)$ and from here we can determine $U(\beta)$ by inversion. The problem of finding $U(\beta)$ is thus reduced to the determination of $\sigma_E^2(U)$. From here it follows, according to what has been said immediately after Eq. (8.11), that the sole determination of $\sigma_E^2(U)$ is sufficient to determine $W_g(E)$ and therefore the complete statististics of the system of oscillators.

8.3. Planck's Law Follows Necessarily

Let us now sketch the main ideas developed in Ref. 1 for determining the functional relation $\sigma_E^2(U)$, in preparation for the present derivation. We start by making a series expansion of $\sigma_E^2(U)$ in the form

$$\sigma_E^2(U) = \sum_{k=0}^{\infty} a_{2k} U^{2k} = a_0 + a_2 U^2 + \varphi(U), \tag{8.15}$$

where only even powers are included since, as shown in Ref. 1, $\sigma_E^2(U)$ is an even function of U.[a]

Because the constants a_{2k} do not depend on T (but at most on the parameter ω), they can be partially fixed by resorting to the high and low temperature limits. As $T \to \infty$, the zero-point energy becomes negligible

[a]The result follows by noticing that $\sigma_E^2 \equiv \langle (E - U)^2 \rangle$ is invariant under the formal inversion $E \to -E$, which leads to the simultaneous substitution $U \to -U$.

and all moments must reduce to their classical values. Using the classical distribution (8.6a) for this limit we find $\langle E^2 \rangle = 2U^2$ (with $U = k_B T$), so that σ_E^2 must satisfy

$$\sigma_E^2(U) \to U^2 \quad (U \to \infty). \tag{8.16}$$

On the other hand, for $T \to 0$, Eqs. (8.2) and (8.13) imply

$$\sigma_E^2(\mathcal{E}_0) \to 0, \quad U \to \mathcal{E}_0. \tag{8.17}$$

The last two equations can be satisfied simultaneously by writing $a_0 = -\mathcal{E}_0^2$, $a_2 = 1$ and $\varphi(U)$ subject to the conditions

$$\varphi(U) \to 0, \quad U \to \mathcal{E}_0, \infty. \tag{8.18}$$

Accordingly, we can write

$$\sigma_E^2(U) = U^2 - \mathcal{E}_0^2 + \varphi(U), \tag{8.19}$$

where the function $\varphi(U)$ vanishes both in the high and low temperature limits. At this point several arguments were given in Ref. 1 to conclude that $\varphi(U)$ is zero at *all* temperatures. Equation (8.19) reduced then to

$$\sigma_E^2 = U^2 - \mathcal{E}_0^2. \tag{8.20}$$

As was stated above, $U(\beta)$ can now be obtained by inserting this result into Eq. (8.14), integrating and then inverting, a procedure that leads directly to

$$U(\beta) = \begin{cases} \dfrac{1}{\beta}, & \text{if } \mathcal{E}_0 = 0; \\ \mathcal{E}_0 \coth \mathcal{E}_0 \beta, & \text{if } \mathcal{E}_0 \neq 0. \end{cases} \tag{8.21}$$

By setting the constant A in Eq. (8.2) equal to $A = \hbar/2$, the last line of this expression shows that indeed Planck's law (with the zero-point energy $\frac{1}{2}\hbar\omega$ included) is an *unavoidable* consequence of the existence of a non-thermal energy $\mathcal{E}_0$ (in the absence of this energy we obtain the classical equipartition result, as seen in the first line of Eq. (8.21)).

Even if not necessary for the purposes of the present work, it is pertinent to somehow complete the exposition by noting that from Planck's law one gets for the intrinsic probability function $g(E)$ the expression

$$g(E) = 2\mathcal{E}_0 \sum_{n=0}^{\infty} \delta(E - E_n), \tag{8.22}$$

where the energy levels E_n are given by

$$E_n = \mathcal{E}_0(2n+1) = \hbar\omega(n + \tfrac{1}{2}), \tag{8.23}$$

in full agreement with those of the quantum harmonic oscillator. This result shows how the discrete features of the theory come about, since the probability density (8.4) turns out to be

$$W_g(E) = \frac{1}{Z}\sum_{n=0}^{\infty}\delta(E - E_n)e^{-\beta E} \qquad \left(Z = \frac{1}{2\sinh \mathcal{E}_0\beta}\right). \tag{8.24}$$

The energy of the stationary states has now a *discrete* distribution of values, even though the energy E is a *continuous* variable. A detailed discussion of these important matters is given in Ref. 1.

As is clear from the results of the present section, and recalling the discussion at the end of section 8.2, the determination of $\sigma_E^2(U)$ in its form (8.20) is enough (together with the selection $\mathcal{E}_0 \neq 0$) to derive Planck's law and therefore sufficient to arrive at the quantum description of the system of oscillators. Our aim in the next section is to convincingly demonstrate that indeed the function $\varphi(U)$ in Eq. (8.19) is identically zero, and thereby provide a cogent derivation of the Planck distribution from the hypothesis of the existence of the (fixed) zero-point energy of value $\mathcal{E}_0 = \hbar\omega/2$.

8.4. Wien's Law and the Variance of the Energy

8.4.1. *First derivation*

Let us now proceed to demonstrate that $\varphi(U)$ is indeed equal to zero. We start by rewriting Wien's law in the form (from now on we put $k_B = 1$)

$$U(\omega, T) = \omega f_1(z), \quad z \equiv \frac{\omega}{T} = \omega\beta. \tag{8.25}$$

Combining this equation with (8.10) we obtain

$$\sigma_E^2 = -\frac{\partial U}{\partial z}\frac{dz}{d\beta} = -\omega\frac{\partial U}{\partial z} = -\omega^2\frac{df_1}{dz} \tag{8.26}$$

or

$$\sigma_E^2 = \omega^2 f_2(z), \tag{8.27}$$

where we have defined $f_2(z) = -df_1/dz$.

Further, when writing the variance σ_E^2 as a function of U we must resort once more to Wien's law (Eq. (8.25)) and write

$$\sigma_E^2(U) = \sigma_E^2(\omega f_1(z)). \tag{8.28}$$

Equations (8.27) and (8.28) together lead to

$$\omega^2 f_2(z) = \sigma_E^2(\omega f_1(z)). \tag{8.29}$$

Since f_1 and f_2 depend on the frequency ω only through z, the fact that the left hand side of Eq. (8.29) must be a function of the variable $\omega f_1(z)$ means that the nontrivial solution for f_2 has the form

$$f_2(z) = Bf_1^2(z) + C, \tag{8.30}$$

with B and C constant. Thus,

$$\sigma_E^2(\omega f_1) = \omega^2 \left(Bf_1^2 + C\right). \tag{8.31}$$

Using Eq. (8.25) we get

$$\sigma_E^2(U) = BU^2 + C\omega^2. \tag{8.32}$$

Now we impose the demand $\sigma_E^2(\mathcal{E}_0) = 0$ (Eq. (8.13)) which, using Eq. (8.2), fixes the constant $C = -BA^2$, thus leading to

$$\sigma_E^2(U) = B\left(U^2 - A^2\omega^2\right) = B\left(U^2 - \mathcal{E}_0^2\right). \tag{8.33}$$

The classical limit (8.16) fixes $B = 1$, so that we finally recover Eq. (8.20). In other words, the function $\varphi(U)$ vanishes for all values of U, hence for all temperatures. We thus see that Wien's law is enough to determine the functional form of the variance $\sigma_E^2(U)$.

8.4.2. *Second derivation*

Let us now present an alternative, more revealing derivation of Eq. (8.20). Our point of departure is the even parity of σ_E^2 as a function of U (as stated in Sec. 8.3),

$$\sigma_E^2(U) = \sigma_E^2(-U). \tag{8.34}$$

We consider now the condition (8.13),

$$\sigma_E^2(\mathcal{E}_0) = 0. \tag{8.35}$$

Equations (8.34) and (8.35) imply that since $\mathcal{E}_0$ is a root of the function $\sigma_E^2(U)$, then also $-\mathcal{E}_0$ is a root. Thus we can expand $\sigma_E^2(U)$ in the general form

$$\sigma_E^2(U) = (U + \mathcal{E}_0)^m \,(U - \mathcal{E}_0)^n \, f_3(U), \tag{8.36}$$

where $m, n = 1, 2, ...$ and $f_3(U)$ is a function (with no zeros) to be determined, for which we write

$$f_3(U) = \sum_{s=0}^{\infty} b_s U^s. \tag{8.37}$$

The coefficients b_s are independent of ω and T (otherwise f_3 could not be written as a function of the argument $\omega f_1(z)$, as demanded by Wien's law). Substituting Eq. (8.37) into Eq. (8.36) we get (with $U = \omega f_1$ and $\mathcal{E}_0 = A\omega$)

$$\sigma_E^2(\omega, T) = \sum_{s=0}^{\infty} b_s \omega^{m+n+s} f_4(z) f_1^s(z), \tag{8.38}$$

where we have defined

$$f_4 = (f_1 + A)^m (f_1 - A)^n . \tag{8.39}$$

From condition (8.27) it follows that Eq. (8.38) is such that it should be possible to write

$$\omega^2 f_2(z) = \sum_{s=0}^{\infty} b_s \omega^{m+n+s} f_4(z) f_1^s(z). \tag{8.40}$$

Since the coefficients b_s do not depend on ω nor T, Eq. (8.40) will be satisfied provided

$$b_s = \begin{cases} 0, & \text{if } s \neq 2 - (m+n); \\ B, & \text{if } s = 2 - (m+n). \end{cases} \tag{8.41}$$

Further, since s can take the values $0, 1, 2, \ldots$ the condition $s = 2 - (m+n)$ implies

$$m + n \leq 2. \tag{8.42a}$$

Also, given that $m, n = 1, 2, \ldots$, it always happens that

$$2 \leq m + n. \tag{8.42b}$$

Thus $m = n = 1$ (the result $m = n$ should be expected from symmetry considerations applied to Eq. (8.36)), and therefore, according to Eq. (8.41), only the term corresponding to $s = 0$ contributes to the sum (8.37). This reduces $f_3(U)$ to the constant value B and consequently Eq. (8.36) becomes

$$\sigma_E^2(U) = B\,(U + \mathcal{E}_0)\,(U - \mathcal{E}_0) = B(U^2 - \mathcal{E}_0^2). \tag{8.43}$$

This result coincides with Eq. (8.33), so that by using the classical limit to set $B = 1$ we once again obtain Eq. (8.20).

8.5. Concluding Remarks

By proving definitely that the variance of the energy can be written in the form of Eq. (8.20), we complete the demonstration that Planck's law follows necessarily and naturally from the assumption of a zero-point energy, without the need to introduce any discrete postulate. The role played by Wien's law along the derivation can hardly be overestimated. It is the entry point for the zero-point energy into the theory, but it is also the law that ultimately leads us to write the energy variance in the form (8.20). In other words, Wien's law is sufficient to determine the functional form of the variance $\sigma_E^2(U)$. Given that $\sigma_E^2(U)$ completely determines the mean energy of the oscillators, we conclude that Wien's law (together with the selection $\mathcal{E}_0 \neq 0$, of course) is all that is needed to derive Planck's law , whence the former seems to stand as a more fundamental law than the latter. Moreover, the fact that Wien's law can be deduced exclusively from dimensional considerations[5] forces us to find the *physical* root of Planck's law —and with it the physical root of the discontinuous (quantum) description of the oscillators— in the selection $\mathcal{E}_0 \neq 0$, that is, in the existence of the zero-point energy. Given the experimental evidence of Planck's law, this result exhibits the zero-point field as a *real* physical entity, rather than *virtual* (as is normally considered).[b]

Acknowledgments

The authors gratefully acknowledge a comment by Professor T. H. Boyer that prompted the present work. One of the authors (AVH) acknowledges financial support from the Instituto de Ciencia y Tecnología del Distrito Federal.

References

1. L. de la Peña, A. Valdés-Hernández and A. M. Cetto, *Statistical consequences of the zero-point energy of the harmonic oscillator*, Am. J. Phys. **76** (10), 947 (2008)
2. T. H. Boyer, *Thermodynamics of the harmonic oscillator: Wien's displacement law and the Planck spectrum*, Am. J. Phys. **71**, 866–870 (2003)

[b]What is commonly considered as real are the *fluctuations* of the zpf, not the zero-point field itself. The non-thermal fluctuations of the zpf are discussed in Ref. 1.

3. E. W. Montroll and M. F. Shlesinger, *Maximum entropy formalism, fractals, scaling phenomena, and 1/f noise: A tale of tails*, J. Stat. Phys. **32**, 209 (1983)
4. K. Huang, *Statistical Mechanics*, John Wiley & Sons, New York (1963)
5. A. Sommerfeld, *Thermodynamics and statistical mechanics. Lectures on theoretical Physics,* vol. 5, Section II.20, New York Academic Press, New York (1956)

Chapter 9

A Functional Study for Newton's Endoreversible Engines

José Luis del Río-Correa

Departamento de Física,
Universidad Autónoma Metropolitana Unidad Iztapalapa,
jlrc@xanum.uam.mx

In this work a theoretical method is presented instead of the usual numerical method to analyze the Newton's endoreversible engines.The power output function and the ecological function are expressed in dimensionless form, a functional relationship between these functions is found. It essentially shows that the ecological function is proportional to the re-scaled power output function. Using this result two additional functional relationships are found, the first between the upper operational temperatures: T_a^P and T_a^E ,and the second between the efficiencies η_P and η_E.

Using two physical facts on the heat flux and on the efficiency, the general analytical behavior of the dimensionless power output function is obtained, and combining it with the functional relationship the analytical behavior of the ecological function is obtained. Using these analytical results and the functional relationships between the upper operational temperatures, and between the efficiencies, it is shown that for Newton's endoreversible engines the following inequalities: $T_a^E > T_a^P$ and $\eta_E > \eta_P$ are always satisfied.

It is shown that from the operational point of view, the functional relationship reduces the amount of numerical calculation. A unified procedure to obtain the upper operational temperature and the efficiency for a Newton's endoreversible engine, upon using either the maximum power output or the maximum ecological function criteria is presented.

Contents

9.1. Introduction

We begin with a briefly description of an endoreversible engine, this is essentially an internally reversible engine making a Carnot cycle and coupled in an irreversible way with external baths, i.e. the interchange of the heat between the baths and the internally reversible engine takes a finite time and is irreversible. An endoreversible engine (EE) is defined as one where the two processes of heat transfer (from and to the heat reservoirs) are the only irreversible processes in the cycle.

The problem is to fix the upper and lower operational temperatures (OT) of the Carnot engine using one criterion of the "best performance" for the EE. In 1975 Curzon and Ahlborn,[1] proposed that the best performance is obtained when the power output function (POF) is maximum, they expressed it in terms of the OT of the internally reversible Carnot engine, they introduced the endoreversible assumption which relates the ratio of the OT with the ratio of the average heat fluxes between the external baths and the Carnot engine, and determined the values of the upper operational temperature (UOT) and lower operational temperature (LOT) which maximize the POF under the restriction given by the endoreversible assumption, with these OT they obtained the efficiency of the Carnot engine at maximum power output function criteria ($MPOFC$). A pedagogical presentation of the Curzon and Ahlborn's paper is given in the reference.[2] After the appearance of Curzon and Ahlborn's paper, a strong interest was developed for the so called thermodynamics of finite time, because with their not very complicated model a reasonable agreement with the efficiency of real engines was obtained; but of course, to obtain a best agreement with the real systems, it was necessary to take into account additional factors, in fact one important factor is the entropy production of the of the cyclic process, due to its relation with the loss of available work.[2] This last fact introduced another criteria for the best performance of the EE. In 1981, Angulo-Brown[3] introduced the ecological function (EF), defined as the POF minus the entropy production σ times T_c the temperature of the external cold bath, this quantity is related to the useful power available in a process, because $T_c\sigma$ is the power dissipated by irreversibilities.[2] He determined the OT of the Carnot cycle maximizing the EF, and evaluated the efficiency, he found that this efficiency is greater than the maximum power efficiency.

In both papers[1] and,[3] it was considered that the endoreversible assumption is valid, the engine is a Carnot one, and the heat fluxes between the external baths and the Carnot engine were modeled by the Newton's law of heat transport. In 2000, Sahin[4] published a paper about the optimum operating conditions of solar driven heat engines, where he relaxed the condition that the heat flux from the hot bath to the Carnot engine was given by the Newton's law. He took into account that the heat is transferred by radiation and convection modes, he proposed that the heat flux from the hot bath to the Carnot engine was model by the Stephan-Boltzmann's radiation heat law. To obtain the optimal OT of the Carnot cycle, Sahin used the $MPOFC$, but he introduced a more simple procedure than the Curzon and Ahlborn's method. In Sahin's procedure the endoreversible assumption is introduced since the beginning into the POF, thus it only depends of one of the OT of the Carnot cycle, and the process to maximize the POF is simplified. He found that the efficiency of this EE is greater than the Curzon and Alhborn's efficiency (CAE) and he concluded that the CAE is not a fundamental upper limit on the efficiency of a cycle heat engine operating at $MPOFC$.

In 2008, Barranco et al[5] used Sahin's procedure with the maximum ecological function criterium $MEFC$. They considered that although the heat flux from the Carnot engine to the cold bath satisfies the Newton's law, the heat flux from the hot bath to the Carnot engine is governed by different heat transfer laws, in fact they considered the Stephan-Boltzmann's law, the Dulong-Petit's law and a linear combination of the Stephan-Boltzmann's radiation law with Newton's conduction law. They compared the results obtained with the $MPOFC$ with the corresponding results obtained when is used the $MEFC$, and they found that for all cases the efficiency with the $MEFC$ was greater than the efficiency with the $MPOFC$, and also corroborated the Sahin's result that the CAE doesn't represent a upper limit for the efficiency of the endoreversible engine.

As is usual in papers about EE, Sahin and Barranco obtain their results and conclusions using numerical calculations, however a theoretical framework for EE is lacking which will allow to distinguish between particular and general results, and which will explain the reasons of some results.

In this work a functional study of the EE is presented. The focus of the analysis is to unify the $MPOFC$ with the $MEFC$. This work is restricted to the case where the heat flux from the Carnot engine to the cold external bath is modeled by the Newton's law, but the analytical form of the heat flux from the hot bath to the Carnot engine is arbitrary. When it is the case the family of EE is known as Newton's endoreversible engines (NEE). The structure of the work is the following:

In Section 9.2, Sahin's procedure is applied to the NEE, the POF and the EF are obtained in terms of the UOT, these are the expressions that are used for the numerical calculations,reduced the problem to numerically find the maximum of these functions, and usually at this point all theoretical consideration is finish.

In Section 9.3, dimensionless variables are introduced, the dimensionless power output function ($DPOF$) and the dimensionless ecological function (DEF) are expressed in terms of them, With these expressions for $DPOF$ and DEF a functional relationship (FR) is obtained between them, one finds essentially that the DEF is proportional to the re-scaled $DPOF$. This relationship reflects the fact that the EF is the useful power available in the cycle. Using this result one established FR between the UOT obtained correspondingly with both criteria. Afterwards a FR between the efficiencies is obtained.

In Section 9.4, one obtains for any hot flux the general analytical behavior of the $DPOF$, as an implication of two simple facts: the dimensionless hot flux (DHF) is a decreasing function, and the efficiency of the NEE is positive but less than one. In particular it is shown that the $DPOF$ is a convex function, which obtains its maximum value for the dimensionless quantity $\theta_P(\alpha)$ associated with the $UOT : T_a^P$ determined when the $MPOFC$ is used. It is also established that $\theta_P(\alpha)$ is always an increasing function of the ratio of the temperatures of the external baths.

In section 9.5, the FR between the DEF and the $DPOF$ is used to obtain the general properties of the DEF, and it is shown that always the $UOT : T_a^E$ of the Carnot cycle, and the efficiency η_E obtained maximizing the DEF are greater than the $UOT : T_a^P$ and the efficiency η_P obtained by maximization of the $DPOF$ respectively.

In section 9.6, the $MPOFC$ and the $MEFC$ are unified with a simple procedure to obtain the UOT and the efficiency for a NEE.

In section 9.7 results and conclusions are presented.

9.2. Sahin's Procedure for Newton's Endoreversible Engines

In this section the analysis of the EE following the Sahin's procedure[4] is presented. It generalizes for arbitrary hot fluxes the Curzon and Ahlborn's treatment to find the endoreversible engines OT and the efficiency when the $MPOFC$ is used. And, it can also be extended to find the same quantities when the $MEFC$ is selected. In the next section we follow this procedure to construct a unified treatment to find these quantities when use is made of the $MPOFC$ or the $MEFC$.

Let us suppose that two thermal reservoirs exist, at temperatures T_h and T_c, and that we extract a quantity of heat $Q_h > 0$ from the high temperature reservoir, to feed it to a Carnot engine working between the $OT : T_a$ and T_b ($T_h > T_a > T_b > T_c$). After inserting a quantity of heat $Q_c < 0$ to the cold reservoir at T_c, and complete the cycle work is deliver to a work sink.

Let it be t_h the time required to transfer an amount of heat Q_h from the thermal bath at temperature T_h to the isothermal branch of the engine cycle at the $UOT : T_a$, and let it be t_c the time required to transfer the amount of heat Q_c from the isothermal branch of the Carnot cycle at the LOT : T_b to the cold bath. Let us assume the time required for the two adiabatic branches is negligible relative to $(t_h + t_c)$; therefore the time required for a complete cycle is $t = (t_h + t_c)$.

When we use the first law for the Carnot cycle, we have that the delivered work is given by $|W| = Q_h + Q_c$ and upon introducing the endoreversible assumption: $\frac{Q_c}{Q_h} = -\frac{T_b}{T_a}$, then the delivered work is given by $|W| = \eta Q_h$ with $\eta = 1 - \frac{T_b}{T_a}$. The mean heat fluxes are given by:

$$J_h = \frac{Q_h}{t_h}; \quad J_c = \frac{-Q_c}{t_c}; \tag{9.1}$$

The endoreversible assumption then implies the next relation between them,

$$\frac{J_h(T_h, T_a)}{T_a} = \tau \frac{J_c(T_b, T_c)}{T_b} \tag{9.2}$$

with $\tau = t_c/t_h$. For a NEE the heat flux to the cold bath obeys the Newton's law:

$$J_c = \kappa_c(T_b - T_c) \tag{9.3}$$

we introduce this fact into the Eq. (9.2), and obtain a relationship between the LOT and UOT,

$$T_b = \frac{T_c}{1 - \frac{J_h(T_h,T_a)}{\tau\kappa_c T_a}} \tag{9.4}$$

Now, we find the efficiency η of the NEE in terms of T_a, to be given by

$$\eta = 1 - \frac{T_b}{T_a} = 1 - \frac{T_c}{T_a - \frac{1}{\tau\kappa_c} J_h(T_h,T_a)} \tag{9.5}$$

The engine power output P is given by $P = \frac{|W|}{t_h+t_c} = \frac{\eta}{1+\tau} J_h$, therefore it can be expressed in terms of T_a, as

$$P = \frac{\eta}{1+\tau} J_h(T_h,T_a) = \frac{1}{1+\tau}\left[1 - \frac{T_c}{T_a - \frac{1}{\tau\kappa_c} J_h(T_h,T_a)}\right] J_h(T_h,T_a) \tag{9.6}$$

We note from Eqs.(9.5) and (9.6) that the efficiency and the POF depend of the UOT and not on the LOT; this is the advantage of Sahin's procedure upon Curzon and Ahlborn's procedure, because in the last one the POF and the efficiency depend on both OT and afterwards, the POF is maximized with the additional constriction given by the endoreversible assumption. In contrast in Sahin's procedure this constriction was incorporated since the beginning to obtain the efficiency and the POF as functions of the UOT, which is the independent variable

Curzon and Ahlborn[1] selected the UOT as the value T_a^P which maximizes the POF. On the other hand Angulo-Brown[3] choose the UOT as the value T_a^E which maximizes the EF which takes into account both the POF and the entropy production function. Of course, there might be other criteria to select T_a, but in the present work only the $MPOFC$ and the $MEFC$ and their relationship will be consider. The EF is defined by:

$$E = P - T_c\sigma \tag{9.7}$$

where σ denotes the entropy production of the NEE given by:

$$\sigma = \frac{1}{t_h+t_c}\left(\frac{-Q_h}{T_h} + \frac{-Q_c}{T_c}\right) = \frac{1}{1+\tau}\left(\tau\frac{J_c}{T_c} - \frac{J_h}{T_h}\right) \tag{9.8}$$

here the second equality follows from Eq. (9.1), for the NEE one has that:

$$\tau\frac{J_c}{T_c} = \tau\kappa_c\left(\frac{T_b}{T_c} - 1\right) = \left(\frac{1}{1 - \frac{J_h}{\tau\kappa_c T_a}}\right)\frac{J_h}{T_a} \tag{9.9}$$

where use has been in the second equality of Eq. (9.4). With this result the entropy production can be written as:

$$\sigma = \frac{1}{1+\tau}\left(\frac{1}{\frac{T_a}{T_h} - \frac{J_h}{\tau\kappa_c T_h}} - 1\right)\frac{J_h}{T_h} \tag{9.10}$$

When one substitutes Eqs. (9.6) and (9.10) into Eq. (9.7) one obtains the following expression for the EF:

$$E = \frac{1}{1+\tau}\left[1 + \frac{T_c}{T_h} - \frac{2\frac{T_c}{T_h}}{\frac{T_a}{T_h} - \frac{J_h}{\tau\kappa_c T_h}}\right] J_h \tag{9.11}$$

The value of T_a is selected as the value T_a^E which maximizes the EF, and the efficiency η_E is obtained afterwards using Eq. (9.5).

This way for determine the efficiency with the $MPOFC$ or with the $MEFC$ is from theoretical point of view very simple, however its practical implementation requires in general numerical analysis to determine the values T_a^P and T_a^E, which depend on the explicit form of the heat flux J_h, and only in a few cases an analytical treatment is possible, one such case ocurrs when this flux satisfies Newton's law.

This is the usual method to analyze the EE performance with the $MPOFC$ and the $MEFC$,[5] upon using this method one cannot know how the OT: T_a^P and T_a^E are related, neither know beforehand which criterion gives the best value for the efficiency.

9.3. Functional Relations for Newton's Endoreversible Engines

In this section we proceed to find a functional relation between the EF and the POF for NEE. Subsequently using this result two additional functional relationships will be found, the first between T_a^P and T_a^E and the second one between the efficiencies η_P and η_E. With this goal in mind let us introduce the following dimensionless quantities:

$$\widetilde{P} = \frac{1+\tau}{\tau}\frac{P}{\kappa_c T_h}; \quad \theta = \frac{T_a}{T_h}; \quad \alpha = \frac{T_c}{T_h}$$
$$F_h(\theta) = \frac{J_h}{\tau\kappa_c T_h}; \quad \widetilde{E} = \frac{1+\tau}{\tau}\frac{E}{\kappa_c T_h}; \quad \widetilde{\sigma} = \frac{1+\tau}{\tau}\frac{\sigma}{\kappa_c}; \tag{9.12}$$

the POF given by Eq. (9.6) takes the form:

$$\widetilde{P}(\theta, \alpha) = [1 - \alpha H(\theta)]F_h(\theta), \tag{9.13}$$

where

$$H(\theta) = \frac{1}{\theta - F_h(\theta)} \tag{9.14}$$

the dimensionless entropy production given in Eq. (9.10) becomes now,

$$\widetilde{\sigma} = (H(\theta) - 1)F_h(\theta) \tag{9.15}$$

the DEF $\widetilde{E}$, from Eq. (9.7) and Eqs. (9.12), is now given by

$$\widetilde{E} = \widetilde{P} - \alpha\widetilde{\sigma} = [1 + \alpha - 2\alpha H(\theta)]F_h(\theta) \tag{9.16}$$

where the second equality comes from Eq. (9.13) and Eq. (9.15); after rearranging terms, one obtains

$$\widetilde{E}(\theta, \alpha) = (1+\alpha)\left[1 - \frac{2\alpha}{1+\alpha}H(\theta)\right]F_h(\theta) \tag{9.17}$$

Upon comparison of Eqs. (9.13) for the $DPOF$ and (9.17) for the DEF one finds that they are related in thc following way:

$$\frac{\widetilde{E}(\theta,\alpha)}{2\alpha} = \frac{\widetilde{P}(\theta, f(\alpha))}{f(\alpha)} \qquad \text{with} \qquad f(\alpha) = \frac{2\alpha}{1+\alpha} \tag{9.18}$$

This functional relationship (FR) establishes that the EF and POF of a NEE have the same analytical properties. In fact the EF for a given value of $\alpha = T_c/T_h$ is proportional to the POF at a different ratio of the temperatures of the external baths, ratio given by $f(\alpha)$ which can be read as a different set of external baths $f(\alpha) = T_c'/T_h'$. Then the EF has the same behavior that the POF of a NEE at different temperatures of the external baths, Eq. (9.18) connects these two operational modes of the NEE.

In the usual procedure the values $\theta_P(\alpha)$ and $\theta_E(\alpha)$ which maximize the POF and the EF respectively, are determined independently. However, due to the FR between the POF and the EF shown above, the procedure is reduced to the determination of only one of these two quantities.

The values $\theta_P(\alpha)$ and $\theta_E(\alpha)$ are determined by the conditions:

$$\left.\frac{\partial\widetilde{P}(\theta,\alpha)}{\partial\theta}\right|_{\theta_P(\alpha)} = 0, \qquad \left.\frac{\partial\widetilde{E}(\theta,\alpha)}{\partial\theta}\right|_{\theta_E(\alpha)} = 0; \tag{9.19}$$

the relation 9.18 implies that these critical values of θ are related one to another.

Their relationship is found upon taking the derivative of Eq. (9.18) with respect to θ, and using the definition of the critical values given by Eq. (9.19), thus one have that:

$$\theta_E(\alpha) = \theta_P(f(\alpha)) \tag{9.20}$$

This result comes about because of the fact that $\widetilde{E}(\theta, \alpha)$ is proportional to $\widetilde{P}(\theta, f(\alpha))$ then both functions obtain their maximum for the same value of θ . Clearly, Eq. (9.20) can be expressed by the equivalent relation:

$$\theta_E(g(\alpha)) = \theta_P(\alpha) \text{ with } g(\alpha) = f^{-1}(\alpha) = \frac{\alpha}{2-\alpha} \tag{9.21}$$

and using this relation, we will obtain the curve θ_E vs α when the curve θ_P vs α is known.

Now, using the FR (9.20) we will obtain the efficiency for the $MEFC$ in terms of the efficiency for the $MPOFC$. The efficiency of the NEE in terms of the dimensionless quantities, takes the form:

$$\eta(\theta, \alpha) = 1 - \frac{\alpha}{\theta - F_h(\theta)} = 1 - \alpha H(\theta) \tag{9.22}$$

The value of the efficiency when $\widetilde{E}$ is maximum, is given by

$$\eta_E(\alpha) = \eta(\theta_E(\alpha), \alpha) = 1 - \alpha H\big(\theta_E(\alpha)\big), \tag{9.23}$$

then we have that

$$H\big(\theta_E(\alpha)\big) = \frac{1 - \eta_E(\alpha)}{\alpha} \tag{9.24}$$

In a similar way, the relationship between the H function and the efficiency when the $DPOF$ is maximum is given by

$$H\big(\theta_P(\alpha)\big) = \frac{1 - \eta_P(\alpha)}{\alpha} \tag{9.25}$$

the FR between θ_E and θ_P expressed by Eq. (9.20) implies that the H function satisfies the following functional relationship:

$$H\big(\theta_E(\alpha)\big) = H\left(\theta_P\big(f(\alpha)\big)\right), \tag{9.26}$$

the combination of this last expression with Eqs. (9.24) and (9.25) leads to

$$\frac{1 - \eta_E(\alpha)}{\alpha} = \frac{1 - \eta_P\big(f(\alpha)\big)}{f(\alpha)}. \tag{9.27}$$

With this result, the FR between the efficiency when the EF is maximum and the efficiency when the POF is maximum becomes:

$$\eta_E(\alpha) = 1 - \frac{\alpha}{f(\alpha)} + \frac{\alpha}{f(\alpha)} \eta_P\big(f(\alpha)\big). \tag{9.28}$$

Through the introduction of the explicit form of $f(\alpha)$ Eq. (9.21), we get:

$$\eta_E(\alpha) = \frac{1-\alpha}{2} + \frac{1+\alpha}{2} \eta_P\left(\frac{2\alpha}{1+\alpha}\right) \tag{9.29}$$

and since the the efficiency of a Carnot engine operating between the external baths at temperatures T_h and T_c is given by

$$\eta_{Carnot} = 1 - \alpha \tag{9.30}$$

the relation (9.29) can be written in the following way:

$$\eta_E(\alpha) = \frac{1}{2}\left[\eta_{Carnot} + (2 - \eta_{Carnot})\eta_P\left(\frac{2\alpha}{1+\alpha}\right)\right] \tag{9.31}$$

The FR Eq. (9.18), establishes that the EF for a NEE with a ratio of external baths temperatures $\frac{T_c}{T_h}$, is proportional to the POF of the same NEE with a ratio of the external baths temperatures $\frac{2T_c}{T_c+T_h}$, therefore it implies that the results of the $MEFC$: T_a^E and η_E, can be evaluated by re-scaling the corresponding results obtained with the $MPOFC$. It is only necessary to focus our attention to the $MPOFC$, since using Eq. (9.18) one obtains the results for the $MEFC$.

9.4. On the Analytical Behavior of the Power Output Function

Given that the $DPOF$ Eq. (9.13), is the product between the efficiency times the dimensionless hot flux $F_h(\theta)$ (DHF), the analytical behavior of the $DPOF$ is determined by the analytical properties of their factor functions. The efficiency $\eta(\theta,\alpha)$ is an increasing function of θ with the property that $0 < \eta < 1$, on the other hand $F_h(\theta)$ is a decreasing function of θ. In this section the analytical behavior of $\widetilde{P}(\theta,\alpha)$ using these facts is obtained.

Let us now show the physical argument which supports that $F_h(\theta)$ is a decreasing function of θ. We notice that for given values of the temperatures of the baths, when the UOT increases θ increases, and the difference $T_h - T_a$ decreases, then the heat flux from the hot bath to the engine decreases, therefore the DHF is a decreasing function of θ. Given that the range of values of θ with physical meaning is the unitary interval, the DHF is maximum for $\theta = 0$ and is null for $\theta = 1$, in this last case because $T_h = T_a$. Now one sees that in general $F_h(\theta)$ is a positive decreasing function of its argument, and that $F_h(\theta = 1) = 0$.

The analytical behavior of the efficiency is found from its expression given by Eq. (9.22). Upon taking the derivative with respect to θ one obtains:

$$\frac{\partial\eta(\theta,\alpha)}{\partial\theta} = \frac{\alpha(1 - F_h'(\theta))}{(\theta - F_h(\theta))^2} > 0. \tag{9.32}$$

The last inequality follows from the fact that as $F_h(\theta)$ is a decreasing function of θ, then its derivative is always negative, and therefore the numerator of Eq. (9.32) would be always positive, implying that the efficiency is an increasing function of θ.

When the physical fact that the efficiency satisfies the inequality $0 < \eta(\theta, \alpha) < 1$ is introduced into Eq. (9.22), one has

$$\theta - F_h(\theta) > 0 \text{ and } \theta - F_h(\theta) > \alpha, \tag{9.33}$$

these inequalities restrict the values θ can take into the unitary interval, in fact the first inequality implies that θ must be greater than Θ which is the fixed point of $F_h(\theta)$, i.e. $F_h(\Theta) = \Theta$. The second inequality imposes a stronger condition, since for a given value of α, it implies that $\theta > \theta_{\min}(\alpha)$, where $\theta_{\min}(\alpha)$ is the solution of the equation $F_h(\theta) = \theta - \alpha$. The graphic solution of this expression is given by the abscissa of the intersection point of $z = F_h(\theta)$ with the straight line $z = \theta - \alpha$. Due to $F_h(\theta)$ is being a decreasing function, $\theta_{\min}(\alpha) > \Theta$ and therefore the physical values of θ are restricted to the interval $(\theta_{\min}(\alpha), 1)$.

The efficiency takes values in the extreme points of the interval $[\theta_{\min}(\alpha), 1]$ given by:

$$\eta(\theta_{\min}(\alpha), \alpha) = 0 \text{ and } \eta(\theta = 1, \alpha) = 1 - \alpha \tag{9.34}$$

Thus, the efficiency as a function of θ has the following behavior, it is an increasing function of θ, which is null for $\theta = \theta_{\min}$, and takes its maximum value in $\theta = 1$; the first result in (9.34) corresponds physically to the value of the efficiency when the two operational temperatures of the Carnot cycle are the same, i.e. $T_a = T_b$. The second one corresponds to the efficiency of the Carnot engine working between the external baths, i.e. $T_a = T_h$ and $T_b = T_c$.

The analytical behavior of the function $\theta_{\min}(\alpha)$ is determined by the fact that $F_h(\theta)$ is a decreasing function. Given that $\theta_{\min}(\alpha)$ is the abscissa of the intersection point between the straight line $z = \theta - \alpha$ and the decreasing curve $z = F_h(\theta)$, when α increases, the intersection point moves down and to the right, i.e. its ordinate decreases and its abscissa increases, therefore $\theta_{\min}(\alpha)$ is an increasing function of α .

With the knowledge of the properties of $F_h(\theta)$ and $\eta(\theta, \alpha)$ discussed above, and the fact that the *POF* is given by the product of these functions, the general analytical properties of $\widetilde{P}(\theta, \alpha)$ can be found. They are the following:

a) The surface $\widetilde{P}(\theta,\alpha)$ is only defined for the points in the plane (α,θ) which are below of the line $\theta = 1$ and above of the curve $\theta = \theta_{\min}(\alpha)$, which is an increasing function of α that goes from $(0,\Theta)$ to $(1,1)$. The coordinates of these points are determined by the fact that $\Theta = \theta_{\min}(\alpha = 0)$ and $\theta_{\min}(\alpha = 1) = 1$. See Figures (9.1) for $F_h(\theta) = (1-\theta)^{5/4}$ and Fig. (9.4) (a) for $F_h(\theta) = (1-\theta)$.

b) $\widetilde{P}(\theta,\alpha)$ is below of the surface $z = F_h(\theta)$. This follows directly from the fact that the *DPOF* is the efficiency times the DHF and $\eta(\theta,\alpha) < 1$.
In the limit when $\alpha \to 0$, the efficiency goes to one, therefore in this limit the *DPOF* is identical with the *DHF* in the interval $\theta \in (\Theta, 1)$.

c) The surface $\widetilde{P}(\theta,\alpha)$ cuts the plane (α,θ), on the curve $\theta_{\min}(\alpha)$ and on the line $\theta = 1$. This result comes from the efficiency being zero on the curve $\theta = \theta_{\min}(\alpha)$, and from the *DHF* being null on the line $\theta = 1$. Therefore the *DPOF* is null on these curves.

d) The level curve $\widetilde{P}(\theta,\alpha_i)$ is a concave function of θ with a single maximum. As $\eta(\theta,\alpha_i)$, with $i = 1,2$ is an increasing function of θ with zero value on $\theta = \theta_{\min}(\alpha_i)$ and $F_h(\theta)$ is a decreasing function of θ which is null on $\theta = 1$, their product $\widetilde{P}(\theta,\alpha_i) = \eta(\theta,\alpha_i)F_h(\theta)$ is a concave function of θ in the interval $[\theta_{\min}(\alpha_i), 1]$, and it is zero on the extreme values of the interval. (See Fig.(9.4)). It implies that $\widetilde{P}(\theta,\alpha_i)$ has a maximum for some value $\theta_P(\alpha_i)$ which is necessarily greater than $\theta_{\min}(\alpha_i)$.

e) The curve $\theta_P(\alpha)$ is an increasing function of α . When α increases $\theta_{\min}(\alpha)$ increases, the level curve of $\eta(\theta,\alpha)$ moves toward the right, and for $\alpha_2 > \alpha_1$ the level curve $\eta(\theta,\alpha_2)$ is below of the level curve $\eta(\theta,\alpha_1)$, when the product of these functions is taken with the decreasing function $F_h(\theta)$, the level curve $\widetilde{P}(\theta,\alpha_2)$ occurs below $\widetilde{P}(\theta,\alpha_1)$, and its maximum moves to the right, then $\theta_P(\alpha_2) > \theta_P(\alpha_1)$, and the curve $\theta_P(\alpha)$ is an increasing function of α. See Fig. (9.2).

f) The curves $\theta_{\min}(\alpha)$ vs α and $\theta_P(\alpha)$ vs α take the same values at $\alpha = 0$ and $\alpha = 1$.
When $\alpha = 0$ the *DPOF* is identical with the $F_h(\theta)$ in the interval $(\Theta, 1)$; this implies that $\mathrm{Lim}_{\alpha\to 0}\theta_P(\alpha) = \Theta$. Now consider the inequality $\Theta < \theta_{\min}(\alpha) < \theta_P(\alpha)$, when one takes the limit $\alpha \to 0$ one obtains that $\mathrm{Lim}_{\alpha\to 0}\theta_{\min}(\alpha) = \Theta$. On the other hand,

Fig. 9.1. The surface $\widetilde{P}(\theta,\alpha)$ for the flux $F_h(\theta)=(1-\theta)^{5/4}$. The definition domain in the plane (α,θ) is showed.

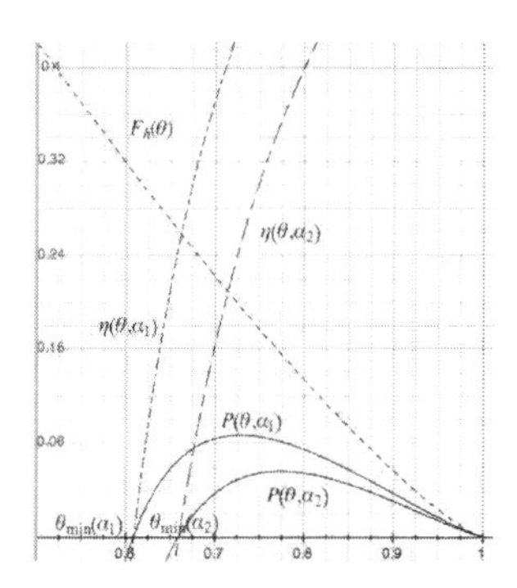

Fig. 9.2. The level curve $\widetilde{P}(\theta,\alpha_i)$ is a concave function of θ with a single maximum. Their function factors: a) $\eta(\theta,\alpha_i)$, with $i=1,2$ is an increasing function of θ and b) $F_h(\theta)=(1-\theta)^{5/4}$ a decreasing function of θ, are shown.

using the fact that $\mathrm{Lim}_{\alpha\to 1}\theta_{\min}(\alpha)=1$, and considering the inequality $\theta_{\min}(\alpha)<\theta_P(\alpha)<1$, a similar argument implies that $\mathrm{Lim}_{\alpha\to 1}\theta_P(\alpha)=1$.

9.5. The Analytical Behavior of the Dimensionless Ecological Function

The FR given by Eq. (9.18), implies that the analytical behavior of the DEF is essentially the same that the behavior of the $DPOF$, to obtain its behavior the relation (9.18) is rewritten in the form:

$$\widetilde{E}(\theta,\alpha)=(1+\alpha)\widetilde{P}(\theta,f(\alpha)) \tag{9.35}$$

it shows that the DEF for a value of α, is a scaled version of the $DPOF$ for a value $f(\alpha) = \frac{2\alpha}{1+\alpha}$, then the DEF behavior is obtained using the results of the last section. The DEF has the following properties:

a) It is defined into the region of the plane (α, θ), below of the line $\theta = 1$ and above of the curve $\theta = \theta_{\text{min}}(f(\alpha))$. As $f(\alpha) > \alpha$ and $\theta_{\text{min}}(\alpha)$ is an increasing function of its argument the curve $\theta_{\text{min}}(f(\alpha))$ is above to the curve $\theta_{\text{min}}(\alpha)$.
b) For any value of $\alpha > 0$, the surface $\widetilde{E}(\theta, \alpha)$ is below of the hot flux $F_h(\theta)$. When $\alpha = 0, f(\alpha = 0) = 0$ then from Eq. (9.35) follows that the DEF is identical with the $DPOF$, and therefore the the DEF is identical with $F_h(\theta)$ in the interval $\theta \in (\Theta, 1)$.
c) The surface $\widetilde{E}(\theta, \alpha)$ cuts the plane (α, θ) on the curve $\theta_{\text{min}}(f(\alpha))$ and on the line $\theta = 1$, i.e.

$$\widetilde{E}\left(\theta_{\text{min}}(f(\alpha)), \alpha\right) = \widetilde{E}(\theta = 1, \alpha) = 0 \tag{9.36}$$

d) The level curve $\widetilde{E}(\theta, \alpha_0)$ is a concave function of θ which is maximum in the same value of θ that the level curve $\widetilde{P}(\theta, f(\alpha_0))$ has its maximum, i.e. $\theta_E(\alpha) = \theta_P(f(\alpha))$, this fact is expressed in the FR given by Eq. (9.20). See Fig. (9.3).

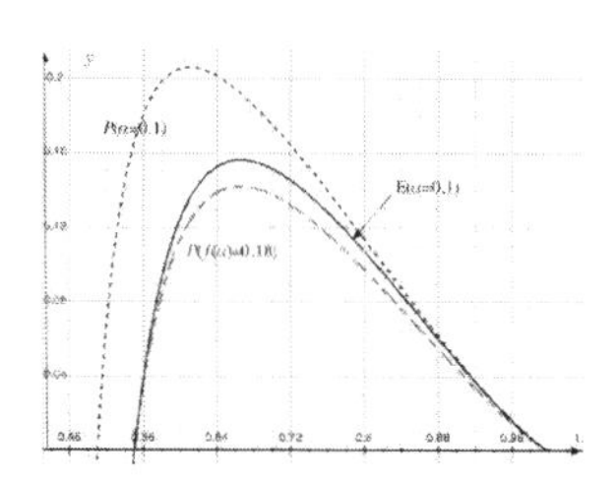

Fig. 9.3. The curve $\widetilde{E}(\theta, \alpha_0)$ is maximum in the same value of θ that the curve $\widetilde{P}(\theta, f(\alpha_0))$ has its maximum, i.e., $\theta_E(\alpha) = \theta_P(f(\alpha))$.

e) The curve $\theta_E(\alpha)$ vs. α is always above of the curve $\theta_P(\alpha)$ vs. α in the interval (0,1), they are increasing function of α; and both curves meet in the extreme values of the interval.
Taken into account: first that $\theta_P(\alpha)$ is an increasing function of α, second that $f(\alpha) > \alpha$, and third the $FR : \theta_E(\alpha) = \theta_P(f(\alpha))$, one obtains the inequality:

$$\theta_E(\alpha) > \theta_P(\alpha), \tag{9.37}$$

from Eq. (9.20) one has that for the fixed points of $f(\alpha)$ satisfy that $\theta_E(\alpha^*) = \theta_P(\alpha^*)$), as $\alpha = 0$ and $\alpha = 1$ are the fixed points of $f(\alpha)$ both functions take the same value for these values of α.

f) The efficiencies obtained with the $MEFC$ and with the $MPOFC$ satisfies the inequality:

$$\eta_E(\alpha) > \eta_P(\alpha) \tag{9.38}$$

The explicit expressions for these efficiencies are given by:

$$\eta_E(\alpha) = \eta(\theta_E(\alpha), \alpha); \qquad \eta_P(\alpha) = \eta(\theta_P(\alpha), \alpha); \tag{9.39}$$

one knows that the efficiency is a increasing function of θ, and that $\theta_E(\alpha) > \theta_P(\alpha)$, taking into account both facts the inequality (9.38) is obtained.

The inequality given by Eq. (9.37) implies that for given values of the temperatures of the baths, the operational temperature T_a^E obtained when the $MEFC$ is used, is greater than the operational temperature T_a^P obtained with the $MPOFC$. On the other hand the inequality given by Eq. (9.38) says that when the $MEFC$ is used the efficiency is greater than when the $MPOFC$ is used. These results are completely general for a NEE. In Fig. (9.4), these facts for newtonian flux $F_h(\theta) = (1 - \theta)$ are showed.

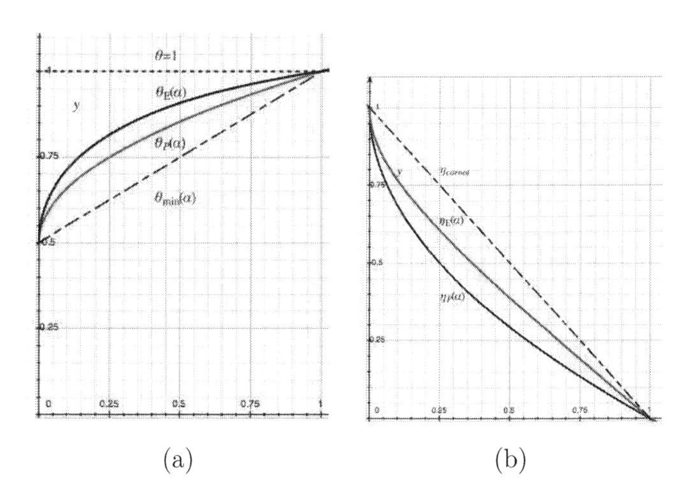

Fig. 9.4. The curve $\theta_E(\alpha)$ is always above of the curve $\theta_P(\alpha)$, left panel. The curve of the efficiency $\eta_E(\alpha)$ is always above of the curve $\eta_P(\alpha)$, right panel.

9.6. A Unified Procedure to Analyze the Newton's Endoreversible Engine

In this section an algorithm based on the FR is found. This algorithm gives a unified procedure to obtain the dimensionless UOT and the efficiency for a NEE when $MPOFC$ or $MEFC$ is used.

An overview of the procedure is the following: given a set of values $\{\alpha_k\}$, a new set $\{g_k\}$ with $g_k = g(\alpha_k) = \alpha_k/(2-\alpha_k)$ is found. Solving numerically Eq. (9.19) the set $\{\theta_k = \theta_P(\alpha_k)\}$ is obtained. As a consequence of the FR expressed by Eq. (9.21), these values also correspond to $\{\theta_E(g_k)\}$, and the curve $\theta_E(\alpha)$ is obtained.

Once, the set $\{H_k = H(\theta_P(\alpha_k))\}$ with $H(\theta) = [\theta - F_h(\theta)]^{-1}$ is constructed, the use of Eq. (9.22) leads to the efficiency $\{\eta_P(\alpha_k)\}$; finally, combining the sets $\{H_k\}$ and $\{g_k\}$ and using again the Eq. (9.22), the efficiency $\{\eta_E(g_k)\}$ is obtained. The details are given below.

The first step of the procedure is to find the dimensionless UOT with the $MPOFC$.

The partial derivative of Eq. (9.13) with respect to θ, is given by

$$\frac{\partial P}{\partial \theta} = \frac{1}{(\theta - F_h(\theta))}\left[F_h'(\theta)\left\{(\theta - F_h(\theta))^2 - \alpha\theta\right\} + \alpha F_h(\theta)\right] \tag{9.40}$$

thus $\theta_P(\alpha)$ is the solution of the following equation:

$$\alpha\left[F_h(\theta) - \theta F_h'(\theta)\right] + F_h'(\theta)\left(\theta - F_h(\theta)\right)^2 = 0 \tag{9.41}$$

in the interval $(\theta_{\min}(\alpha), 1)$, where $\theta_{\min}(\alpha)$ is the solution of the Eq. (9.33).

For a set of values $\{\alpha_k\}$ on the unitary interval, solving numerically Eq. (9.41) the set $\{\theta_k = \theta_P(\alpha_k)\}$ is obtained. Up this point, this is the usual procedure for the analysis of the NEE. In order to obtain $\theta_E(\alpha)$ Eq. (9.21) will be called for.

The functional relation (9.21) establishes that:

$$\theta_P(\alpha) = \theta_E\big(g(\alpha)\big) \quad \text{with} \quad g(\alpha) = f^{-1}(\alpha) = \frac{\alpha}{2-\alpha}$$

it means that the set $\{\theta_k = \theta_P(\alpha_k)\}$ also corresponds to values of the function $\theta_E(\alpha)$ but evaluated at different values of its argument, in fact these values conform the set $g_k = g(\alpha_k)$. Then in order to find $\theta_E(\alpha)$, instead of numerically solving Eq. (9.19) for the DEF, one evaluates the function $g(\alpha)$. Also, with the use of this procedure it is a very simple matter to find the curve $\theta_E(\alpha)$ with the previous knowledge of the curve $\theta_P(\alpha)$.

The procedure just described can be expressed in compact with the following array:

$$\begin{array}{ccccc} \theta_1 & \theta_2 & \theta_3 & \cdots & \theta_n \\ \alpha_1 & \alpha_2 & \alpha_3 & \cdots & \alpha_n \\ g(\alpha_1) & g(\alpha_2) & g(\alpha_3) & \cdots & g(\alpha_n) \end{array} \tag{9.42}$$

where $\theta_k = \theta_P(\alpha_k)$. The numerical values $\theta_P(\alpha)$ vs. α are shown in the first and second rows of this array and the numerical values of $\theta_E(\alpha)$ vs. α are shown in the first and third row due to the fact that $\theta_k = \theta_E\left(g(\alpha_k)\right)$.

In the second part of the procedure the efficiencies are obtained. This is done by evaluating for each value of $\theta_P(\alpha)$ the H function defined by Eq. (9.14), i.e.

$$H\left(\theta_P(\alpha)\right) = \frac{1}{\theta_P(\alpha) - F_h\left(\theta_P(\alpha)\right)}$$

With the resulting set $\{H_k = H(\theta_k)\}$ and the use of Eq. (9.22), which relates the efficiency with the H function,

$$\eta_P(\alpha) = 1 - \alpha H\left(\theta_P(\alpha)\right)$$

the set $\{\eta_P(\alpha_k) = 1 - \alpha_k H_k\}$ is obtained and the efficiency of the $MPOFC$ is thus obtained.

As a consequence of the FR (9.21) the set $\{H_k\}$ also corresponds with the values of the H function for $\{\theta_E(g(\alpha_k))\}$.

The efficiency for the $MEFC$ for $g(\alpha)$ is found using Eqs. (9.22) and (9.21), i.e.

$$\eta_E\left(g(\alpha)\right) = 1 - g(\alpha)H\left(\theta_E(g(\alpha))\right) = 1 - g(\alpha)H\left(\theta_P(\alpha)\right)$$

then the set $\{\eta_E(g_k) = 1 - g_k H_k\}$ in its turn gives the efficiency of the MEFC. We express the above results in a compact form using the following array:

$$\begin{array}{cccc} H_1 & H_2 & \cdots & H_n \\ 1 - \alpha_1 H_1 & 1 - \alpha_2 H_2 & \cdots & 1 - \alpha_n H_n \\ 1 - g(\alpha_1)H_1 & 1 - g/\alpha_2)H_2 & \cdots & 1 - g(\alpha_n)H_n \end{array} \tag{9.43}$$

In this array the first row is $H_k = H(\theta_k)$, the second row is the efficiency $\eta_P(\alpha)$ obtained with the MPOFC and the third row is the efficiency $\eta_E(g(\alpha))$ obtained with the MEFC. Due to $H_k = H(\theta_P(\alpha_k)) = H(\theta_E(g(\alpha_k)))$, the efficiencies are related in the following way: $H_k =$

$H\left(\theta_P(\alpha_k)\right) = H\left(\theta_E\big(g(\alpha_k)\big)\right)$, the efficiencies are related in the following way:

$$\eta_E\left(g(\alpha_k)\right) = 1 - g(\alpha_k)\,\frac{1-\eta_P(\alpha_k)}{\alpha_k} \tag{9.44}$$

The algorithm presented in this section only requires the numerical solution of Eq. (9.41) to obtain the dimensionless operational temperature and the efficiency for a NEE when MPOFC or MEFC is selected. Notice that relation (9.44) is also obtained when Eq. (9.28) is evaluated with $g(\alpha) = f^{-1}(\alpha)$.

9.7. Results and Conclusions

A first important result of this work is the existence of a functional relation between the POF and the EF expressed in Eq. (9.18), because it implies that all the results obtained with the $MPOFC$ also contains the results of the $MEFC$.

The consequences of the FR were analyzed in detail; one of them is the FR between the functions $\theta_P(\alpha)$ and $\theta_E(\alpha)$ given in Eq. (9.20) another is that the efficiencies $\eta_P(\alpha)$ and $\eta_E(\alpha)$ are related by the FR expressed in the Eq. (9.22).

A second important result is that only two physical properties of the NEE are enough to determine the general behavior of the POF. These properties are: that the dimensionless heat flux from the hot bath to the Carnot engine is always a decreasing function of θ, and that the efficiency of the NEE satisfies $0 < \eta(\theta, \alpha) < 1$.

In Sec. (9.4), the POF analytical properties were analyzed. The functional relationship (9.18) implies that EF has a similar analytical behavior to the POF, all of its properties were given in Sec. (9.5).

A third result is the two showed general properties of the POF, first that the value $\theta_P(\alpha)$ where the POF has its maximum, is an increasing function of α, and second that the efficiency of the NEE, $\eta(\theta, \alpha)$ is an increasing function of θ. The combination of these properties with the FR between $\theta_P(\alpha)$ and $\theta_E(\alpha)$, has showed that the inequalities: $\theta_E(\alpha) > \theta_P(\alpha)$ and $\eta_E(\alpha) > \eta_P(\alpha)$ are satisfied for any analytical form whatsoever of the heat flux from the hot bath to the Carnot engine.

These are the type of results which were mentioned at the introduction, in the sense that they are frequently found numerically, but never before it has been recognized that they have a general validity. For example, when the hot flux is given the Stefan-Boltzmann's law, Sahin obtained that the curve $\theta_P(\alpha)$ vs. α is an increasing function, but he doesn't give any importance to this fact. Another example that shows the need to have a theoretical framework for the NEE, appears when Barranco et al paper[5] is analyzed; they obtained a lot of numerical results, however it is very difficult to separate the general results from the particular ones. In that paper they found the curves $\theta_P(\alpha)$ vs. α and $\theta_E(\alpha)$ vs. α when the hot flux satisfies different empirical laws, in all the cases their curves satisfy the inequality $\theta_E(\alpha) > \theta_P(\alpha)$, but they didn't realize the generality of these results[6] .They also obtained the curves $\eta_P(\alpha)$ vs. α and $\eta_E(\alpha)$ vs. α, for several hot fluxes,[7] in all cases they found $\eta_E(\alpha) > \eta_P(\alpha)$, and they reported: "This result has been systematically observed in all kind of thermal engine models operating under ecological conditions. This property is considered to be concomitant with ecological goals from a long-term energy conversion point of view", however they didn't give a theoretical support of this fact. On the other hand with the theoretical framework presented here, both inequalities are consequence of: a) the heat flux is a decreasing function of its argument, b) the efficiency satisfies that $0 < \eta(\theta, \alpha) < 1$, and c) there is a FR between the POF and the EF, namely they one shows in Sec. (9.4).

From the operational point of view, the FR (9.21) reduces the numerical calculation since after numerically solving Eq. (9.41), the function $\theta_P(\alpha)$ is obtained, and afterward $H(\theta_P(\alpha))$ can be evaluated. With the use of these functions it is very simple to obtain $\eta_P(\alpha)$ and the corresponding quantities for the $MEFC$, i.e. the functions $\theta_E(\alpha)$ and $\eta_E(\alpha)$, with the simple algorithm proposed in the Section 9.6.

All the results found in this work, are valid for an arbitrary heat flux between the hot bath and the Carnot engine, and also can be used to check the numerical results of the NEE.

Acknowledgments

The author wishes to thank the reviewer for his careful, unbiased and constructive suggestions, which led to this revised manuscript.

References

1. F. Curzon and B. Ahlborn, Am. J. of Phys. 43 (1975) 22.
2. G. Lebon, D. Jou and J. Casas-Vasquez. See Chapter 5, Understanding Non-equilibrium thermodynamics, Springer-Verlag 2008.

3. F. Angulo-Brown J. Appl. Phys. 69 (1991) 7465.
4. A. Z. Sahin, Energy Conversion and Management 41 (2000) 1335.
5. M. A. Barranco-Jiménez, N. Sanchez-Salas, F. Angulo-Brown, Rev. Mex. Fis. 54 (2008) 284
6. See Ref.,[5] Figs. (8) and (14), showed the curves $\theta_P(\alpha)$ vs. α and $\theta_E(\alpha)$ vs. α for the Stefan-Boltzmann's law, and for the Dulong-Petit's law.
7. See Ref.,[5] Figs. (4), (9) and (15), showed the curves $\eta_P(\alpha)$ vs. α and $\eta_E(\alpha)$ vs. α, for the linear combination of Stephan-Boltzmann's law and Newton's law, Stephan-Boltzmann's law and Dulong-Petit's law.

PART 2

Statistical Physics and Biological Physics

Chapter 10

Diffusion between Two Chambers Connected by a Conical Capillary

I. Pineda, M.V. Vázquez, and L. Dagdug

Deptartamento de Física, Universidad Autonóma Metropolitana-Iztapalapa, Distrito Federal, 09340, México

dll@xanum.uam.mx

The concentration change due to diffusion between two chambers connected by a conical capillary tube is approached by a microscopical model, by means of the use of propagator functions, G_{ji}, which state the probability of found a particle at time t in chamber j, given that the particle was in chamber i at time $t = 0$. The description of the propagator functions involved the equilibrium probabilities and relaxation functions which describe the way the system reach the equilibrium state. The latter functions had to be written down in terms of the fluxes of particles escaping from the capillary, $f_{\mathrm{tr},i}$ and $f_{\mathrm{r},i}$, named the translating and returning fluxes from chamber i. To obtain the fluxes, the Fick-Jacobs' equation was solved inside the tube, subject to radiative boundary conditions, in the Laplace's space, then used these solutions to write down the Laplace transforms of the propagator functions. Monte Carlo computational simulations were performed to complement the analytical solution previously obtained by finding the appropriate model to describe the effective diffusion in the capillary tube, in the interval of geometrical parameters accessible to actual biological systems. Our results show that this model could has serious applications in modelling the diffusion between the cell's surroundings and its interior trough a varying cross-section channel, as those specific to the potassium ions, and contribute to understand key features of those systems. Additionally, the simulated data yielded two noticeably results: the non-equivalent behavior of diffusion trough the conical tube went trough in opposite directions, and the inability of present theoretical models to completely describe the observed patterns.

We want to dedicate this work to our friend, colleague and professor, Leopoldo García Colín.

Contents

10.1. Introduction

Brownian motion was discovered early in the nineteenth century and explained in the early years of the twentieth century. Now, at the beginning of the twenty-first century, it is still a subject of considerable interest and current research.

In 1827, Brown was investigating the way in which pollen acted during impregnation. He wanted to use non-spherical grains, in order to be able to observe their orientation. The first plant he studied under the microscope was *Clarkia pulchella*, whose pollen contains granules varying from about five to six microns in linear dimension. It is these granules, not the whole pollen grains, upon which Brown made his observations.

This inherent, incessant motion of small particles suspended in a fluid is nowadays called *Brownian motion* in honor of Robert Brown. Similar observations had, in fact, been made early by other workers. Brown however, was the first to give them serious scientific study, and to show that, in fact, they were not due to the living origin of the moving particles. It can be said that Brown, showed that the phenomenon was not one of biology, but one of physics.

The first paper by Albert Einstein on Brownian motion was published in 1905. The paper is a remarkable document. He first argues that particles suspended in a fluid must exert the same osmotic pressure as particles dissolved in the fluid, the concentrations being the same. Thus the molecular-kinetic theory of heat requires that suspended particles posses a motion of the same character as that of dissolved ones. He thus realizes that the theoretical problem is truly one of probability theory.

Einstein assumes that the position of a suspended particle as a function of time, in a time-coarse-grained sense, is a Markov process. He then writes down the Chapman-Kolmogorov equation for the transition probability, and expands it, assuming that the partial range of the transition probability is small in short time intervals. The final result is a diffusion equation for the probability density of the position of the suspended particle.

The solution of the diffusion equation was well known, and Einstein could immediately deduce that the mean square displacement of the suspended particle grows linearly with the time.

The diffusion equation has universal validity and satisfactorily describes the features of the mass transport in any geometry. However, up to now this equation can not be analytically solved for a set of boundary conditions describing complex geometries, like interconnected spatial regions.

Our interest in complex geometries arise from the study of biological systems as the ion channels. Ion channels are transmembranal proteins specifically designed to transport charged species across the cellular membrane, between the cell and its surroundings, and exhibit high degree of selectivity. The physical model used trough this work represents an approach to the problem of mass transport between the cell's interior and its environment trough a channel of varying cross-section.

This paper deals with the kinetics of equilibration of diffusing particles between two chambers connected by a conical capillary tube. This complex geometry is shown in Figure 10.1. This two chambers model is a simplified version of the more general problem of describing the kinetics of solute transport between several chambers; a problem common to biophysics, chemical engineering, and physiology. We derive general solutions for the Laplace transforms of the relaxation functions describing the equilibration of particles between the two chambers and the capillary. These

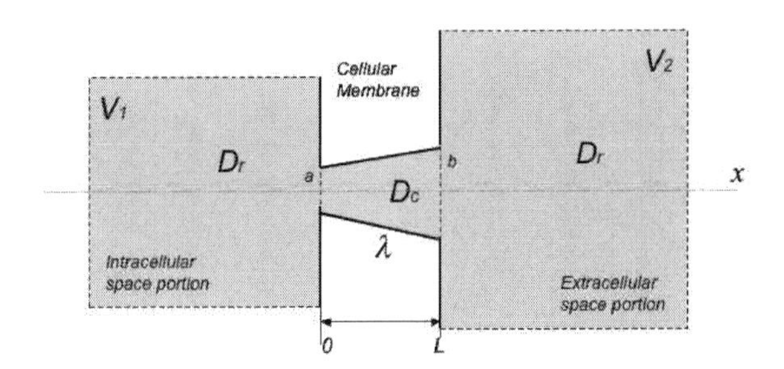

Fig. 10.1. Schematic diagram of two chambers connected by a conical capillary. In the sake of simplicity the diffusion constants in both volumes were taken to be the same, but a distinct value were assigned to the diffusion constant in the channel.

solutions show how the relaxation functions depend on geometric parameters. The cavities' volumes, V_1 and V_2, the length of the tube, L, and its radii a (the minimum radius) and b (the maximum radius), or its slope $\lambda = (b - a)/L$, as well as the diffusion constants in the two chambers, D_r, and in the conical capillary, D_c.

10.2. The Diffusion of Particles between Two Chambers Connected by a Capillary Tube is Well Described by Propagator Functions

To find the way that particles inside the chambers diffuse across a conical capillary we used a pair of propagator functions[2]. A propagator, denoted by G_{ij}, where i, j can take on the values 1, 2 (the volumes V_1 or V_2) or c (the capillary), and is defined as the probability that a particle in chamber i at time $t = 0$ will be found in chamber j at time $t > 0$. So, the equilibrium probabilities are found to be

$$\lim_{t\to\infty} G_{ji} = P_j^{eq} = \frac{V_j}{V_1 + V_2 + V_c}, \qquad j = 1, 2, c \tag{10.1}$$

where $V_c = \pi L(a^2 + ab + b^2)/3$ is the volume of the conical capillary. To describe the propagator functions of the system, we can use a couple of relaxation functions $R_{ji}(t)$, which indicate how the particles in chamber j reach the equilibrium if a fixed concentration of particles is placed initially in chamber i. The propagators can be related with the relaxation functions and equilibrium probabilities. If all the particles are initially in chamber 1, we have,

$$G_{11}(t) = P_1^{eq} + (1 - P_1^{eq})R_{11}(t) \tag{10.2}$$

and

$$G_{21}(t) = P_2^{eq}(1 - R_{21}(t)) \tag{10.3}$$

The domain and the range of the relaxation functions $R_{ji}(t)$ are $[0, \infty)$ and $[0, 1]$ respectively.

10.3. Calculation of the Relaxation Functions Involves Fluxes of Particles Escaping from the Capillary

In order to calculate the relaxation functions it will be necessary to introduce new functions that describe the probability fluxes escaping from the two ends of the conical capillary at time t, under the condition that the particles go into the capillary from chamber i at $t = 0$, where $i = 1, 2$. These fluxes are produced by trajectories that escape from the capillary for the first time at time t. We denote fluxes due to translocating and returning particle trajectories by $f_{tr,i}(t)$ and $f_{r,i}$, respectively. Accordingly, the integral of the fluxes are the translation and return probabilities:

$$P_{tr,i} = \int_0^\infty f_{tr,i}(t)dt \tag{10.4}$$

$$P_{r,i} = \int_0^\infty f_{r,i}(t)dt \tag{10.5}$$

The temporal derivatives of propagators satisfy a pair of non-Markovian integro-differential equations. For example, for particles initially in chamber 1, these equations are

$$\begin{aligned}\dot{G}_{11}(t) &= -k_1 G_{11}(t) + k_1 \int_0^t f_{r,1}(t-\tau)G_{11}(\tau)d\tau \\ &\quad + k_2 \int_0^t f_{tr,2}(t-\tau)G_{21}(\tau)d\tau \end{aligned} \tag{10.6}$$

$$\begin{aligned}\dot{G}_{21}(t) &= -k_2 G_{21}(t) + k_2 \int_0^t f_{r,2}(t-\tau)G_{21}(\tau)d\tau \\ &\quad + k_1 \int_0^t f_{tr,1}(t-\tau)G_{11}(\tau)d\tau \end{aligned} \tag{10.7}$$

which are to be solved subject to the initial conditions $G_{11}(0) = 1$ and $G_{21}(0) = 0$. On the other hand, the constants $k_i = 4Da_i/V_i$, $i = 1, 2$, are the rate constants which satisfy the propagator functions, $G_{ij}(t)$, Eqs. (10.2) and (10.3), in a two-state kinetics.[3]

Now, focusing on the right-hand side of equation (10.6), we see that the first term accounts for those realizations of particle's trajectory that leave chamber 1 at time t. The second term accounts for those realizations that leave chamber 1 at time $\tau < t$, returning to chamber 1 at time t without entering chamber 2. The third one accounts for those realizations that go into the capillary from chamber 2 at time $\tau < t$, spend a time $t - \tau$ in the capillary, and then escape into chamber 1 at time t (see Fig. (10.2)). The terms on the right-hand side of the equation (10.7) can be interpreted in a similar way. The set of equations (10.6) and (10.7), with the initial conditions $G_{11}(0) = 1$ and $G_{21}(0) = 0$, can be solved in Laplace space. Let the Laplace transform of a generic function $h(t)$ be denoted by $\hat{h}(s) = \int_0^\infty e^{-st}h(t)dt$. Taking the Laplace transform of (10.6) and (10.7) changes

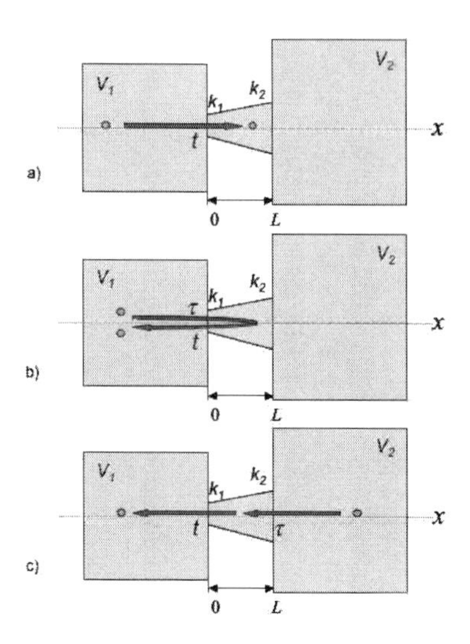

Fig. 10.2. Representation of the trajectories that are related to each term on the right-hand side in the equation (10.6).

our integro-differential equations problem into a pair of easily solvable linear equations,

$$s\hat{G}_{11} - 1 = -k_1\hat{G}_{11} + k_1\hat{f}_{r,1}\hat{G}_{11} + k_2\hat{f}_{tr,2}\hat{G}_{21} \tag{10.8}$$

$$s\hat{G}_{21} = -k_2\hat{G}_{21} + k_2\hat{f}_{r,2}\hat{G}_{21} + k_1\hat{f}_{tr,1}\hat{G}_{11} \tag{10.9}$$

Solving the system for $\hat{G}_{11}(s)$ and $\hat{G}_{21}(s)$,

$$\hat{G}_{11}(s) = \frac{s + k_2[1 - \hat{f}_{r,2}(s)]}{\Delta(s)} \tag{10.10}$$

$$\hat{G}_{21}(s) = \frac{k_1\hat{f}_{tr,1}(s)}{\Delta(s)} \tag{10.11}$$

where $\Delta(s)$ is given as

$$\Delta(s) = \{s + k_1[1 - \hat{f}_{r,1}(s)]\}\{s + k_2[1 - \hat{f}_{r,2}(s)]\} - k_1k_2\hat{f}_{tr,1}\hat{f}_{tr,2} \tag{10.12}$$

Then, using the equilibrium probabilities given by equations (10.2) and (10.3), we can write the Laplace transform of the R_{i1}, $i = 1, 2$, as functions of the Laplace transforms of the translating and returning fluxes from the capillary, and is given by,

$$\hat{R}_{i1}(s) = \frac{\hat{G}_{i1}(s) - \frac{1}{s}P_i^{eq}}{\delta_{i1} - P_i^{eq}}, \qquad i = 1, 2 \tag{10.13}$$

in which δ_{i1} is the Kronecker delta.

10.4. The Fluxes from the Capillary are Obtained Solving the Fick-Jacobs Equation in the Capillary

The calculation of fluxes from the capillary is based on the reduction of a three dimension conical capillary, by an effective diffusion on one dimension. This mapping was justified theoretically by Zwanzig[4] as follows. Starting with the Smoluchowsky equation in two dimensions with a general potential,

$$\begin{aligned}\frac{\partial C(x,y,t)}{\partial t} &= D\frac{\partial}{\partial x}e^{-\beta U(x,y)}\frac{\partial}{\partial x}e^{\beta U(x,y)}C(x,y,t)\\ &\quad + D\frac{\partial}{\partial y}e^{-\beta U(x,y)}\frac{\partial}{\partial y}e^{\beta U(x,y)}C(x,y,t)\end{aligned} \tag{10.14}$$

where $\beta = 1/k_BT$, k_B is the Boltzmann constant and T the absolute temperature. For small deviations from equilibrium, integrating in one variable one can obtain,

$$\frac{\partial c(x,t)}{\partial t} = D\frac{\partial}{\partial x}\int e^{-\beta U(x,y)}\frac{\partial}{\partial x}e^{\beta U(x,y)}C(x,y,t)dy \tag{10.15}$$

where the linear concentration $c(x,t)$ is defined by,

$$c(x,t) = \int C(x,y,t)dy \tag{10.16}$$

Then, introducing the free energy $F(x)$, we can write,

$$e^{-\beta F(x)} = \int e^{-\beta U(x,y)}dy \tag{10.17}$$

Defining

$$\rho(x|y) = \frac{e^{-\beta U(x,y)}}{e^{-\beta F(x)}}, \tag{10.18}$$

under the local equilibrium hypothesis, we can define,

$$C(x, y, t) \cong c(x, t)\rho(x|y) \tag{10.19}$$

Substituting the two last expressions in equation (10.15) we obtain the Fick-Jacobs' equation, where $F(x) = A(x)$

$$\frac{\partial c(x,t)}{\partial t} = D\frac{\partial}{\partial x}\left\{A(x)\frac{\partial}{\partial x}\left[\frac{c(x,t)}{A(x)}\right]\right\} \tag{10.20}$$

To calculate the fluxes from the capillary we use the reduction of the original three-dimensional problem to one dimension as given by equation (10.20). For the sake of simplicity we take the length of the tube as a unit length, then the position dependent radius is given by,

$$R(x) = 1 + \lambda x, \qquad 0 < x < L = 1 \tag{10.21}$$

so $\lambda = b - a$, and the position dependent area is given by,

$$A(x) = \pi R(x)^2 \tag{10.22}$$

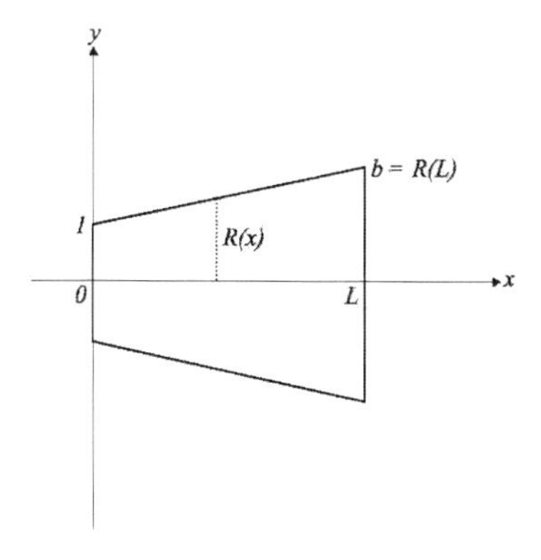

Fig. 10.3. The conical capillary tube. This geometry is described as a varying radius $R(x)$ with fixed lengths at one end, $a = R(0) = 1$, and the other end determined by the tube's axial length, $b = R(L)$.

Making the pertinent substitutions, for the conical capillary, the Fick-Jacobs' equation, Eq. (10.20) will be,

$$\frac{\partial c(x,t)}{\partial t} = D_c\left\{\frac{\partial^2 c(x,t)}{\partial x^2} - \frac{2\lambda}{1+\lambda x}\frac{\partial c(x,t)}{\partial x} + \frac{2\lambda^2}{(1+\lambda x)^2}c(x,t)\right\} \tag{10.23}$$

This equation must be solved with the initial condition,

$$c(x, 0) = \delta(x - x_0) \tag{10.24}$$

The solution of equation (10.23) in the Laplace space is,

$$\hat{c}(x,s) = \begin{cases} (1+\lambda x)A\left[e^{\sqrt{s/D_c}x} - \alpha e^{-\sqrt{s/D_c}x}\right] & \\ & \text{for } 0 \leqslant x < x_0, \\ [1+\lambda(L-x)]B\left[e^{\sqrt{s/D_c}(L-x)} - \beta e^{-\sqrt{s/D_c}(L-x)}\right] & \\ & \text{for } x_0 < x \leqslant L. \end{cases} \tag{10.25}$$

In the limit when $\lambda = 0$, the solution for the cylindrical capillary is obtained.[2]

Boundary conditions for this geometry, in Laplace space, have the following form,

$$D_c \hat{c}'(x,s)\Big|_{x=0} = \kappa_1 \hat{c}(0,s) \tag{10.26}$$

$$D_c \hat{c}'(x,s)\Big|_{x=L} = -\kappa_2 \hat{c}(L,s) \tag{10.27}$$

Where $\kappa_i = 4D_i/(\pi b_i)$, $i = 1,2$ with $b_1 = R(0)$ and $b_2 = R(L)$ accordingly to Eq. (10.21). The constant κ_i is the efficiency of particles to arrive from the chamber i to the correspondent conical aperture.[2] Introducing these conditions in Eq. (10.25) we get that,

$$\alpha = \frac{\kappa_1 - \sqrt{sD_c} - D_c\lambda}{\kappa_1 + \sqrt{sD_c} - D_c\lambda} \tag{10.28}$$

$$\beta = \frac{\kappa_2 - \sqrt{sD_c} - D_c\lambda}{\kappa_2 + \sqrt{sD_c} - D_c\lambda} \tag{10.29}$$

Using the continuity condition in the Laplace transform of equation (10.20), we have that,

$$\int_{x_0-\epsilon}^{x_0+\epsilon} [s\hat{c}(x,s) - \delta(x-x_0)]dx$$
$$= D_c \int_{x_0-\epsilon}^{x_0+\epsilon} \left\{ \hat{c}''(x,s) - \frac{2\lambda}{1+\lambda x}\hat{c}'(x,s) + \frac{2\lambda^2}{(1+\lambda x)^2}\hat{c}(x,s) \right\} dx \tag{10.30}$$

After some algebraic manipulation is obtained that,

$$A = \frac{(1+\lambda L)(1-\beta e^{-2\widetilde{L}})}{2\sqrt{sD_c}(1+\lambda L)(1-\alpha\beta e^{-2\widetilde{L}}) + (1-\beta e^{-2\widetilde{L}})(1-\alpha)(2+\lambda L)D_c\lambda} \tag{10.31}$$

$$B = \frac{(1-\alpha)e^{-\widetilde{L}}}{2\sqrt{sD_c}(1+\lambda L)(1-\alpha\beta e^{-2\widetilde{L}}) + (1-\beta e^{-2\widetilde{L}})(1-\alpha)(2+\lambda L)D_c\lambda} \tag{10.32}$$

where

$$\widetilde{L} = \sqrt{s/D_c}L \tag{10.33}$$

To find the fluxes, we use the following relations,

$$\hat{f}_{r,1}(s) = \kappa_1 \hat{c}(0, s|0) \tag{10.34}$$

$$\hat{f}_{tr,1}(s) = \kappa_2 \hat{c}(L, s|0) \tag{10.35}$$

$$\hat{f}_{r,2}(s) = \kappa_2 \hat{c}(L, s|L) \tag{10.36}$$

$$\hat{f}_{tr,2}(s) = \kappa_1 \hat{c}(0, s|L) \tag{10.37}$$

For the sake of simplicity, taking $D_r = D_c = D$, and using the previous results, we can write the Laplace transform of the fluxes as,

$$\hat{f}_{tr,1}(s) = \frac{\kappa_2(\alpha-1)(1-\beta)e^{2L\sqrt{s/D}}}{D\bar{\Delta}(s)} \tag{10.38}$$

$$\hat{f}_{r,1}(s) = \frac{\kappa_1(\alpha-1)(1+\lambda L)(e^{2L\sqrt{s/D}}-\beta))}{D\bar{\Delta}(s)} \tag{10.39}$$

where

$$\begin{aligned}\bar{\Delta}(s) = \beta\Big\{&\lambda(2+\lambda L) + \alpha\{2\sqrt{s/D}(1+\lambda L) - \lambda(2+\lambda L)\}\Big\} \\ &- e^{2L\sqrt{s/D}}\Big\{(1-\alpha)L\lambda^2 + 2(\sqrt{s/D}L - \alpha + 1)\lambda \\ &+ 2\sqrt{s/D}\Big\}.\end{aligned} \tag{10.40}$$

And,

$$\hat{f}_{tr,2}(s) = \frac{\kappa_2(\beta-1)(1-\alpha)e^{2L\sqrt{s/D}}}{D\bar{\Delta}'(s)} \tag{10.41}$$

$$\hat{f}_{r,2}(s) = \frac{\kappa_1(\beta-1)(1+\lambda L)(e^{2L\sqrt{s/D}}-\alpha)}{D\bar{\Delta}'(s)} \tag{10.42}$$

where,

$$\begin{aligned}\bar{\Delta}'(s) = \alpha\Big\{&\lambda(2+\lambda L) + \beta\{2\sqrt{s/D}(1+\lambda L) - \lambda(2+\lambda L)\}\Big\} \\ &- e^{2L\sqrt{s/D}}\Big\{(1-\beta)L\lambda^2 + 2(\sqrt{s/D}L - \beta + 1)\lambda \\ &+ 2\sqrt{s/D}\Big\}.\end{aligned} \tag{10.43}$$

Finally, we can use the expressions in the given form, Eqs. (10.38)–(10.43), to obtain the Laplace transforms of the propagators, Eq. (10.10) and Eq. (10.11) which are the main results of this paper.

Up to now, we have derived analytical expressions to describe the diffusion of particles in a complex geometry. To illustrate these results, we calculated the relaxation functions, according the equation (10.13), in a system with the geometrical parameters: $\lambda = 0.1$, $L = 30$ nm, $P_1^{eq} + P_2^{eq} = 0.98$ y $P_c^{eq} = 0.2$. These parameters were taken from the geometry of an ionic channel KscA,[1,6] see Fig. (10.4).

In Fig. (10.4) we can observe a small difference between the two relaxation functions calculated. The first one, R_{11}, is slightly below to the second one, R_{21}. This is explained by the relation of the propagators with the tube's fluxes. While relaxation in chamber 2 depends on the time that particles spend escaping from chamber 1 and going across all the channel, relaxation in chamber 1 do not have to. So, the equilibrium in chamber 1 is reached faster than in chamber 2, depending on the size of the channel.

Given the dependence on the fluxes, one can predict the influence of the capillary's geometrical parameters on the effective diffusion. To determine the range of applicability of the reduction to the one-dimensional description of the problem, we performed Monte Carlo computational simulations, and are treated in the next section.

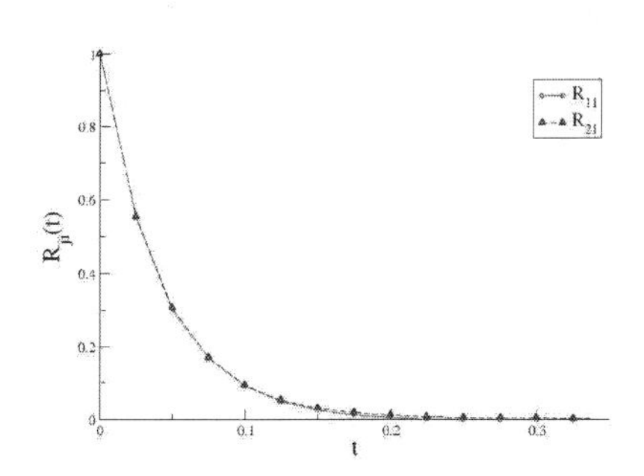

Fig. 10.4. Relaxation functions for a system of two chambers connected by a conical capillary. The number of particles in chamber 1 reaches the equilibrium slightly faster than in chamber 2, due to the initial conditions and the relation with the fluxes from the capillary.

10.5. Computational Monte Carlo Simulations Validate Assumptions Made in our Model

As we discuss in the previous section, if one asumes that the distribution of the solute in any cross section of the tube is uniform as it is at equilibrium, directing the x-axis along the center line of a tube, one can write an approximate one-dimensional effective diffusion as the Fick-Jacob's equation. As Zwanzig pointed out,[4] Fick-Jacobs' equation can be considered as the Smoluchowski equation for diffusion in the entropy potential $U(x)$ defined as,

$$U(x) = -k_B T \ln \frac{A(x)}{A(x_0)}, \tag{10.44}$$

where k_B is the Boltzmann constant and T the temperature, and $U(x)$ at $x = x_0$ is taken to be zero, $U(x_0) = 0$.

Equation (10.20) with position-independent diffusion coefficient, $D(x) = D$, is known as the Fick-Jacobs' (FJ) equation.[10] To improve FJ's reduction, Zwanzig (Zw) derived one-dimension diffusion equation assuming that the tube radius $r(x)$ is a slowly varying function, $|r'(x)| \ll 1$.[4] He showed that $c(x,t)$ satisfies the conservation probability equation

$$\frac{\partial c(x,t)}{\partial t} = -\frac{\partial j(x,t)}{\partial x} \tag{10.45}$$

where the flux, $j(x,t)$, is given by,

$$j(x,t) = -A(x)D(x)\frac{\partial}{\partial x}\left[\frac{c(x)}{A(x)}\right] \tag{10.46}$$

The expression for $D(x)$ deducted by Zwanzing is as follows,[4]

$$D_{Zw}(x) = \frac{D}{1 + r'(x)^2/2}. \tag{10.47}$$

Recently, in the same spirit of rectifying the FJ's reduction, Reguera and Rubí (RR) suggested, based on heuristic arguments, the following expression for $D(x)$,[8]

$$D_{RR}(x) = D\left[1 - \frac{1}{2}R'(x)^2\right] \simeq \frac{D}{\sqrt{1 + r'(x)^2}}. \tag{10.48}$$

Now, we want to discuss the range of applicability of these diffusion coefficients in a conical channel. To this end we ran computational simulations of random walks in a conical tube as the one shown in Fig. (reffig:cone). Each trajectory starts from a point uniformly distributed over the reflecting wall, and ends in an absorbent one, both perpendicular to the tube's axial length. The side-walls were taken as reflective, and therefore the contact between the particle and the side-walls were treated classically as an elastic collision. In simulations we find the mean first-passage time τ (MFPT), defined as the time that particle takes to reach the absorbing wall for the first time. When running simulations we take $D_0 = 1$ and the time step $\Delta t = 10^{-4}$, so that $\sqrt{2D_0\Delta t} = \sqrt{2} \times 10^{-2} \ll 1$. When the particle's initial position is fixed at the wide wall and the absorbing at the narrow one, the MFPT will be denoted $\tau_{w\to n}(\lambda)$, where n and w stand for the narrow and wide ends of the tube. On the other hand we have $\tau_{n\to w}(\lambda)$ when the starting point and the absorbing wall are reversed

Thereafter, both MFPT were related to the correspondent effective diffusion coefficients trough the following equations,

$$\tau_{n\to w}(\lambda) = \frac{L^2}{6D_{n\to w}}\left(\frac{3+\lambda L}{1+\lambda L}\right) \tag{10.49}$$

$$\tau_{w\to n}(\lambda) = \frac{L^2}{6D_{w\to n}}(3+2\lambda L) \tag{10.50}$$

as reported by Berezhkovskii et al.[9]

Comparison between theory and experimental results are shown in Fig. (10.5). Curves show theoretical values using the effective diffusion given by Zwanzig (bold-dashed line, equation 10.47), Reguera-Rubí (bold-solid line, equation 10.48), and Fick-Jacobs (light-dotted line, where $D(x) = D_0$) depending on direction, τ_i, $i = n \to w, w \to n$.

First of all, It should be noted that transport along each direction in the conical tube is not equivalent, revealing a strong influence of the entropic potential on the MFPT. In Fig. (10.5) (*a*) ($w \to n$), the behavior of the MFPT remains well close to the Reguera-Rubí model in the whole range of values for λ, while in the opposite direction ($n \to w$), Fig. (10.5) (*b*), for $\lambda > 0.2$, neither model satisfactorily explains the observed pattern of τ. It is important to point it out that a huge effort should be done to obtain a diffusion coefficient expression capable to reproduce the whole set of experimental data. Even so, the Fick-Jacobs approximation is justified in the interval $\lambda \leq 0.2$, that suitably fits the characteristic dimensions of actual ionic channels (see example shown in Fig. (10.1)).

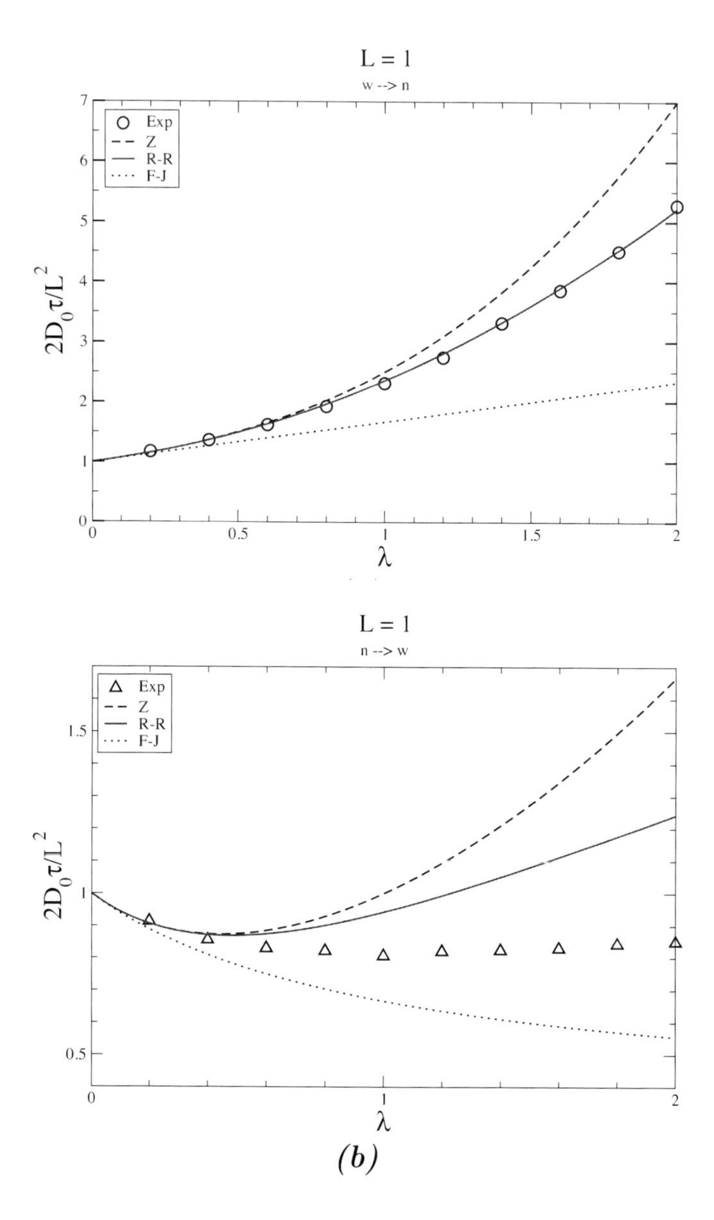

Fig. 10.5. Normalized mean-survival times of a Brownian particle diffusing trough a conical channel. This trajectory can be accomplished in two distinct and opposite directions, denoted (***a***) ($w \to n$), and (***b***) ($n \to w$), where circles and triangles denote experimental data, respectively. Lines correspond to equations (10.50) and (10.49), respectively, with three theoretical models of diffusion: Zwanzig (bold dashed, Z), Reguera-Rubí (solid, R-R), and Fick-Jacobs (dotted, F-J).

10.6. Conclusions

The method used, based on the propagator functions, enabled us to obtain analytic expressions —in the Laplace space— for the translating and returning fluxes. Eventually, after a simple substitution, those fluxes lead to the relaxation functions. From parameters α, Eq. (10.28), and β, Eq. (10.29), both of which contain geometrical information of the system that make a clear distinction between the fluxes (equations (10.38), (10.39); and equations (10.41), (10.42)) and this result have important biological implications. This noticeable behavior is also confirmed by numerical simulations of Brownian dynamics performed to asses the range of application of our assumptions. The experimental data fit well with theoretical predictions in the interval $\lambda \leq 0.2$ showing that the use of the Fick-Jacobs' approximation is well justified. Additionally, the experimental results also show that the behavior of the system in the $n \to w$ direction is not well explained by the available models so far, leaving this problem open to further research.

Acknowledgments

I. Pineda wishes to thank the National Council of Science and Technology (CONACyT) the financial support trough his Postgraduate studies.

References

1. M. B. Jackson, *Molecular and Cellular Biophysics*, Cambridge University Press, UK (2006).
2. L. Dagdug, A. M. Berezhkovskii, S. Y. Shvartsman, G. H. Weiss, *J. Chem. Phys.*, 119(23) 12473-12478 (2003).
3. I. V. Grigoriev, Yu. A. Makhnovskii, A. M. Berezhkovskii, and V. Yu. Zitserman, *J. Phys. Chem.*, 116,9574 (2002).
4. R. Zwanzig, *J. Phys. Chem*, 96, 3926-3930 (1992).
5. S. M. Bezrukov, A. M. Berezhkovskii, M. A. Pustovoit, A. Szabo, *J. Chem. Phys.*, 113, 8206 (2000).
6. I. Pineda, M. V. Vázquez, L. Dagdug, *Difusión a través de un canal cónico: Reducción efectiva a una dimensión*, in La Física Biológica en México: Temas Selectos (Vol. 2), L. García-Colín *et al.* (Coords.), El Colegio Nacional, México (2008).
7. L. Dagdug, A. M. Berezhkovskii, S. M. Bezrukov, G. H. Weiss, *J. Chem. Phys.*, 118(5) 2367-2373 (2003).
8. M. Reguera, and J. M. Rubí, *Phys. Rev. E*, 64, 061106 (2001).
9. A. M. Berezhkovskii, M. A. Pustovoit, S. M. Bezrukov, *J. Chem. Phys.*, 126, 134706 (2007).
10. M. H. Jacobs, *Diffusion Processes* (Springer, New York, 1967).

Chapter 11

Non-Equilibrium Thermodynamics of Transcriptional Bursts

Enrique Hernández-Lemus*

Departamento de Genómica Computacional
Instituto Nacional de Medicina Genómica
Periférico Sur No. 4124, Torre Zafiro II, Piso 6 Col. Ex Rancho de Anzaldo, Álvaro Obregón México, D.F. C.P. 01900, Phone. +52 (55) 5350-1970, Fax 5350-1999

Gene transcription or Gene Expression (GE) is the process which transforms the information encoded in DNA into a *functional* RNA message. It is known that GE can occur in bursts or pulses. Transcription is irregular, with strong periods of activity, interspersed by long periods of inactivity. If we consider the average behavior over millions of cells, this process appears to be continuous. But at the individual cell level, there is considerable variability, and for most genes, very little activity at any one time. Some have claimed that GE bursting can account for the high variability in gene expression occurring between cells in isogenic populations. This variability has a big impact on cell behavior and thus on phenotypic conditions and disease. In view of these facts, the development of a thermodynamic framework to study gene expression and transcriptional regulation to integrate the vast amount of molecular biophysical GE data is appealing. Application of such thermodynamic formalism is useful to observe various dissipative phenomena in GE regulatory dynamics. In this chapter we will examine at some detail the complex phenomena of transcriptional bursts (specially of a certain class of anomalous bursts) in the context of a non-equilibrium thermodynamics formalism and will make some initial comments on the relevance of some irreversible processes that may be connected to anomalous transcriptional bursts.

Contents

*Also at the Centro de Ciencias de la Complejidad, UNAM

11.1. Introduction

The forms and functions of living cells are affected by a wide variety of factors. Between these are, physiological constraints related to the cells function within the organisms, interplay of the cell with environmental factors and also the potential interaction of many genes and proteins. Hence, one important tool to analyze diversity at the cellular level is gene expression analysis (GEA). One particular tool to perform GEA is microarray (MA) technology, in particular the use of high density oligonucleotide arrays has become widely used in several instances in the molecular biomedical research community. The system, also known as GeneChip-technology made use of oligonucleotides, usually of 25 base-pairs in longitude that are used to probe genes. Messenger RNA (mRNA) experimental samples are prepared by RNA extraction from the cells culture or tissue. mRNA in the extract is then labeled with a fluorescent dye and hybridized to the arrays. Afterwards the chips are scanned with a laser and images are produced and analyzed to obtain an intensity value associated to each probe. The intensity of the fluorescent signal of a probe is related to the concentration of the mRNA molecule corresponding (tagged) by this probe.[1]

Since the process of gene expressionis complex, from the biochemical and thermodynamical points of view,[2] there are many shortcomings related to technicalities in the application and interpretation of GEAs: The transcription mRNA for a certain gene from its DNA template is frequently regulated by different processes between genes and their products, this set of various biophysico-chemical mechanisms is commonly called gene regulation; also the measurement process produces extremely noisy signals and, for practical and economic limitations, the number of experimental samples is limited whilst the number of variables is in the order of tens of thousands, namely, the expression level of each mRNA transcript within a cell. It is expected thus, that all these complex phenomena will require a phenomenological description of transcriptional regulation since concentrations (expression levels) and chemical potentials of mRNA transcripts are combinatorially correlated in a non-equilibrium environment within the living cell. This points out to the necessity of an irreversible thermodynamical description of the gene regulatory processes.

On the other hand, experimental data on the fluctuations in genetic activity both between individual cells and within the same cell over time have showed that gene expression is itself, a noisy process. Analyzing gene expression at the single-cell level has provided insight into oscillatory or nonlinear behavior in asynchronous cells and has revealed the cell-to-cell variability that arises owing to the complex nature of gene expression.[3,4] Transcriptional network analysis have showed that, instead of being independent, different levels of gene regulation are strongly coupled. This cooperativity-complexity relationship take special importance with regards to pathologies strongly related to gene de-regulation such as cancer. An important case of such cooperativity phenomena is that of transcriptional bursts (TBs) which are experimentally verifiable outcomes of this phenomena, as it could be observed that protein production often occurs in bursts, each due to a single promoter or transcription factor binding event.[6] TBs are ubiquitous and fundamental phenomena in living matter;[5] however anomalous TBs have been observed in a variety of biological settings (both normal and diseased): ulcerative colitis,[7] endocrine system,[8] cellular differentiation in developmental processes.[9] It has been stated that positive feedback between mRNA and regulatory-protein production may result in bistability and seemingly stochastic bursts in gene transcription.[10,11] In fact, when explicit mRNA diffusion is considered [a], the bursts may be much more irregular; the periods between bursts may be much shorter; and, depending on the circumstances, the burst window may be reduced or extended.[12] Some researchers even suggest that TBs are responsible for whole genome expression coordination.[13]

11.2. A Two-State Model for Hybridization Thermodynamics

We are interested in the thermodynamics associated with gene expression quantification and profiling in high throughput experiments (that is, microarray experiments, for example Affymetrix GeneChips)[16,17] since this is

[a] Diffusive phenomena are not explicitly considered in the model presented in this paper, however this could be implemented if experimental data on whole-genome diffusion is available.

the ultimate and more accurate laboratory tool to study the mechanism of genetic transcription and regulation. According with a Langmuir adsorption model of oligonucleotide hybridization the specific-hybridization intensity (or gene expression signal) for a gene probe i within a set of M gene-probes ($i = 1, ..., M$) as measured by (for example) an Affymetrix-type gene chip[18] is given by:[19,20]

$$\varphi_i(c_i, \Delta G_i) = \frac{A_i c_i e^{-\beta \Delta G_i}}{1 + c_i e^{-\beta \Delta G_i}} \tag{11.1}$$

where $\beta = \frac{1}{RT}$, T is the local temperature, R is the gas constant, c is the mRNA concentration for this species, ΔG is the free energy of hybridization, and A is a parameter that sets the scale of intensity corresponding to the saturation limit $c >> e^{\beta \Delta G}$. The local chemical potential μ_i of species i due the hybridization process could be calculated as follows:[20]

$$\mu_i = \left(\frac{\partial \Delta G_i}{\partial c_i} \right)_{T,P,c_j} = \left(\frac{\partial \Delta G_i}{\partial \varphi_i} \right)_{T,P,c_j} \left(\frac{\partial \varphi_i}{\partial c_i} \right)_{T,P,c_j} \tag{11.2}$$

After some straightforward calculations we finally get:

$$\mu_i = \frac{\frac{\varphi_i}{c_i}\left(1 - \frac{\varphi_i}{A_i}\right)}{\beta \varphi_i \left(\frac{\varphi_i}{A_i} - 1\right)} = \frac{-1}{\beta c_i} = \frac{-RT}{c_i} \tag{11.3}$$

This level of description (two state Langmuir adsorption model) gives an expression for the chemical potential that mathematically resembles that of an *ideal gas*. Given the low concentrations of mRNA transcripts in solution and also the fact that current technologies are very efficient in reducing the rate of unspecific hybridization this model is appropriate.[18]

11.3. Irreversible Thermodynamics of Transcriptional Regulation

The process of gene regulation within a cell is extremely complex from the bio-physicochemical standpoint. For the reasons mentioned above, an irreversible thermodynamics formalism has been recently proposed[20] to try to capture some of these complex phenomena within a general scheme. Here we will explore some of its main assumptions and consequences, as well as its limitations and range of applicability within the present state of molecular biomedical research, in particular concerning bursts of transcriptional regulation. As one may notice, an important source of complexity in the thermodynamical characterization of such systems lies in the fact that a cell [b] is a *small system*, in the sense that its dimensions are so reduced that the application of the thermodynamic limit is not valid due to the role of fluctuations and stochasticity in the description . Yet, small systems thermodynamics *for equilibrium systems* has been studied in the past[22,23] and some results were even expected to extend the local equilibrium settings within cellular sized biosystems.[24] Theories have been developed to explain several results. Recently, attempts have been made to include mesoscopic thermodynamical approaches[25,28] to irreversible phenomena as is done by the formalism of Mesoscopic Non-Equilibrium Thermodynamics (MNET).[25] MNET is an extension of the equilibrium thermodynamics of small systems developed by Hill and co-workers in the 1990s.[22,24] Within MNET it is recognized that scaling down the description of a physical system brings up energy contributions that are usually neglected in thermodynamical descriptions. MNET was developed to characterize non-equilibrium small systems MNET could be a good choice to model activated processes, like a system crossing a potential barrier, *provided one has a suitable model* or microscopic means to infer the probability distribution for the non-equilibrated degrees of freedom.

Whole-genome transcriptional regulation consists on a (huge) series of biochemical reactions, and many of these has unexplored chemical kinetics, thus a detailed MNET analysis is un attainable at the present moment. On what follows, we will explore a phenomenologically based approach that nevertheless takes into account (although in a more intuitive, less explicit way) similar considerations as the MNET framework. This phenomenological approach[20] is based on the Extended Irreversible Thermodynamics assumption of enlargement of the thermodynamical variables space.[34,35]

11.3.1. *Extended irreversible thermodynamics*

We shall start our discussion by assuming that a generalized entropy-like function Ψ exists, which may be written in the form:[36,37]

[b]The model presented here deals with information of mRNA levels within the cells as reflected by indirect measurements through the use of mRNA extracted from a pool of cells and hybridized on GeneChips. For a comprehensive description of MA experiments see section 2 of reference 1.

$$d_t\Psi = T^{-1}\left[d_tU + pd_tv - \sum_i \mu_i d_tC_i - \sum_j \chi_j \odot d_t\Phi_j\right] \tag{11.4}$$

T is the local temperature, p and V the pressure and volume, etc. X_j and Φ_j are the extended thermodynamical fluxes and forces. In the case of a multicomponent mRNA mixture at fixed volume and pressure, we will take as the set of relevant variables, the temperature T(r,t) and concentrations of mRNA species $C_i(\vec{r},t)$ and also the transcription *mass flux* of these species $\Phi_i(\vec{r},t)$ as *fast* variables that will take into account the presence of inhomogeneous regions and so will correct the predictions based on the local equilibrium hypothesis. The non-equilibrium Gibbs free energy for a mixture of $i = 1, ..., M$, mRNA transcripts at constant pressure, then reads:

$$d_tG = -\Psi d_tT + \sum_i \mu_i d_tC_i + \sum_j \Phi_j \odot d_t\chi_j \tag{11.5}$$

In order to take into account the gene regulatory interactions underlying transcriptional bursts, we retain the generalized force-flux terms. We then propose a form for the extended fluxes and forces within this highly fluctuating regime, that at the same time allow for experimental verification, is simple enough to be solved and it is compatible with the axioms of extended irreversible thermodynamics: A system of linear (in the forces) coupled fluxes with memory. This type of form has been used to successfully characterize another highly fluctuating system, a fluid mixture near the critical point.[38]

The constitutive equations are

$$\vec{\Phi}_j(\vec{r},t) = \sum_k \int_{-\infty}^{t} \lambda_{j,k}^{\Phi}\vec{u}e^{\frac{(t'-t)}{\tau_j^{\Phi}}}\mu_{j.k}(\vec{r},t')dt' \tag{11.6}$$

$$\vec{X}_j(\vec{r},t) = \int_{-\infty}^{t} \lambda_j^{X} e^{\frac{(t'-t)}{\tau_j^{X}}}\vec{\Phi}_j(\vec{r},t')dt' \tag{11.7}$$

The λ's are time-independent amplitudes, $\vec{u}$ is a unit vector in the direction of mass flow and τ's are the associated relaxation time. It is possible to take the limits $\tau_j^{\Phi} \to 0$ and $\tau_j^{X} \to 0$ since most GE studies nowadays are made either on steady state experiments or on time-courses with a small number of *distant* time-points, then the integrals become evaluated delta functions to give:

$$\vec{\Phi}_j(\vec{r},t) = \vec{u}\sum_k \lambda^{\Phi}_{j,k}\mu_{j.k}(\vec{r},t) \tag{11.8}$$

$$\vec{X}_j(\vec{r},t) = \lambda^X_j \vec{\Phi}_j(\vec{r},t) \tag{11.9}$$

In the future, it is expected that it will turn possible to measure gene expression in shorter time intervals (maybe even in real time). In that case, the appropriate theoretical setting will be given by Eqs. (11.6) and (11.7) that represent the dynamic nature of the coupling better than Eqs. (11.8 and (11.9). Also due to the spatial nature of the experimental measurements (either RNA blots or DNA/RNA chips measure space-averaged mRNA concentrations) it is possible to work with the related scalar quantities instead, to give:

$$\Phi_j(\vec{r},t) = \sum_k \lambda^{\Phi}_{j,k}\mu_{j.k}(\vec{r},t) \tag{11.10}$$

$$X_j(\vec{r},t) = \lambda^X_j \Phi_j(\vec{r},t) \tag{11.11}$$

Substituting Eqs. (11.10) and (11.11) into Eq. (11.5) and assuming the generalized transport coefficient λ^X_j to be independent of the flux Φ_j we are able to write[20] (in terms of the transcription regulation *chemical potentials* $\mu_{j,k}$):

$$d_tG = -\Psi d_tT + \sum_i \mu_i d_tC_i + \sum_j\sum_k \left(\lambda^{\Phi}_{j,k}\mu_{j.k}\right)\lambda^X_j\, d_t\Phi_j \tag{11.12}$$

Defining $L_{j,k} = \frac{\left(\lambda^{\Phi}_{j,k}\right)^2\lambda^X_j}{2}$

$$d_tG = -\Psi d_tT + \sum_i \mu_i d_tC_i + \sum_j\sum_k L_{j,k}d_t\mu^2_{j.k} \tag{11.13}$$

As we have stated, fluorescence intensity signals as measured by, for example, Microarray experiments (i.e. gene chips) are the usual technique to acquire information about the concentration of a given gene under certain cellular conditions. From Eq. (11.1), the concentration of a given gene-probe (with hybridization energy ΔG_i) is a function of the intensity as follows:

$$c_i = \frac{\varphi_i}{A_i e^{-\beta \Delta G_i} - \varphi_i e^{-\beta \Delta G_i}} \tag{11.14}$$

$\mu_i = +RT/c_i$ could be used in the thermodynamical characterization of gene expression as given by Eq. (11.13). If we insert Eq. (11.14) into Eq. (11.3) we get:

$$\mu_i = \frac{RT\left(A_i e^{-\beta \Delta G_i} - \varphi_i e^{-\beta \Delta G_i}\right)}{\varphi_i} \tag{11.15}$$

to give:

$$d_t G = -\Psi d_t T + \sum_i \frac{A_i e^{-\beta \Delta G_i}}{\beta \varphi_i \left(A_i e^{-\beta \Delta G_i} - \varphi_i e^{-\beta \Delta G_i}\right)} d_t \varphi_i + \sum_j \sum_k L_{j,k} d_t \mu^2_{j.k} \tag{11.16}$$

If we define $\Gamma_i = \frac{A_i e^{-\beta \Delta G_i}}{\beta \varphi_i \left(A_i e^{-\beta \Delta G_i} - \varphi_i e^{-\beta \Delta G_i}\right)}$ as the *thermodynamic* conjugate variable to the probe intensity φ_i we obtain:

$$d_t G = -\Psi d_t T + \sum_i \Gamma_i d_t \varphi_i + \sum_j \sum_k L_{j,k} d_t \mu^2_{j.k} \tag{11.17}$$

11.3.2. *Maxwell-like relations*

Recalling Eq. (11.17) it is possible to further extract additional thermodynamical insight in the form of Maxwell-like relations. These relations arise from the assumption that $d_t G$ has the structure of an exact differential form.[37] From this assumption one could notice that:

$$\left(\frac{\partial \Gamma_i}{\partial T}\right)_{\varphi_j, \mu_{j,k}} = -\left(\frac{\partial \Psi}{\partial \varphi_i}\right)_{\varphi_j, \mu_{j,k}} \tag{11.18}$$

taking the definition of Γ_i we get

$$\left(\frac{\partial \Gamma_i}{\partial T}\right)_{\varphi_j, \mu_{j,k}} = \left(\frac{\partial}{\partial T}\left[\frac{A_i e^{-\beta \Delta G_i}}{\beta \varphi_i A_i e^{-\beta \Delta G_i} - \varphi_i e^{-\beta \Delta G_i}}\right]\right)_{\varphi_j, \mu_{j,k}} \tag{11.19}$$

After performing derivatives and re-arrange terms one gets:

$$\left(\frac{\partial \Gamma_i}{\partial T}\right)_{\varphi_j,\mu_{j,k}} = \frac{A_i e^{-\beta \Delta G_i}}{\beta(A_i e^{-\beta \Delta G_i} - \varphi_i e^{-\beta \Delta G_i})} \tag{11.20}$$

By substitution of the definition of Γ_i Eq. (11.20) gives:

$$\left(\frac{\partial \Gamma_i}{\partial T}\right)_{\varphi_j,\mu_{j,k}} = \Gamma_i \times \varphi_i \tag{11.21}$$

That substituted into Equation (11.18) gives as a result that:

$$\left(\frac{\partial \Psi}{\partial \varphi_i}\right)_{\varphi_j,\mu_{j,k}} = -\Gamma_i \times \varphi_i \tag{11.22}$$

Eq. (11.22) could be parametrically integrated to give:

$$\Delta\Psi_{\varphi_i} = -\frac{A_i}{\beta} \ln\left[A_i e^{-\beta \Delta G_i} - \varphi_i e^{-\beta \Delta G_i}\right] + \Theta(\varphi_j, \mu_{j,k}) \tag{11.23}$$

Here $\Theta(\varphi_j, \mu_j, k)$ is an *integration constant* that could be function of all other independent variables, namely $\varphi_j \forall j \neq i$, and $\mu_{j,k} \forall j, \forall k$. This non-equilibrium entropy contribution takes into account the fact that individual gene expression levels (in this case of gene i) modify the whole *entropic landscape* of the system.

From the integrability conditions[37] of Eq. (11.17) one can also notice that:

$$\left(\frac{\partial \Gamma_i}{\partial \varphi_j}\right)_{T,\varphi_k,\mu_{j,k}} = -\left(\frac{\partial \Gamma_j}{\partial \varphi_i}\right)_{T,\varphi_k,\mu_{j,k}} \tag{11.24}$$

In the absolutely non-regulated case (i.e. the expression levels of all genes are statistically independent), Eq. (11.24) trivially holds. This fact, however implies a far from trivial consequence, critical for the whole concept of systems level biology: *Every change in the expression level of a single gene (i) perturbs the expression levels for the whole genome* (every other gene j). In this way Eq. (11.24) represents a *systematic interconectedness* constraint for whole genome expression.

The third set of Maxwell like relations derived from Eq. (11.21) refers to the thermal coupling of transcription factor activity. Specifically we have that:

$$\left(\frac{\partial L_{j,k}}{\partial T}\right) = -\left(\frac{\partial \Psi}{\partial \mu_{j,k}^2}\right) \tag{11.25}$$

Eq. (11.25) holds for the transcription-factor-regulated setting. According to the definition of $L_{j,k}$ previously given we obtain:

$$\left(\frac{\partial L_{j,k}}{\partial T}\right) = \frac{1}{2}\frac{\partial\left[(\lambda_{j,k}^{\Phi})^2\lambda_j^X\right]}{\partial T} \tag{11.26}$$

Which after some calculations and substitutions gives:

$$\left(\frac{\partial L_{j,k}}{\partial T}\right) = L_{j,k}\frac{\partial\left[\ln\left(\sqrt{\lambda_{j,k}^{\Phi}}\lambda_j^X\right)\right]}{\partial T} \tag{11.27}$$

Or, given the Maxwell-like relations just mentioned:

$$\left(\frac{\partial \Psi}{\partial \mu_{j,k}^2}\right) = -L_{j,k}\frac{\partial\left[\ln\left(\sqrt{\lambda_{j,k}^{\Phi}}\lambda_j^X\right)\right]}{\partial T} \tag{11.28}$$

Given a model of experimental data, that accounts for the dependence of the $\lambda_{j,k}^{\Psi}$'s and the λ_j^X's in the thermodynamical parameters, Eq. (11.32 could be formally integrated to give:

$$\Delta\Psi_{\mu_{j,k}} = -\int L_{j,k}\frac{\partial\left[\ln\left(\sqrt{\lambda_{j,k}^{\Phi}}\lambda_j^X\right)\right]}{\partial T}d\mu_{j,k}^2 \tag{11.29}$$

Eq. (11.29) or its analogues could play an important role in the experimental characterization of such transcription-regulation *chemical potentials* ($\mu_{j,k}$) since, in principle, it is possible to develop techniques (e.g. capillary micro-calorimetry inside the cell) to measure intra-cellular thermal dissipation (let us recall that the non-equilibrium entropy-like function Ψ is closely related with uncompensated heat phenomena) and via a model for the λ-parameters study these $\mu_{j,k}$'s.

11.4. Gene Regulation

Now let us examine in some detail the structure of Eq. (11.21). In the isothermic, non-regulated steady state (i.e. $d_tG = 0$, $d_tT = 0$, $\mu_{j,k}^2 = 0$),

Eq. (11.21) is nothing but a formal non-equilibrium extension of the Gibbs-Duhem relation $\sum_i \Gamma_i d_t \varphi_i = 0$. Without any gene regulatory mechanism, and without explicit dissipation, the energetics of gene expression within a cell are just the ones of a non-interacting dilute mixture of its components (in this case the different mRNA transcripts).[20] A more realistic case is the regulated, isothermal steady state given by: $d_t G = 0$, $d_t T = 0$ and at least some $\mu^2_{j,k} \neq 0$. Let us then consider the regulated isothermal steady-state version of Eq. (11.21), namely:

$$\sum_i \Gamma_i d_t \varphi_i + \sum_i \sum_k L_{j,k} d_t \mu^2_{j,k} = 0 \tag{11.30}$$

One could see that changes in the mRNA concentration of gene i as measured by its probe intensity φ_i could depend *not only in their own* characteristic thermodynamical parameters (A_i, ΔG_i, and T) but also on other mRNA transcript (say n) via a coupling given by a term $L_{n,i}\mu^2_{n,i}$. In that case one says that the n-th gene regulates the i-th gene, or that n is a *transcription factor* for i (conversely i is a *transcriptional target* of n).

11.4.1. *Gene-switching model*

Consider the irreversible thermodynamic coupling that sets the process of transcriptional regulation between two genes. In this case, let gene number 1 be a transcription factor for gene number 2 and that gene 1 is not regulated. This means that $\mu_{1,2} \neq 0$ whereas $\mu_{1,1} = \mu_{2,1} = \mu_{2,2} = 0$. Then Eq. (11.30) will read:

$$\Gamma_1 d_t \varphi_1 + \Gamma_2 d_{tt} \varphi_2 + L_{1,2} d_t \mu^2_{1,2} = 0 \tag{11.31}$$

We will consider SYK, the transcript responsible for the synthesis of *spleen tyrosine kinase* as gene 1 and IL2RB (*interleukin 2 receptor, beta*) as gene 2. SYK is known for being an inducer of gene transcription, specially in the case of the beta domain interleukin 2 receptor.[39] Also, there is evidence indicating the possible role of these two genes in the course of the so-called C-MYC network of reactions, a very important, cancer-related biochemical pathway. It is known that bursts of C-MYC expression are correlated with passage of cells from $\mathcal{G}_0$ into the cell cycle[40] and with the presence of differentiation of *human burst-forming unit, erythroid (BFU-E)*[41] also important in cell's cycle control and thus potentially influencing cancer outcomes.[42]

The value of the parameters could be calculated:[20] ΔG_1=483.55 kcal/mol and ΔG_2=463.05 kcal/mol. From Eq. (11.1 we obtain A_i as follows:

$$A_i = \lim_{\varphi_i \to \varphi_i^{sat}} \frac{\varphi_i(1 + c_i e^{-\beta \Delta G_i})}{c_i e^{-\beta \Delta G_i}} \tag{11.32}$$

To obtain: A_1=5513 intensity units/mol and A_2=1105 intensity units/mol. Given these parameters, it is possible to calculate both $\Gamma_1 = \Gamma_1(\varphi_1)$ and $\Gamma_2 = \Gamma_2(\varphi_2)$, and if dynamic data is available via $\varphi_1(t)$ y $\varphi_2(t)$ we could as well obtain the time evolution for $\mu_{1,2}$, hence characterizing in a complete form the transcriptional regulation for this *gene switch*. We have then, the following expressions for the thermodynamic functions in terms of the experimentally measurable intensities (in all cases physiological temperature of $T = 37^o$ C is assumed), hence $\beta = 1.622507 \times 10^{-6}$ mol kcal $^{-1}$, $e^{-\beta \Delta G_1}$=0.99922, $A_1 \times e^{-\beta \Delta G_1} = 5508.67950$ intensity units/mol; also $e^{-\beta \Delta G_2} = 0.99925$, and $A_2 \times e^{-\beta \Delta G_2} = 1104.17014$ intensity units/mol.

Calculating the intensity-dependent chemical potentials we obtain, from Eq. (11.15) (in units of kcal/mol):

$$\mu_1 = \frac{3.395 \times 10^9 - 6.158 \times 10^5 \varphi_1}{\varphi_1} \tag{11.33}$$

and

$$\mu_2 = \frac{6.805 \times 10^8 - 6.159 \times 10^5 \varphi_2}{\varphi_2} \tag{11.34}$$

There is a noticeable difference in transcriptional behavior for gene 1 (SYK) which is a transcription factor and gene 2 (IL2RB) which is not (and, in fact is a transcriptional target). The maximum intensity (related to a maximum concentration peak) attainable in both cases in the spontaneous regime (i.e. solving Eqs.(11.33 and (11.34 for null-values of the chemical potential) is of 5513 intensity units for SYK, whereas in the case of IL2RB is of just 1105 intensity units. If we recall the values of the saturation constants A_i, an immediate physical interpretation is given in terms of the respective gene expression chemical potentials. This means that, in order for IL2RB to be produced at higher rates, the presence of chemical environment modifications (e.g via transcription factors) is needed.

We are now able to calculate expressions for Γ_1 and Γ_2 as follows.

$$\Gamma_1 = \frac{5508.6795}{0.008938\varphi_1 - 1.6098 \times 10^{-6}\varphi_1^2} \tag{11.35}$$

$$\Gamma_2 = \frac{1104.17014}{0.001795\varphi_2 - 1.6243 \times 10^{-6}\varphi_2^2} \tag{11.36}$$

Substitution of Eqs.(11.35 and (11.36) into Eq. (11.31) we get (after integration):

$$-L_{1,2}\mu_{1,2}^2 = 616321.2687 \ln\left|\frac{-1.6098 \times 10^{-6}\ \varphi_1}{0.008938 - 1.6098 \times 10^{-6}\ \varphi_1}\right| + 615,136.5683 \ln\left|\frac{-1.6243 \times 10^{-6}\ \varphi_2}{0.001795 - 1.6243 \times 10^{-6}\ \varphi_2}\right| \tag{11.37}$$

Taking experimental values of φ_1 and φ_2, Eq. (11.37) could be solved for $\mu_{1,2}$.

11.4.2. *Thermodynamic differences between transcription factors (TF) and target genes (TG)*

We have found an interesting trend in the *transcriptional activity coefficients* Γ_i for other well characterized gene probes[18] (see table (11.4.2)). It could be seen that the values of Γ_i are in general lower for genes that act as transcription factors (such as RHOH, SYK, SMAD3, POU2AF1) and high for genes with no-known TF-activity (such as IL2RB, CD69, TNFRSF1B and TNFRSF14). One possible exception is GLDC which codes for a glycinedehydrogenase enzyme that is anchored to the mithochondrion and has no reported evidence of transcription factor activity yet is grouped with the TF genes in the set of low Γ's.

This trend in transcriptional activity coefficients could be traced-back to the values of the saturation constants A_i: TFs have shown higher values of the saturation constants that represent differences in transcriptional energetics between two genesets (TFs vs TGs) as it could be seen in Fig.(11.1) (For visualization bold lines are TFs whereas slim lines are TGs). These differences have shown to be statistically significant (p-value=0.007967076 under a two-tailed heteroskedastic regularized t-test). From the definition of Γ_i it is possible to see that $\frac{d\Gamma_i}{dA_i} < 0 \forall i$. Hence high values of saturation limits imply low activity coefficients. This fact supports the **TF → low** Γ

hypothesis. The physicochemical meaning of this finding seem to point-out to transcription factors as genes whose expression is regulated by lower activation-energy barriers. Since TF's are involved in the transcriptional activation of other genes, it is expected that they are synthesized first when energy is started to being released by metabolic processes in the cell. Transcriptional targets should, in general be synthesized later and with higher activation energies. It is important to stress as it is shown in Fig.(11.2) that the chemical potentials μ_i for *spontaneous* transcription do not show a big difference between TFs and TGs at low gene expression intensity levels (φ_i) but show a significant difference at high values of φ_i (recall that the values of φ_i at which the chemical potential becomes negative are the saturation constants A_i and these show statistical significant differences between TFs and TGs). Thus, higher saturation limits for the chemical potentials of TFs suggest both stability and spontaneity in the expression of these as compared to TGs.

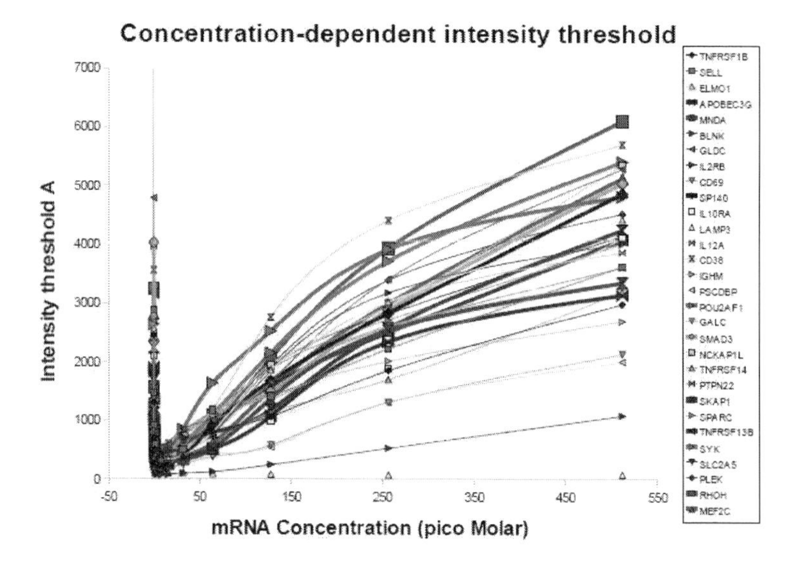

Fig. 11.1. Gene expression intensity saturation limit.

Another interesting feature of transcriptional dynamics is observed by considering the generalized-entropy contribution (i.e. *uncompensated* heat generation by irreversible processes) coming from individual gene expression

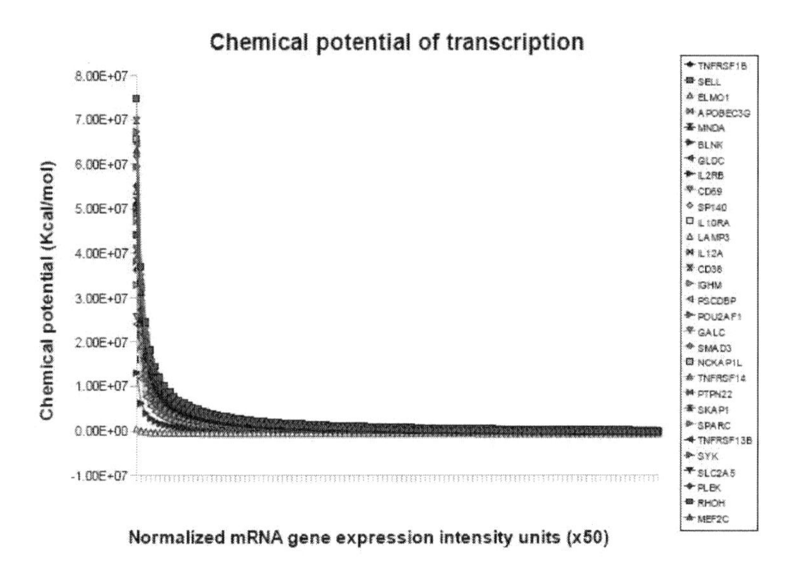

Fig. 11.2. Intensity-dependent chemical potentials of transcription.

as given by equation. As we have previously stated, a marked difference is seen in the behavior of a transcription factor and a transcriptional target at the level of *energy dissipation* In Fig.(11.3) we show some experimentally-derived evidence of regulatory phenomena. As we could see in panel A the expression levels of SYK (gene 1) and IL2RB (gene 2) in two cellular conditions (epithelial cell lines a549 and beas2b from human lung[43]) are highly flucuating, specially in the case of the TF (SYK). In panel B we found a possible energetic explanation since the related chemical potentials of individual transcription are also highly variable, whereas in panel C we observe that the regulatory chemical potentials (related to transcriptional *coupling*) are somewhat stable during this time course (of 48 hours) giving rise to almost constant (and very low) values of dissipation (in the form of uncompensated heat release) that nevertheless show differences between TFs and TGS. This probably means that not only TF's are produced easier than transcriptional targets (that is, before-hand and with lower activation energies) during the non-equilibrium process of mRNA transcription, but also they could be produced *spontaneously* during a longer period and could thus attain a higher expression level . This facilitates the role of TF's as *master regulators* of the whole-genome expression phenomena.

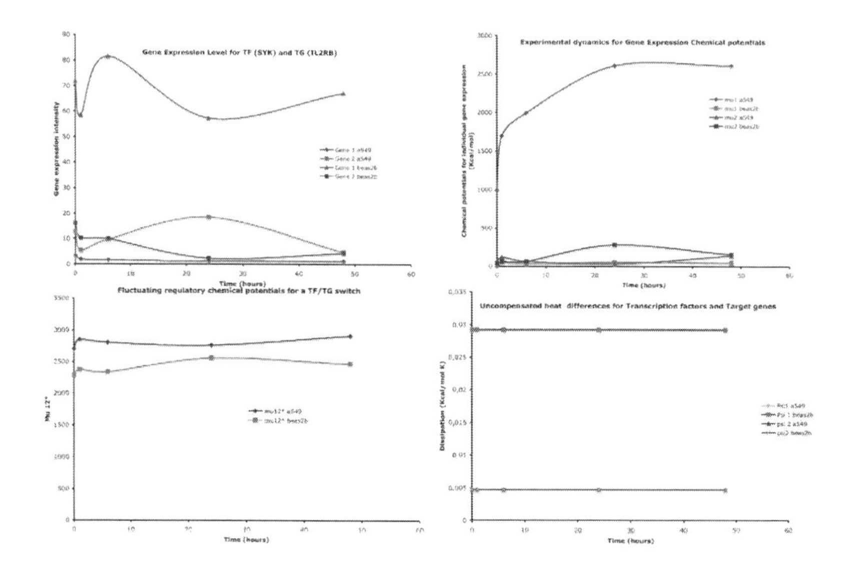

Fig. 11.3. Experimental Dynamics of Gene Regulation for SYK (TF) and IL2RB (TG): a) Gene expression levels b) Chemical potentials of transcription (see Eqs. (11.33 and (11.34)) c) Regulatory Chemical potential (Eq. (11.37)) d) Uncompensated heat dissipation (Eq. (11.23)).

All these pieces put-together point out to TF-thermodynamics as responsible for the phenomena of TBs: low activation barriers in TFs imply that stochastic events [c] within the cell energetics could lead to TF expression, since TFs also have higher saturation limits they could reach their TGs for longer times, thus enabling the propagation of the *transcriptional cascade*. The fact that TGs have higher energetic barriers and lower saturation limits may lead to *counter-regulation* of TBs and hence homeostatic steady states. Thus, as we have discussed bursts in the transcription of TFs could lead via transcriptional regulation to TBs in overall gene activity,[13] nevertheless in non-diseased cellular conditions these bursts are expected to decay for thermodynamic reasons.

In conclusion we have showed here that a non-equilibrium thermodynamics description of cell level transcriptional regulation could be formulated in terms of experimentally measurable quantities, and that essential

[c] In terms of a statistical biophysics description, stochasticity refers to a set of extremely complex biological phenomena that interact in a highly non-linear fashion, e.g. a series of hidden biochemical events within the living cell.

Table 1: Genes in the Latin Square Experiments[19,20]

Gene symbol	Transcription Factor activity	$\Delta G_{t,r}$ (Kcal/mol)
APOBEC3G	Reverse TF	477.38
MNDA	TF Regulation	433.97
SP140	TF activity	700.57
POU2AF1	TF Regulation	473.5
SMAD3	TF activity + Binding	465.08
SKAP1	TF	452.57
TNFRSF13B	TF inducer	481.8
SYK	TF inducer	483.55
RHOH	TF regulation	467.94
MEF2C	TFact, RNA Pol ind	472.81
TNFRSF1B		502.52
SELL		446.34
ELMO1		471.35
BLNK		436.55
GLDC		468.99
IL2RB		463.05
CD69		398.72
IL10RA		474.56
LAMP3		636.01
IL12A		453.31
CD38		569.56
IGHM		482.07
PSCDBP		458.19
GALC		410.95
NCKAP1L		716.67
TNFRSF14		782.09
PTPN22		414.77
SPARC		443.24
SLC2A5		488.56
PLEK		459.23

features in gene regulatory dynamics could be studied with it. In particular, regulatory differences between transcription factors and transcriptional targets are shown to occur in TB-related settings, such as the ones already discussed. Transcriptional bursts in gene expression are extremely complex phenomena, here we presented a highly idealized analysis that shed some light on the origins and possible explanation of TBs within the context of non-equilibrium thermodynamics of cell energetics. The description just sketched is far from completeness and some fundamental issues could still being missed. The problem of transcriptional regulation (and the related TB phenomena) although of enormous biological and biomedical relevance, remains largely unknown and many basic features of it are still not well-understood. One final statement of this paper is then, that a lot of work in this field remains to be done, before we reach a complete understanding of transcriptional regulation.

Acknowledgments

The author wishes to thank Karol Baca López for technical support in manuscript editing.

References

1. Baca-López, K., Hernández-Lemus, E., Mayorga. M.; *Information-theoretical analysis of gene expression data to infer transcriptional interactions*, Revista *Mexicana de Física*, (en prensa)(2009).
2. Rockmanm M. V., Kruglyak, L., *Genetics of global gene expression*, Nat. Rev. Genet. 7, 862- 872, (2006).
3. Longo D., Hasty, J. *Dynamics of single-cell gene expression*, Molecular Systems Biology 2, 64, (2006) doi:10.1038/msb4100110.
4. Attila Becskei, A., Kaufmann, B.B., van Oudenaarden, A., *Contributions of low molecule number and chromosomal positioning to stochastic gene expression*, Nature Genetics 37, 937 - 944 (2005).
5. See for example, Golding, I; Paulsson, J; Zawilski, SM; Cox, EC (2005). *Real-time kinetics of gene activity in individual bacteria*, *Cell* 123 (6): 1025-36. Also Chubb, JR; Trcek, T; Shenoy, SM; Singer, RH (2006). *Transcriptional pulsing of a developmental gene*, *Current biology* 16 (10): 1018-25, and Raj, A; Peskin, CS; Tranchina, D; Vargas, DY; Tyagi, S (2006); *Stochastic mRNA synthesis in mammalian cells*, *PLoS Biology* 4 (10): e309.
6. Ingram P.J., Stumpf, M.P.H, Stark, J. *Nonidentifiability of the Source of Intrinsic Noise in Gene Expression from Single-Burst Data.* PLoS Comput Biol 4(10): e1000192, (2008).
7. Jin, X., Malykhina, A.P., Lupu, F. and Akbarali, H.I., *Altered gene expression and increased bursting activity of colonic smooth muscle ATP-sensitive K+ channels in experimental colitis*, Am J Physiol Gastrointest Liver Physiol 287: G274-G285, 2004.
8. Shorte, S.L., Leclerc, G.M., Vazquez-Martinez, R., Leaumont, D.C., Faught, W.J., Frawley, L.S. and Boockfor, F.R., *PRL Gene Expression in Individual Living Mammotropes Displays Distinct Functional Pulses That Oscillate in a Noncircadian Temporal Pattern*, Endocrinology Vol. 143, No. 3 1126-1133, 2002.
9. Niwa, Y., Masamizu, Y., Liu, T. Nakayama, R., Deng, C-X, and Kageyama, R., *The Initiation and Propagation of Hes7 Oscillation Are Cooperatively Regulated by Fgf and Notch Signaling in the Somite Segmentation Clock*, Developmental Cell 13, 298-304, August 2007.

10. Raj, A., van Oudenaarden, A., *Single-Molecule Approaches to Stochastic Gene Expression*, Annual Review of Biophysics, June 2009, Vol. 38, Pages 255-270.
11. Tripathi, T., Chowdhury, D., *Transcriptional bursts: A unified model of machines and mechanisms*, 2008 Euro. Phys. Lett. 84 68004.
12. Zhdanov, V., *Effect of mRNA diffusion on stochastic bursts in gene transcription*, JETP Letters, Volume 85, Number 6, May 2007, pp. 302-305(4).
13. Cai, L., Dalal, C. K. , Elowitz, M.B.,*Frequency-modulated nuclear localization bursts coordinate gene regulation* Nature 455, 485-490 (25 September 2008).
14. Komili, S., Silver, S., *Coupling and coordination in gene expression processes: a systems biology view*, Nat. Rev. Genet. 9, 38-48, (2008).
15. Margolin, A.A., Nemenman, I., Basso, K., Wiggins, C., Stolovitzky, G., Dalla Favera, R., Califano, A., *ARACNe: An Algorithm for the Reconstruction of Gene Regulatory Networks in a Mammalian Cellular Context*, BMC Bioinformatics, 7 (Suppl I):S7, (2006) doi:10.1186/1471-2105-7-S1-S7.
16. Lu, Z.J., Matthews, D.H., *Efficient siRNA selection using hybridization thermodynamics*, Nucleic Acids Research, 36, 2, 640-647, (2008).
17. Hekstra, D., Taussig, A. R., Magnasco, M., Naef, F., *Absolute mRNA concentrations from sequence-specific calibration of oligonucleotide arrays*, Nucleic Acids Research, 31, 7, 1962-1968, (2003).
18. http://www.affymetrix.com/ (see especially the section: products_ services/ research_solutions/methods/wt_geneexpression_altsplicing.affx also see for the Latin Square Data the section support/technical/sample_data/ datasets.affx).
19. Carlon, E., Hein, T., *Thermodynamics of RNA/DNA hybridization in high density oligonucleotide arrays*, Physica A, 362, 433-449, (2006).
20. Hernández-Lemus, E., *Non-equilibrium thermodynamics of gene expression and transcriptional regulation*, Journal of Non-equilibrium Thermodynamics, (in press) (2009).
21. Held, G. A., Grinstein, G., Tu, Y., *Modeling of DNA microarray data by using physical properties of hybridization*, PNAS, 100, 13. 7575-7580, (2003).
22. Hill, T.L., and Chamberlin, R.V., *Extension of the thermodynamics of small systems to open metastable states: An example*, Proc Natl Acad Sci. USA 95(22): 12779-12782, (1998).
23. Hill, T.L., *Thermodynamics of Small Systems*, Dover, N. York (2002).
24. Hill, T.L., *Free Energy Transduction and Biochemical Cycle Kinetics*, Dover, N. York (2004).
25. Rubí J.M., *Non-equilibrium thermodynamics of small-scale systems*, Energy, 32, 4, 297-300, (2007).
26. Rubí, J.M., *Mesoscopic Non-equilibrium Thermodynamics*, Atti dellAccademia Peloritana dei Pericolanti, Classes di Scienze Fisiche, Matematiche e Naturali, Vol. LXXXVI, C1S081020, Suppl. 1, (2008).
27. Rubí, J.M., Bedeaux, D., Kjelstrup, S., *Unifying thermodynamic and kinetic descriptions of single molecule processes: RNA unfolding under tension*, J. Phys. Chem. B., 111, 32,9598-9602, (2007).

28. Ritort, F., *The nonequilibrium thermodynamics of small systems*, Comptes Rendus Physique, 8, 5-6, 528-539,(2007).
29. D. J. Evans, D. J. Searles, *Fluctuation theorem for stochastic systems*, Phys. Rev. E 50, 1645 (1994).
30. G. Gallavotti, E. G. D. Cohen, *Dynamical ensembles in nonequilibrium statistical mechanics* , Phys. Rev. Lett. 74, 2694 (1995).
31. C. Jarzynski, *Nonequilibrium equality for free energy differences*, Phys. Rev. Lett. 78, 2690 (1997).
32. G. E. Crooks, *Nonequilibrium measurements of free energy differences for microscopically reversible Markovian systems*, J. Stat. Phys. 90, 1481 (1998).
33. de Groot, S.R., Mazur, P., *Non-Equilibrium Thermodynamics*, Dover, N.York, (1984).
34. Jou, D., Casas-Vazquez, J., Lebon, G., *Extended Irreversible Thermodynamics*, Springer-Verlag, N. York, (1998).
35. Jou, D., Casas-Vazquez, J., Lebon, G., *Extended Irreversible Thermodynamics*, Rep. Prog. Phys. 51 1105-1179, (1988).
36. García-Colín, L.S., Rodríguez, R. F., López de Haro, M., Jou, D., Pérez García, C.; *Generalized hydrodynamics and extended irreversible thermodynamics*, Phys. Rev. A, 31, (1985), 2502-2508.
37. Chen, M.; Eu, B.C., *On the integrability of differential forms related to nonequilibrium entropy and irreversible thermodynamics*, J.Math. Phys. 34, (7), (1993), 3012-29.
38. Hernández-Lemus, E. and García-Colín, L.S., *Non-equilibrium critical behavior: An extended irreversible thermodynamics approach*, Journal of Non-equilibrium Thermodynamics, 31, 4, 397-417, (2006).
39. Minami, Y., Nakagawa, Y., Kawahara, A., Miyazaki, T., Sada, K., Yamamura, H., Taniguchi, T., *Protein tyrosine kinase Syk is associated with and activated by the IL- 2 receptor: possible link with the c-myc induction pathway*, Immunity. 2(1):89-100, (1995).
40. Welch, J.N. and Chrysogelos, S.A., *Positive mediators of cell proliferation in neoplasic transformation* p.73 in Coleman, W.B. and Tsongalis, Q.J. (eds.), The molecular basis of human cancers, Humana Press, New Jersey, (2002).
41. Umemura T, Umene K, Nishimura J, Fukumaki Y, Sakaki Y, Ibayashi H., *Expression of c-myc oncogene during differentiation of human burst-forming unit, erythroid (BFU-E)*, Biochem Biophys Res Commun. 1986 Mar 13;135(2):521-6.
42. Yu, Q., Ciemerych, M. A., Sicinski, P., *Ras and Myc can drive oncogenic cell proliferation through individual D-cyclins*, Oncogene (2005) 24, 7114-7119.
43. Nymark, P., Lindholm, P.M., Korpela, M.V., Lahti, L., Rousaari, S., Kaski, S., Hollmen, J., Anttila, S., Kinnula, V.L., Knuutila, S., *Gene Expression profiles in asbestos-exposed epithelial and mesothelial lung cell lines*, BMC Genomics, 8, 62, (2007). NCBI Online data accession: GSE6013.

Chapter 12

Applications of β-Peptides in Chemistry, Biology, and Medicine

Yamir Bandala and Eusebio Juaristi

Departamento de Química, Centro de Investigación y de Estudios Avanzados del Instituto Politécnico Nacional, Apartado Postal 14-740, México, D. F., México

juaristi@relaq.mx

As a consequence of the fundamental importance of peptides and proteins in physiological events, the interest in peptide synthesis as well as in their characterization has increased exponentially in recent years. Although the vast majority of natural peptides and proteins are constituted by α-amino acids, recent studies have shown that the incorporation of β-amino acids, instead of α-amino acids, in peptides significantly modifies their biological activity and increases their hydrolytic stability. As a consequence, inclusion of β-amino acid residues in peptides increases significantly the potential application of the resulting unnatural peptides. The present compilation is intended to provide an overview of the more significant advances achieved in the synthesis and applications of β-peptides in chemistry, biology, and medicine that it is being carried out in Mexico applying modern synthetic methods.

Contents

12.1. Introduction

In recent years, the synthesis of β-amino acids has been increasingly attracting the interest of scientific community owing to the discovery that their presence as constituents of various peptides results in outstanding pharmacological properties. Thus, the synthesis, characterization, and application of peptides consisting of β-amino acids, called β-peptides, has received great attention, especially since the pioneering studies of Seebach and Gellman which have demonstrated their biological potential.[1] β-Peptides are conformationally more flexible than their natural counterparts, the α-peptides, because of the additional CH_2 group inserted between the C=O group and the nitrogen atom. The substituent can be situated adjacent either to the carbonyl or to the amino function in β^2- or β^3-amino acids, respectively.

12.2. Synthesis of β-Amino Acids and β-Peptides

The potential application of β-peptides in medicine has motivated the development of interesting methodologies and strategies for the preparation of non-racemic β-amino acids[2] and β-peptides.[3] In the case of β-amino acids, the most general methodologies can be separated in several categories: homologation of α-amino acids, chiral pool strategy, use of chiral auxiliaries, synthesis via 1,3-dipolar cycloadditions, stereoselective hydrogenations, addition or substitution reactions, and preparation via radical reactions, among others. There exist several reviews regarding the preparation of β-amino acids,[2;4] therefore, this manuscript illustrates only a few representative methodologies.

In an interesting example, Reyes-Rangel *et al.*[5] reported the asymmetric synthesis of β^2-homovaline, β^2-homoleucine, and β^2-homotryptophan employing *trans*-hexahydrobenzoxazolidinone (hereinafter compound **I**; for convenience, transcendent compounds will be referred with roman numerals) as *chiral auxiliary*, achieving enantiomeric excesses greater than 99%. The methodology here involves N-acylation of **I** followed by a highly diastereoselective alkylation reaction, hydrogenolysis, and Curtius rearrangement. Finally, the hydrolysis and hydrogenation of the alkylated derivative afforded the desired β^2-amino acids, which were N-Fmoc protected for their use in solid-phase synthesis (Fig.(12.1)).[5]

On the other hand, in an application of the *chiral pool* concept and starting from (S)-asparagine, Ávila-Ortiz and co-workers[6] developed the synthesis of enantioenriched β^2-homoDOPA, **II**, a β-amino acid homolog of DOPA, the therapeutic agent most successfully used in the treatment of Parkinson's disease. Compound **II** was obtained via the diastereoselective alkylation of a chiral pyrimidinone with veratryl iodide. The diastereomeric mixture of products was separated and the desired β-amino acid was obtained in 34% enantiomeric excess via hydrolysis and purification by means of silica gel chromatography (Figure 1.2a).

By contrast, essentially enantiopure (R)- and (S)-**II** were obtained by *resolution* of racemic N-benzyloxycarbonyl-2-(3,4-dibenzyloxybenzyl)-3-aminopropionic acid with (R)- or (S)-α-phenylethylamine followed by hydrogenolysis (Fig.(12.2b)).[6]

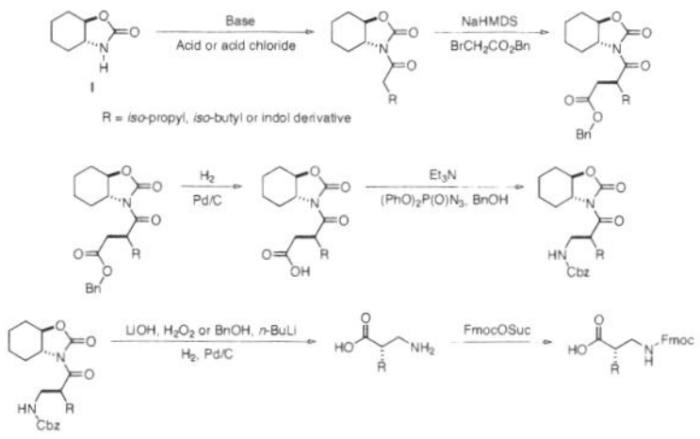

Fig. 12.1. Enantioselective synthesis of β^2-homovaline, β^2-homoleucine, and β^2-homotryptophan from the chiral auxiliary *trans*-hexahydrobenzoxazolidinone **I**.[5]

Guzmán-Mejía and co-workers[7] observed that the use of *microwave irradiation* facilitated the synthesis of (S)-α-benzyl-β-alanine, **III**, with a reduction in reaction times, increased yields, and preservation of the enantiomeric purity in the final products. Similarly, the preparation of *chiral auxiliary* (R,R)-bis(α-phenylethyl)amine, **IV**, was realized under microwave heating following highly diastereoselective alkylation reaction and hydrogenolysis. The critical hydrolysis reaction was carried again under microwave (MW) irradiation. In this form, the desired (S)-α-benzyl-β-alanine **III** was obtained with an enantiomeric excess greater than 98% in only 16 h. By comparison, β-amino acid **III** was obtained with a lower 77% enantiomeric excess and a much longer 5 days reaction time under conventional heating conditions (Fig.(12.3)).[7]

In this context, by means of the selective addition of 'chiral ammonia' to conjugated prochiral acceptors according to the methodology described by Davies,[8] Bandala *et al.*[9] synthesized the β^3-amino acid (S)-β^3-homophenylglycine, **V**. This procedure involved the highly diastereoselective *conjugate addition* of N-benzyl-β-phenylethylamine to *tert*-butylcinnamate. Subsequently, the compound (S)-**V** was N-protected using 9-fluorenylmethoxycarbonyl succinimide prior to their use in solid-phase peptide synthesis (Fig.(12.4)).[9]

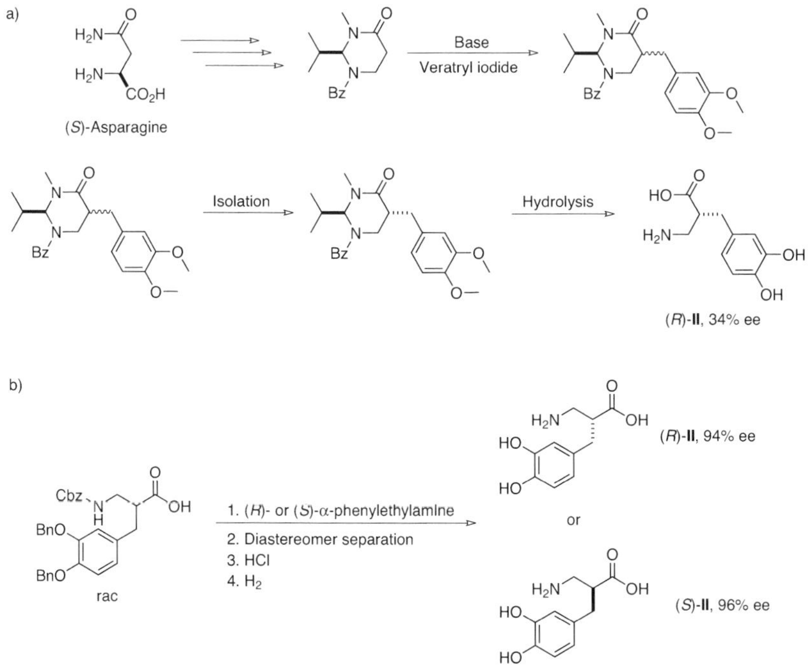

Fig. 12.2. Preparation of β^2-homoDOPA **II** from (S)-asparagine by diastereoselective alkylation (a) and chemical resolution (b).[6]

On the other hand, as in the case of their α-counterparts, the synthesis of β-peptides can be carried out following two principal methodologies: *in solution* or in *solid-phase*. The solution protocol involves the coupling of one amino acid to another in homogeneous phase with subsequent purification under conventional methods. By contrast, solid-phase synthesis implies the use of a functionalized polymeric support (resin) in which the peptide bond is gradually formed. The purification step in this second strategy involves a simple filtration and washing process. The simplicity of solid-phase synthesis has made it the selected choice in most cases, bearing in mind the limitation that the amount of product is generally low. Several illustrative reviews of peptide synthesis in solution and solid-phase describe salient strategies that can be used in small and large scale.[10]

An interesting development in the field of peptide synthesis in solution was described by Escalante and collaborators,[11] who discovered that the

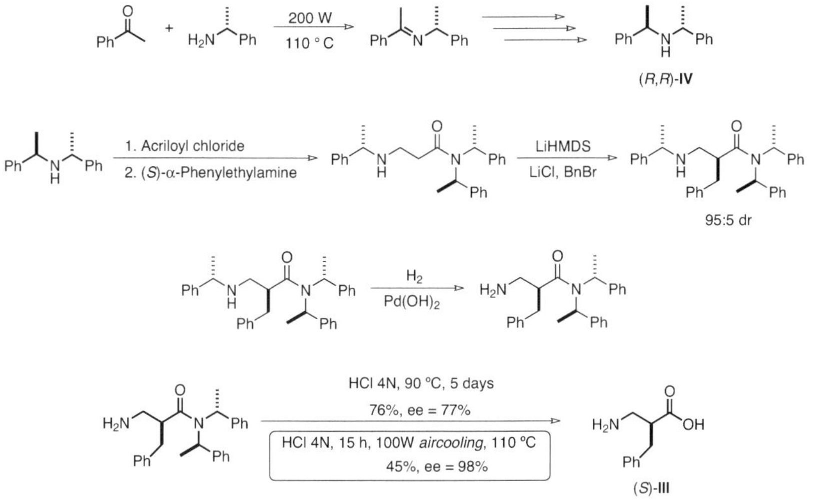

Fig. 12.3. Synthesis of (S)-α-benzyl-β-alanine **III** facilitated by microwave irradiation.[7]

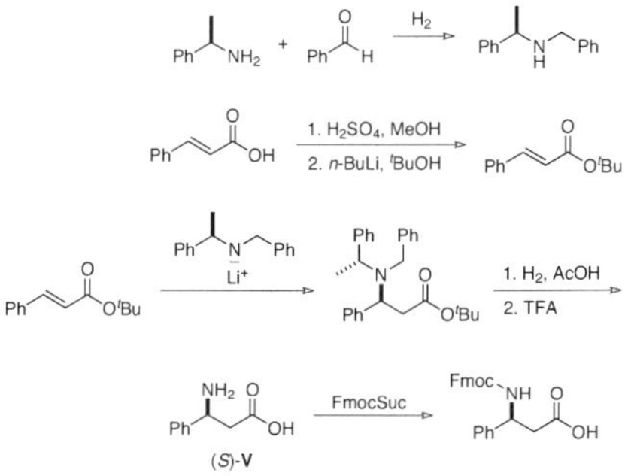

Fig. 12.4. Synthesis of (S)-β^3-homophenylglycine **V** via conjugate addition.[9]

use of $PhP(O)Cl_2$ as activating agent in the cyclization of β-amino acid **VI** affords significant amounts of cyclo β-dipeptide **VIII** (Fig.(12.5)).

This finding opened the door to innovative applications in the synthesis of novel β-amino acids for their incorporation in biologically active peptides. Indeed, cyclo-β-dipeptide **VIII** may serve as scaffold for the incorporation of various chiral residues into polymeric chains. Furthermore, the cyclo-β-dipeptide framework offers the possibility to carry out diastereoselective alkylations due to its rigid conformation (Fig.(12.6)).[12]

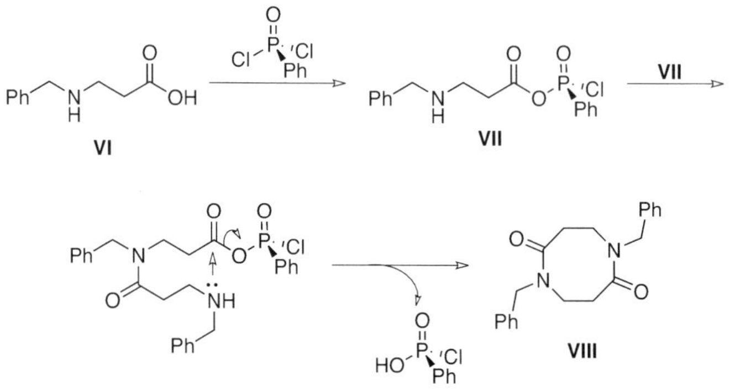

Fig. 12.5. Application of $PhP(O)Cl_2$ in the formation of cyclo β-peptides.[11]

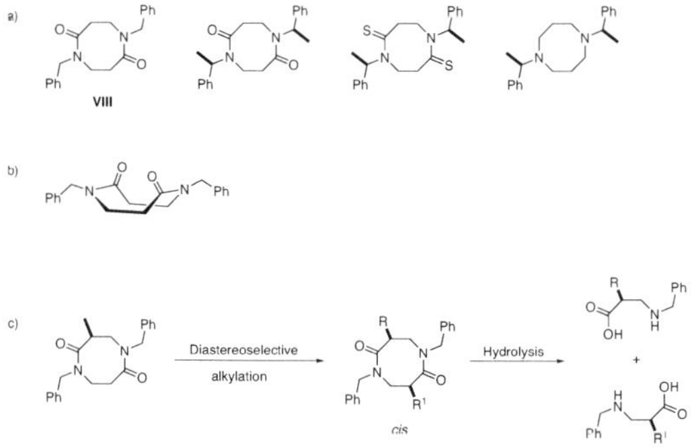

Fig. 12.6. (a) Potential ligands based on cyclo-β-dipeptide **VIII**. The preferred *boat* conformation of **VIII** (b) allows for the enantioselective synthesis of β^2-amino acids (c).[12]

In this context, a typical solid-phase protocol for the synthesis of peptides can be carried out via manual or automated synthesizers, or under microwave irradiation. In 2008, Bandala *et al.*[9] reported the application of these procedures in the preparation of thirty four α/β-tetrapeptides by means of the Fmoc strategy employing Wang resin (Fig.(12.7)).

12.3. Chemical and Biological Properties of β-Peptides

The relevance of peptides as therapeutic agents has been recognized in view of their high biological activity, low cost, and low toxicity. A few

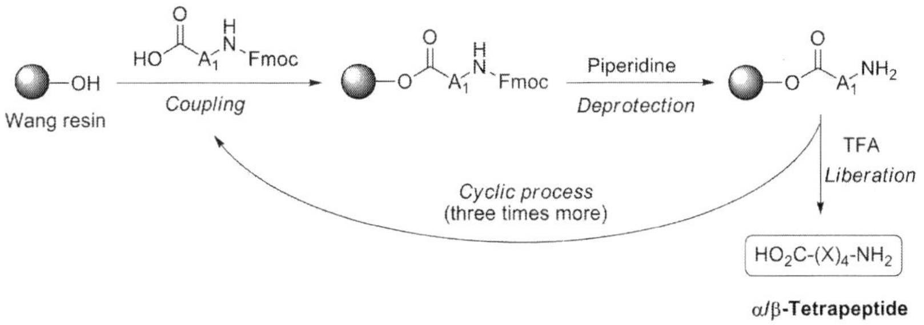

Fig. 12.7. Diagram illustrating the preparation of α/β-tetrapeptides using Fmoc chemistry in solid-phase. Coupling, deprotection, and liberation steps can be realized via manual, on automated synthesizers or under microwave irradiation.[9]

examples showing the diversity of medicines based on peptide include Byetta® (treatment of diabetes mellitus type 2), Cetrorelix® (treatment of hormone-sensitive prostate and breast cancer), Fuzeon® (HIV treatment), Integrelin® (antiplatelet agent), Neoral® (immunosuppressant), and Symlin® (control of blood sugar level).[10] The vulnerability of these drugs to digestive enzymes has motivated the search of novel peptide-analogs including those obtained by incorporation of β-amino acids, which are usually more resistant to enzymatic hydrolysis (Fig.(12.8)).

As previously indicated, the pioneering studies of Seebach and Gellman have shown the wide biological potential of β-peptides.[1] For instance, in a recent publication Seebach presents a number of β-peptidic derivatives that effectively bind to proteins involved in the mechanisms of immune respose.[13] In the same way, Gellman has shown that α/β-peptides play a relevant role in regard of their antimicrobial activity.[14]

12.3.1. *Conformational behavior of β-peptides*

It is suggested that β-peptidic compounds are stable against proteolytic, hydrolytic, and metabolizing enzymes in several organism; so, with the knowledge and understanding of the secondary structures of β-peptides, is possible to design stable structures that conserve their biological functions without loss of their hydrolytic and metabolic properties.[15] Most interesting, Bandala and Juaristi[16] have reported that the inclusion of β-amino acid **V** in a series of lineal α/β-tetrapeptides (α/β-tP) induces folding. By comparison, in the α-analog (α-tetrapeptide) the conformation preference is for an extended structure. For instance, it was observed that the

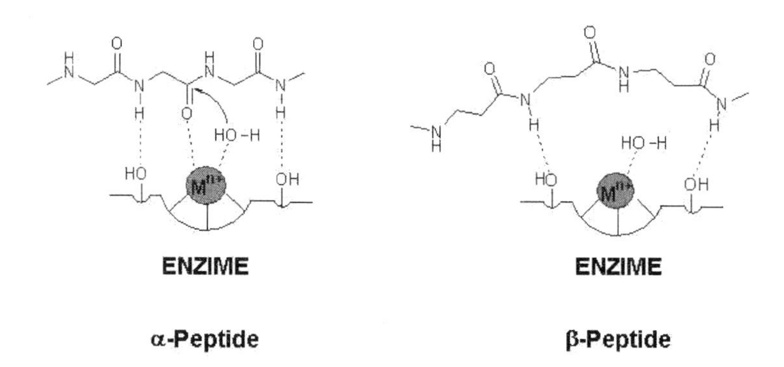

Fig. 12.8. Suggested interaction between α- and β-peptides with the same hydrolytic enzyme. The α-peptide is effectively hydrolyzed whereas a β-peptide is not.

circular dichroism spectrum of α/β-tP **IX** (HO_2C-α-Ile-α-Phe-β^3-**hPhg**-α-Ala-NH_2) presents an intense negative band between 190-200 nm, which can be ascribed to folded conformers, whereas the α-tP **X** (HO_2C-α-Ile-α-Phe^1-α-**Phe**^{2}-α-Ala-NH_2) shows a contrasting behavior corresponding to a linear arrangement (Fig.(12.9)).[17]

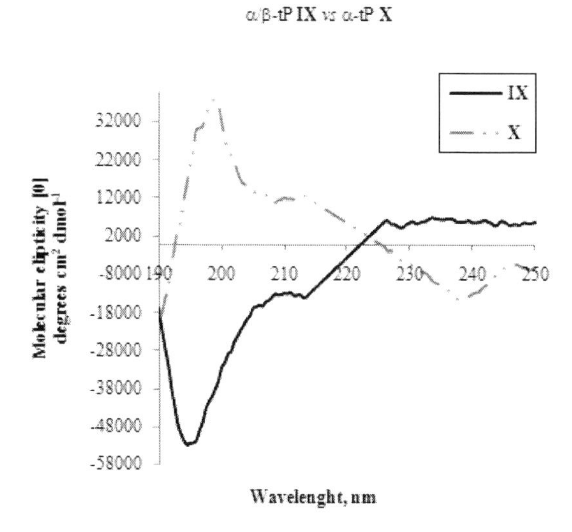

Fig. 12.9. Circular dichroism spectra of compounds α/β-tP **IX** (HO_2C-α-Ile-α-Phe-β^3-**hPhg**-α-Ala-NH_2) and α-tP **X**(HO_2C-α-Ile-α-Phe^1-α-**Phe**^{2}- α-Ala-NH_2), c=2.5 x 10^{-6} M in H_2O, T=23 °C.

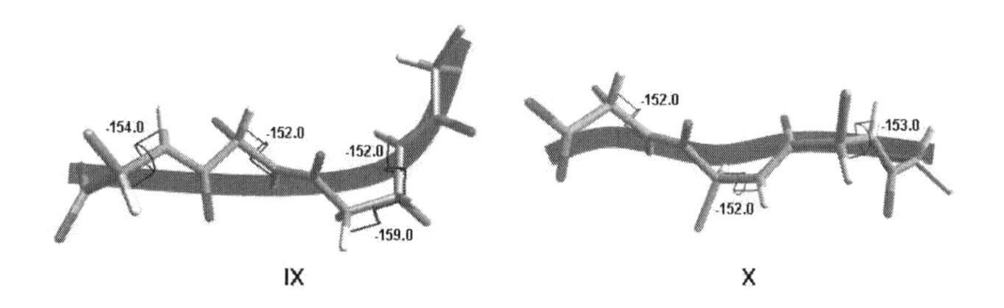

Fig. 12.10. Conformations of α/β-tP **IX** and α-tP **X** as deduced from NMR analysis and molecular modelling.

On the other hand, the analysis of the $^3J_{NH-CHa}$ and $^3J_{CHa-CHb}$ NMR coupling constants for α/β-tP **IX** indicates a folded conformation as a consequence of the additional CH_2 group present on β-amino acid, while in α-tP **X** predominates an extended conformation (Fig.(12.10)), which is supported as well by molecular modelling (DFT, B3LYP 6-311++G//B3LYP 6-31G).[16] DFT indicates the theoretical computational method used to investigate the structure of molecules via their electronic density; B3LYP is a functional that uses Becke's exchange functional (with part of the Hartree-Fock exchange) and the LYP correlation function; 6-31G and 6-311++G describe the set of functions used to create the molecular orbitals.

Finally, X-ray diffraction analysis of α/β-tP **XI** (HO_2C-β^3-**hPhg**-α-Phe-α-Phe-α-Ile-NH_2) shows the folding induced by the residue of β^3-homophenylglycine (Fig.(12.11)).[16]

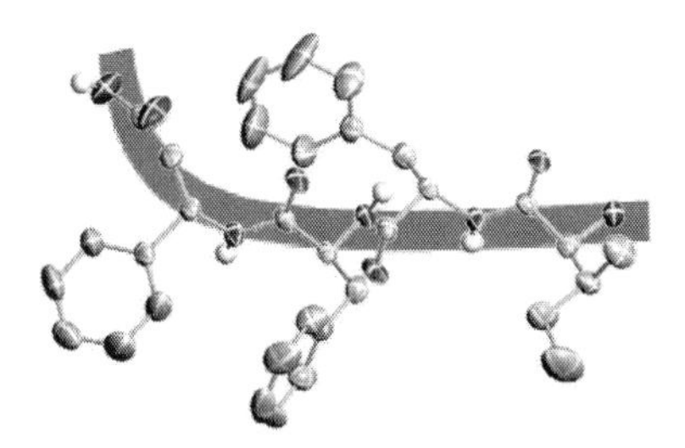

Fig. 12.11. ORTEP diagram of α/β-tP **XI** (HO_2C-β^3-**hPhg**-α-Phe-α-Phe-α-Ile-NH_2) illustrating the folding induced by **V**.

The relevance of this observation lies in the fact that it is possible to use β^3-amino acids to induce the formation of stable secondary structures in proteins.

On the other hand, the fundamental role of metal ions and inorganic salts in several chemical and biological processes, motivated the evaluation of the interaction between selected α/β-tetrapeptides containing the β-amino acid residue and alkali (Li^+, Na^+, K^+), alkaline (Mg^{2+}, Ca^{2+}), and alkaline earth (Cu^{2+}, Zn^{2+}) metal ions by means of electrospray ionization-mass spectrometry (ESI-MS) as well as computational analysis.[9]

Based on the relative peptide/metal ion complex intensities in the resulting ESI-MS spectra,[18] a dominant affinity for cations Na^+ and Mg^{2+} was established. These preferences indicate that α/β-tetrapeptides can be useful as selective chelating agents. Additionally, molecular modelling (DFT, B3LYP 6-311++G//B3LYP 6-31G) shows the substantial conformational changes induced by most of this cations on the initially linear arrangement of the peptide. This study contributes to increase our understanding of substrate-receptor phenomena where conformational changes of the native peptide or protein induced by the substrate may lead to the observed catalytic activity (Fig.(12.12)).[9]

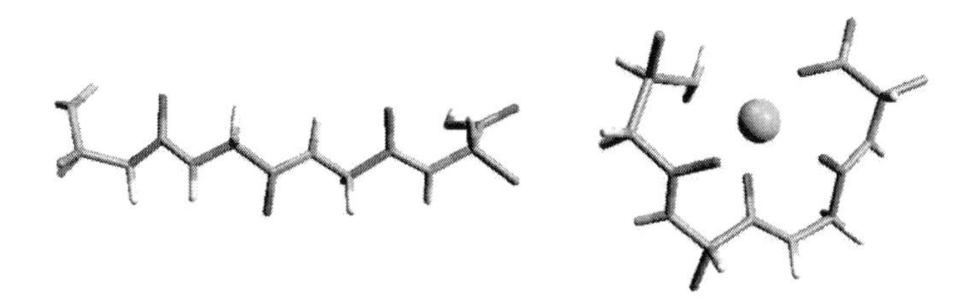

Fig. 12.12. Structure and conformation of α/β-tP **XII** (HO_2C-α-Phe-α-Phe-α-Ile-β^3-**hPhg**-NH_2) in the absence of salts and in the presence of Mg^{2+} illustrating the conformational change generated by the cation. The side chains were removed for clarity.

12.4. Biological Applications of β-Peptides

On the other hand, the relevant goal of pest control, e.g., in the eradication of the mosquito *Aedes aegypti* (the dengue vector), has motivated the exploration of new compounds for the development of effective agents for the control of the population of insects. In this area, the effectiveness of β-peptides has gained great relevance principally because of their potential in environmentally friendly pest control strategies.

In this context, Zubrzak *et al.*[19] reported that the substitution of selected α-amino acids for their β^2- and β^3-homologs in a series of hexa- and hepta kinin peptides display potent diuretic activity in insects (Fig.(12.13)), which die as a consequence of the imbalance generated by the resistance of the unnatural peptides to enzymatic degradation.[19]

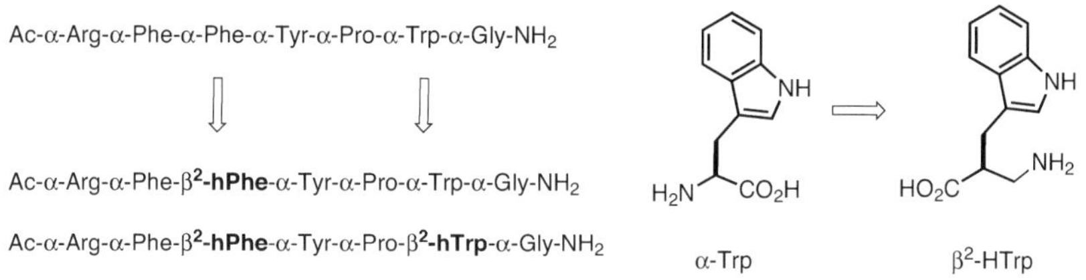

Ac-α-Arg-α-Phe-α-Phe-α-Tyr-α-Pro-α-Trp-α-Gly-NH_2

Ac-α-Arg-α-Phe-β^2-**hPhe**-α-Tyr-α-Pro-α-Trp-α-Gly-NH_2

Ac-α-Arg-α-Phe-β^2-**hPhe**-α-Tyr-α-Pro-β^2-**hTrp**-α-Gly-NH_2

α-Trp ⟹ β^2-HTrp

Fig. 12.13. Replacement of α-amino acids by β-amino acids in kinin peptides in the development of potential pest control agents.

In particular, α/β-hexapeptide Ac-α-Arg-α-Phe-α-Phe-β^3-**Pro**-α-Trp-α-Gly-NH_2, **XIII**, was found to be the most powerful stimulant of fluid secretion. This α/β-hexapeptide presents a low 11% of enzymatic hydrolysis over 2 h of incubation, while the natural kinin is hydrolyzed to the extent of 82% in the same period of time.[19]

The model developed to interpret this effect establishes that the insect kinin peptide adopts a β-turn conformation over the segment α-Phe-α-Phe-α-Pro/α-Ser-α-Trp where the α-Phe and α-Trp are associated with the receptor site. The β^3-Pro in **XIII** is compatible with the β-turn conformation required at this position, and its high activity is probably maintained because modifications at this position have a relatively small impact on the interaction of the analog with the receptor site (Fig.(12.14)).[19]

Following these ideas, other interesting example is presented by Taneja-Bageshwar,[20] who has shown the effectiveness of several α/β-peptides against the enzymatic degradation on recombinant insect kinin receptors from cattle tick and the mosquito *Aedes aegypti*, the dengue vector.[20] Their results show that the substitution of β-amino acids instead α-amino acids led to important modification on activity of kinin receptors from cattle tick, and *A. aegypti*: the single replacement (α-Pro for β^3-Pro) leads to a very powerful activity whereas the replacement of α-Phe and α-Trp residues for β^3-Phe and β^3-Trp, respectively, led to severe losses in activity.[20] These α/β-analogs display as promising pest control agents capable of disrupt insect kinin regulated processes.

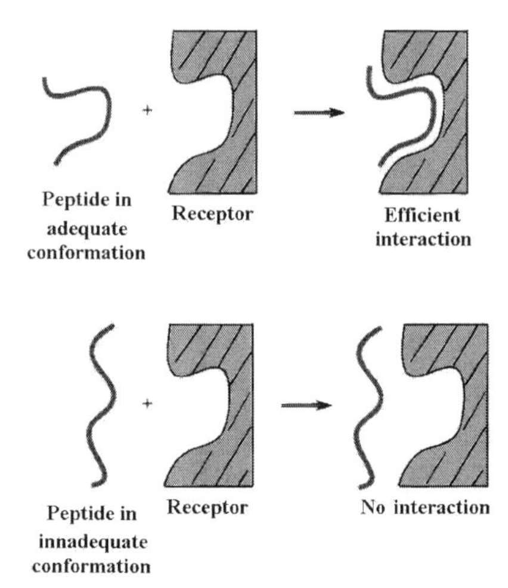

Fig. 12.14. Representative illustration of an effective interaction by adoption of an adequate conformation by a peptide, necessary to realize a productive biological process (top case).

In a recent work, Nachman and co-workers[21] extended the investigation on pest control by the evaluation of pheromonotropic (sex pheromone biosynthesis), melanotropic (defense, melanin biosynthesis), pupariation (development), and hindgut contractile properties (feeding, gut contractions) of several neuropeptides incorporating β-amino acids in various insects.[21]

Nachman showed that the terminal pyrokinin hexapeptide is susceptible to degradation by the peptidases neprilysin and angiotensin converting enzyme, whereas two of the β-amino acid analogs were found to exhibit significantly enhanced resistance to these same enzymes.[21]

12.5. Conclusions and Outlook

This brief account presents several strategies for the synthesis and application of β-peptides that have been developed and applied in Mexico. Since the number of applications involving β-peptides in chemistry, medicine and biology has grown enormously, the search of novel and interesting synthetic methodologies is an active field of research. For example, in chemistry, some β-peptides have found application as organocatalysts in asymmetric reactions.[22] On the other hand, in biology and medicine, unnatural peptides have been designed for their application in genetic therapy. Furthermore, a group of β-peptides has been prepared for their use as antimicrobial agents principally against Gram-positive bacteria. Finally, in a recent collaboration with González-Mariscal, the study of β-peptides to enhance drug delivery across epithelial and endothelial cells has been proposed.[23]

Acknowledgments

The authors are grateful to Consejo Nacional de Ciencia y Tecnología (CONACyT, Mexico) for financial support. Y. Bandala is also indebted to CONACyT for their scholar fellowship.

Glossary

AcOH	Acetic acid
B3LYP	Becke, three-parameter, Lee, Yang, Parr
BnBr	Benzyl bromide
BnOH	Benzyl alcohol
n-BuLi	*n*-Buthyl lithium
^{t}BuOH	*tert*-Butyl alcohol
DIC	Diisopropylcarbodiimide
DFT	Density Functional Theory
DMAP	Dimethylaminopyridine
DOPA	3,4-Dihydroxyphenylalanine
ESI-MS	Electrospray ionization-mass spectrometry
Fmoc	9-Fluorenylmethoxycarbonyl
FmocOSuc	9-Fluorenylmethoxycarbonyl succinimide
HCl	Hydrochloric acid
HOBt	Hydroxybenzotriazole
^3J	Coupling constant at three bonds
LiCl	Lithium chloride
LiHMDS	Lithium hexamethyldisilazide
LiOH	Lithium hydroxide
MeOH	Methanol
MW	Microwave irradiation
NaHMDS	Sodium hexamethyldisilazide
NMR	Nuclear magnetic resonance
$PhP(O)Cl_2$	Dichlorophenylphosphine oxyde
TFA	Trifluoroacetic acid

References

1. See for example: (a) R. P. Cheng, S. H. Gellman, and W. F. DeGrado, *Chem. Rev.*, 3219 (2001). (b) D. Seebach, A. K. Beck, and D. J. Bierbaum, *Chem. Biodiv.*, 111 (2004). (c) D. Seebach, T. Kimmerlin, R. Sebesta, M. Campo, and A. K. Beck, *Tetrahedron*, 7455 (2004).
2. (a) E. Juaristi, D. Quintana, and J. Escalante, *Aldrichimica Acta*, 3 (1994). (b) D. C. Cole, *Tetrahedron*, 9517 (1994). (c) G. Cardillo, and C. Tomasini, *Chem. Soc. Rev.*, 117 (1996). (d) E. Juaristi, Ed., *Enantioselective Synthesis of β-Amino Acids* (Wiley-VCH, New York, 1997). (e) E. Juaristi, and H. López-Ruiz, *Curr. Med. Chem.*, 983 (1999). (f) S. Abele, and D. Seebach, *Eur. J. Org. Chem.*, 1 (2000). (g) F. Fülöp, *Chem. Rev.*, 2181 (2001). (h) M. Liu, and M. P. Sibi, *Tetrahedron*, 7991 (2002). (i) J. A. Ma, *Angew. Chem., Int. Ed.*, 4290 (2003). (j) A. Córdova, *Acc. Chem. Res.*, 102 (2004). (k) E. Juaristi, and V. A. Soloshonok, Eds., *Enantioselective Synthesis of β-Amino Acids* (Second edition, Wiley-VCH, Hoboken, 2005). (l) A. Dondoni, and A. Massi, *Acc. Chem. Res.*, 451 (2006).
3. (a) J. K. Murray, and S. H. Gellman, *Org. Lett.*, 1517 (2005). (b) F. Fülöp, T. A. Martinek, and G. K. Tóth, *Chem . Soc. Rev.*, 323 (2006). (c) J. K. Murray, and S. H. Gellman, *J. Comb. Chem.*, 58 (2006). (d) C. A. Olsen, M. Lambert, M. Witt, H. Franzyk, and J. W. Jaroszewski, *Amino Acids,* 465 (2008).
4. (a) Y. Bandala, and E. Juaristi, *Recent Developments in the Synthesis of Beta Amino Acids.* In *Amino Acids, Peptides, and Proteins in Organic Chemistry,* A. Hughes, Ed., (Wiley-VCH, Weinheim, 2009). (b) D. Seebach, A. K. Beck, S. Capone, G. Deniau, U. Grošelj, and E. Zass, *Synthesis,* 1 (2009). (c) Reference 1, this work.
5. G. Reyes-Rangel, E. Jiménez-González, J. L. Olivares-Romero, and E. Juaristi, *Tetrahedron: Asymmetry,* 2839 (2008).
6. C. G. Ávila-Ortiz, G. Reyes-Rangel, and E. Juaristi, *Tetrahedron,* 8372 (2005).
7. R. Guzmán-Mejía, G. Reyes-Rangel, and E. Juaristi, *Nature Protocols,* 2759 (2007).
8. (a) S. Davies, and O. Ichihara, *Tetrahedron: Asymmetry,* 183 (1991). (b) S. Davies, N. Garrido, O. Ichihara, and I. Walters, *J. Chem. Soc. Chem. Commun.*, 1153 (1993).
9. Y. Bandala, J. Aviña, T. González, I. A. Rivero, and E. Juaristi, *J. Phys. Org. Chem.*, 349 (2008).
10. (a) O. Marder, and F. Albericio *Chimica Oggi,* 6 (2003). (b) T. Bruckdorfer, O. Marder, and F. Albericio, *Curr. Pharm. Biotechnol.*, 29 (2004). (c) S. Y. Han, and Y. A. Kim, *Tetrahedron,* 2447 (2004). (d) C. A. G. N. Montalbetti, and V. Falque, *Tetrahedron,* 10827 (2005). (e) F. Guzmán, S. Barberis, and A. Illanes, *Electronic J. Biotechnol.*, 279 (2007).
11. J. Escalante, M. A. González-Tototzin, J. Aviña, O. Muñoz-Muñiz, and E. Juaristi, *Tetrahedron,* 1883 (2001).
12. (a) E. Juaristi, and J. Aviña, *Pure Appl. Chem.*, 1235 (2005). (b) Y. Bandala,

R. Melgar-Fernández, R. Guzmán-Mejía, J. L. Olivares-Romero, B. R. Díaz-Sánchez, R. González-Olvera, and E. Juaristi, *J. Mex. Chem. Soc.*, (2009) accepted. (c) R. Guzmán-Mejía, and E, Juaristi, *unpublished results.*
13. D. Seebach, and J. Gardiner, *Acc. Chem. Res.*, 1366 (2008).
14. (a) M. A. Schmitt, B. Weisblum, and S. H. Gellman, *J. Am. Chem. Soc.*, 6848 (2004). (b) R. F. Epand, M. A. Schmitt, S. H. Gellman, A. Sen, M. Auger, D. W. Hughes, and R. M. Epand, *Mol. Membrane Biol.*, 457 (2005). (c) M. A. Gelman, and S. H. Gellman, *Using Constrained β-Amino Acid Residues to Control β-Peptide Shape and Function.* In *Enantioselective Synthesis of β-Amino Acids,* E. Juaristi, and V. A. Soloshonok, Eds., (Second edition, Wiley-VCH, Hoboken, 2005). (d) J. D. Sadowsky, W. D. Fairlie, E. B. Hadley, H. S. Lee, N. Umezawa, Z. Nikolovska-Coleska, S. Wang, D. C. S. Huang, Y. Tomita, and S. H. Gellman, *J. Am. Chem. Soc.*, 139 (2007). (e) M. A. Schmitt, B. Weisblum, and S. H. Gellman, *J. Am. Chem. Soc.*, 417 (2007). (f) F. Erinna, J. D. Sadowsky, B. J. Smith, P. E. Czabotar, K. Peterson, J. Kimberly, P. M. Colman, S. H. Gellman, and W. D. Fairlie, *Angew. Chem. Int. Ed.*, 4318 (2009).
15. (a) D. Seebach, M. Rueping, P. I. Arvidsson, T. Kimmerlin, P. Micuch, and C. Noti, *Helv. Chim. Acta*, 3503 (2001). (b) T. L. Raguse, E. A. Porter, B. Weisblum, and S. H. Gellman, *J. Am. Chem. Soc.*, 12774 (2002). (c) A. Mollica, M. P. Paradisi, D. Torino, S. Spisani, and G. Lucente, *Amino Acids*, 453 (2006). References 1 and 14, this work.
16. Y. Bandala, and E. Juaristi, *44° Mexican Congress of Chemistry*, (Sociedad Química de México, Puebla, September 2009). Examples of molecular modelling of peptides can be found in: (a) L. B. Kier, and J. M. George, *J. Med. Chem.*, 384 (1972). (b) Y. Isogai, G. Nmethy, and H. A. Scherega, *Proc. Natl. Acad. Sci. USA*, 414 (1977). (c) G. Loew, G. Hashimoto, L. Williamson, S. Burt, and W. Anderson, *Mol. Pharmacol.*, 667 (1982). (d) M. D. Beachy, D. Chasman, R. B. Murphy, T. A. Halgren, and R. A. Friesner, *J. Am. Chem. Soc.*, 5908 (1997). (e) N. Gresh, S. A. Kafafi, J. F. Truchon, and D. R. Salahub, *J. Comput. Chem.*, 823 (2004).; examples of molecular modelling of amino acids and peptides with metal cations can be found in: (a) S. Hoyau, and G. Ohanessian, *Chem. Eur. J.*, 1561 (1998). (b) T. Marino, N. Russo, and M. Toscano, *Inorg. Chem.*, 6439 (2001). (c) C. H. S. Wong, N. L. Ma, and C. W. Tsang, *Chem. Eur. J.*, 4909 (2002). (d) J. Zhang, E. Dyachokva, T. K. Ha, R. Knochenmuss, and R. Zenobi, *J. Phys. Chem. A*, 6891 (2003). (e) M. Kohtani, M. F. Jarrold, S. Wee, and R. A. J. O'Hair, *J. Phys. Chem. B*, 6093 (2004). (f) F. M. Siu, N. L. Ma, and C. W. Tsang, *Chem. Eur. J.*, 1966 (2004). (g) M. Benzakour, M. Mcharfi, A. Cartier, and A. Daoudi, *J. Mol. Struct. THEOCHEM*, 169 (2004). (h) S. Abirami, Y. M. Xing, C. W. Tsang, and N. L. Ma, *J. Phys. Chem. A*, 500 (2005). (i) A. S. Reddy, and G. N. Sastry, *J. Phys. Chem. A*, 8893 (25). (b) R. W. Woody, *Circular Dichroism of Peptides and Proteins* in *Circular Dichroism: Principles and Applications.* K. Nakanishi, N. Berova, and R.005).
17. (a) G. D. Rose, L. M. Gierasch, and J. A. Smith, *Adv. Prot. Chem.*, 1 (1985). (b) R. W. Woody, *Circular Dichroism of Peptides and Proteins* in *Circular Dichroism: Principles and Applications.* K. Nakanishi, N. Berova, and

R.tCircular Dichroism: Principles and Applications. K. Nakanishi, N. Berova, and R. Woody, Eds., (Wiley-VCH, New York, 1994). (c) W. C. Johnson, *Ann. Rev. Biophys. Chem.*, 145 (1988).
18. (a) W. Hutchens, R. W. Nelson, M. H. Allen, C. M. Li, and T. T. Yip, *Bio8*(a) W. Hutchens, R. W. Nelson, M. H. Allen, C. M. Li, and T. T. Yip, *Biol. Mass Spectrom.*, 151 (1992). (b) K. Wang, X. Han, R. W. Gross, and G. W. Gokel, *J. Am. Chem. Soc.*, 7680 (1995). (c) J. A. Loo, *Mass Spectrom. Rev.*, 1 (1997). (d) T. J. D. Jorgensen, P. Roepstorff, and A. J. R. Heck, *Anal. Chem.*, 4427 (1998). (e) P. A. Brady, and J. K. M. Sanders, *New J. Chem.*, 411 (1998). (f) F. Giorgianni, A. Cappiello, S. Beranova-Giorgianni, P. Palma, H. Trufelli, and D. M. Desiderio, *Anal. Chem.*, 7028 (2004). (g) A. Wortmann, F. Rossi, G. Lelais, and R. Zenobi, *J. Mass Spectrom.*, 777 (2005). (h) Y. Ye, M. Liu, J. Kao, and G. Marshall, *Biopolymers*, 472 (2006). (i) A. Wortmann, M. C. Jecklin, D. Touboul, M. Badertscher, and R. Zenobi, *J. Mass Spectrom.*, 600 (2008).
19. P. Zubrzak, H. Williams, G. M. Coast, R. E. Isaac, G. Reyes-Rangel, E. Juaristi, J. Zabrocki, and R. J. Nachman, *Biopolymers*, 76 (2007).
20. S. Taneja-Bageshwar, A. Strey, P. Zubrzak, H. Williams, G. Reyes-Rangel, E. Juaristi, P. Pietrantonio, and R. Nachman, *Peptides*, 302 (2008).
21. R. J. Nachman, O. B. Aziz, M. Davidovitch, P. Zubrzak, R. E. Isaac, A. Strey, G. Reyes-Rangel, E. Juaristi, H. J. Williams, and M. Altstein, *Peptides*, 608 (2009).
22. See for example: G. Luppi, P. G. Cozzi, M. Monari, B. Kaptein, Q. B. Broxterman, and C. Tomasini, *J. Org. Chem.*, 7418 (2005).
23. This collaboration has generated a first report: D. Chamorro, L. Alarcón, A. Ponce, R. Tapia, H. González-Aguilar, M. Robles-Flores, T. Mejía-Castillo, J. Segovia, Y. Bandala, E. Juaristi, and L. González-Mariscal, *Mol. Biol. Cell*, 4120 (2009).

Chapter 13

Laser-Induced Fluorescence in Mononuclear Cells: Direct Estimate of the NADH Bound/Free Ratio

A. Silva-Pérez, J. R. Godínez-Fernández
Departamento de Ingeniería Eléctrica,
Universidad Autónoma Metropolitana-Iztapalapa,
Apartado Postal, 55-532, México 09340 D.F., México
anilo31@yahoo.com.mx, gfjr@xanum.uam.mx

M. Fernández-Guasti, E. Haro-Poniatowski
Departamento de Física, Universidad Autónoma Metropolitana-Iztapalapa,
Apartado Postal, 55-532, México 09340 D.F., México
mfg@xanum.uam.mx, haro@xanum.uam.mx

C. Campos-Muñiz
Departamento de Ciencias de la Salud,
Universidad Autónoma Metropolitana-Iztapalapa,
Apartado Postal, 55-532, México 09340 D.F., México
caro@xanum.uam.mx

L. Llorente
Departamento de Inmunología, Instituto Nacional de Ciencias Médicas y Nutrición, Vasco de Quiroga 15, Tlalpan 14000 México D.F.
lllyrp@quetzal.innsz.mx

Different laser light sources are used to selectively excite the various natural fluorophores present in mononuclear live cells. The radiation characteristics using a laser-induced fluorescence technique together with an appropriate tissue preparation permitted an improved spectral resolution. The NADH spectrum in phosphate buffered saline solution was very closely fitted with a double Gaussian that was used as the reference spectrum for the intra cellular experiments. The NADH molecules within the cells provide the dominant contribution to the relatively narrow band fluorescence spectrum obtained with this procedure under 355 nm Nd:YAG laser excitation. The deconvolution of the fluorescence curve allowed for the direct calculation of the NADH bound/free ratio. This dimensionless quantity, estimated at 1.35 with this method, is an important indicator of the energetic metabolic state of the cells.

Contents

13.1. Introduction

Laser radiation is becoming an increasingly important tool in various fields of science and technology due to its characteristics of high intensity output, monochromaticity (high spatial and temporal coherence), tunability and either the possibility of continuous or pulsed operation. These properties have allowed increased frequency resolution, which in turn have often revealed the structure of previously unresolved spectroscopic contributions. Suitable laser frequencies (or lines) have extended over an ample range of the electromagnetic spectrum thus generating a rapid progress in fluorescence spectroscopy of atoms and molecules.

There are many research fields where fluorescence spectroscopy of samples excited with laser sources has been successfully implemented. For example, kinetics of chemical reactions; measurement of the lifetime of excited states of atoms and molecules; determination of trace concentrations of elements in gases; fluorometry of condensed media and others.[16] There are various applications specific to medicine and biochemistry that exploit either the high energy content or wavelength resolution of the laser source. In the former case several types of lasers are used in order to generate heat while interacting with living tissue. For example, CO_2 and Nd:YAG laser systems are used in surgery of carcinomas of the upper aerodigestive tract showing some advantages in reepithelization over conventional surgery.[16] In other circumstances, the wavelength resolution capability of lasers plays an important role in the process.

A Xenon Chlorine laser, for example, has been successfully utilized instead of a conventional NB-UVB (290 – 320 nm) lamp in the therapy of some inflammatory and hyperproliferative diseases taking advantage of the laser narrow bandwidth.[18] Another example is the use of a Q-switched Nd:YAG laser together with an argon ion laser. This system has been employed to induce a temperature-jump relaxation in order to examine the kinetics of NADH binding to lactate dehydrogenase over the nanosecond to millisecond time scales.[8] Regarding medical diagnosis and biochemical studies, laser sources are being used as non-invasive techniques. In medical optics, Laser Induced Fluorescence (LIF) has been used with sufficiently small laser fluence so as to avoid any damage of the tissue. In most applications, the photon energy is usually chosen sufficiently below the binding energies of the molecules in order to avoid DNA photodissociation.[28] In this way it has been possible, for example, to detect cervical intraepithelial neoplasia in vivo.[23]

The LIF technique has been successfully implemented in the detection of lesions in cervix,[21,22] bladder[3,13] and colon.[5,28] Tissue samples often exhibit two drawbacks when they are analyzed using this technique. On the one hand, the natural fluorophores present in the tissue have rather broad absorption and emission spectra and they are often partially overlapped.[19] This overlap limits the quantitative measurement of specific fluorophores.

On the other hand, the strong dispersion exhibited by the tissue also distorts these measurements. A long standing problem in biomedical optics has been to find an efficient way of separating the signals coming from the biochemical substances within the cell and the dispersion produced by the tissue.[24] Appropriate haematologic tissue samples together with the adequate laser source may overcome these drawbacks. A suspension of living mononuclear cells of young rats in phosphate-buffered saline solution eliminates the extra cellular material consisting mainly on collagen and elastin. This preparation also eliminates the porphyrins and hemoglobin present in erythrocytes and thus avoids the attenuation of the incident beam and the fluorescence reabsorption.[12] In addition, there are hardly any lipopigments present in these cells because they are extracted from young rats.[23]

Furthermore, light dispersion is reduced since the microscopic heterogeneities of the extra cellular elements are eliminated and the dispersion coefficient of a phosphate-buffered saline solution is also relatively small.[10] Therefore, cells in this suspension reduce the absorption and dispersion of light that limit the spectral resolution of intracellular fluorophores emission bands.[10,24] In order to understand LIF spectra in living cells it is crucial to know the autofluorescence spectra generated by each particular fluorophore. Monici et al.[18] have characterized the autofluorescence of leukocytes. They have measured the emission spectra of granulocytes and agranulocytes in a suspension consisting on Iscoveás modified Dulbeccos medium plus 1% foetal calf serum.

They reported emission peaks associated with tryptophan centered at 340 nm and nicotinic coenzymes (e.g. NADH) at 440 nm using a conventional spectro fluorometer. Heintzelman et al.[11] used the autofluorescence diagnosis method called excitation-emission matrix spectroscopy. Their experiments reveal three maxima of excitation and emission: the tryptophan peak at 290 nm of excitation, 330 nm emission, the NAD(P)H excitation peak at 350 nm, 450 nm emission and the FAD excitation peak at 450 nm, 530 nm emission. The NAD(P)H emission shown in Figure (13.2) (c) in Ref.[11] has an approximate bandwidth of 120 nm. They used a spectro fluorometer with a Xenon lamp of 450 W as the light source and two double monochromators. The data of these two reports[11,17] will be important in order to understand the fluorescence spectra under laser light excitation reported here. It is worth mentioning that conventional incoherent light sources have much larger bandwidths than laser sources even when heavily filtered through double monochromators. A typical bandwidth for 1 mm slits is 1.7 nm.[17]

For this reason, it is expected that broader absorption bands are excited with these incoherent sources compared with laser ones; as a result, emission bands are likely to be also wider in the former case. It is well known that biochemical energetic metabolic processes where the coenzyme NADH interacts are of fundamental importance. The biological function of the electronic transports is to transfer reduced equivalents (electrons) from the NADH to the oxygen, which is the ultimate acceptor of electrons in the aerobic process of oxidation. In between, the NAD+/NADH molecule (in its oxide and reduced forms) interacts in the processes of glycolysis and the citric acid cycle. These processes are present in all living cells except those of anaerobic bacteria. The binding of NADH with enzymes has a catalytic action; without this bond, the production and storage of energy in the form of ATP is halted. It is therefore important to know the NADH bound/free ratio as an indicator that contributes to the understanding of the energetic metabolic activity in different types of cells.[27]

In the present work, we obtain the intracellular NADH bound/free ratio from direct fluorescence observations. To this end, four different laser sources were used to selectively excite the three fluorophores present in the cellular samples, namely pyridoxine, the NADH coenzyme and the flavin adenine dinucleotide FAD. This procedure allows us to obtain a reproducible fluorescent signal with an adequate spectral resolution for the former two fluorophores within the cell. The spectra of the NADH molecule in phosphate-buffered saline solution were also obtained under Nd:YAG, N2 and dye laser excitation. These data were compared with the previous reports mentioned above. The $\lambda = 355nm$ Nd:YAG laser excitation yields the best results for NADH excitation within the cell and permits the deconvolution of the signal into the bound and free contributions.

13.2. Materials and Methods

13.2.1. *Materials*

Nicotinamide adenine dinucleotide reduced form, purity 98% was obtained from Sigma Chemical Co. (St. Louis, MO). Phosphate-buffered saline (PBS) solution was acquired from Gibco BRL (Grand Island N.Y. 14072 U.S.A.). Nycodenz solution for the isolation of mononuclear cells from most mammalian species was obtained from Nycomed Pharma (Oslo Norway). Wistar rat blood was obtained by heart puncture to avoid hemolysis; approximately 2 mL of blood anticoagulated with sodium heparin was collected.

13.2.2. *Sample preparation*

Wistar rat blood was diluted 1:1 with phosphate-buffered saline solution. Mononuclear cells were separated from granulocytes by applying a 1.077 Nicodenz-density gradient at 600 g for 15 min. Cells at the interface, consisting of lymphocytes and monocytes were collected. They were subsequently washed in PBS solution, centrifuged at 400 g for 15 min and resuspended in PBS. All cell suspensions exhibited cell viability around 98% when checked with the trypan blue dye exclusion test. The cells in suspension were placed in a 825 μL quartz cuvette with 5 mm of optical path. For our purposes, it was sufficient to use merely 120 μL of cells suspension in order to obtain adequate fluorescence spectra. The cellular count was made with a Neubauer camera.

13.2.3. *Laser sources*

The fluorescence measurements were made using four different laser systems: i) A molecular Nitrogen laser made in our installations[9] emitting at a wavelength centered in 337.1 nm with 20 μJ energy per pulse. ii) A Nd:YAG commercial laser (HyperYAG Laser System, Lumonics Ltd, model HY1200) with third harmonic at 355 nm with 30 μJ energy per pulse. The laser bandwidth at the 1064 nm fundamental line is 0.47 nm at room temperature; the third harmonic has a somewhat narrower bandwidth of 0.053 nm. The shot to shot intensity variation is within 10% of the total emitted intensity. iii) A Nd:YAG pumped dye laser made in our laboratory; operation with coumarin 460 dye (Exciton Chemical Co. Inc.) yields tunability centered at 440 nm with 5 μJ energy per pulse. iv) An Argon Ion commercial laser (Lexel laser, Inc. model 95) emitting in the 457.9 nm or 472.7 nm lines with 80 mW power continuous wave. This latter system was used together with a mechanical chopper (Stanford Research Systems, Inc. model SR540) in order to obtain light pulses that could be electronically processed in a similar fashion as the pulsed laser systems.

13.2.4. *Optical detection*

A 0.5 m Czerny-Turner (Pacific Instruments) monochromator was used with 0.1 nm resolution together with a photomultiplier (Hamamatsu model 1P21). A perpendicular excitation geometry was used illuminating the cell at its center. The sample fluorescence was collected with a lens located at twice the focal distance providing an image at the monochromator entrance slit with unit magnification thus minimizing aberrations.[4]

13.2.5. *Signal processing*

In order to process the signal electronically a Gated Integrator and Boxcar Averager Module (Stanford Research Systems, Inc. model SR250) was employed. It consists of a gate generator, a fast gated integrator and exponential averaging circuitry. This type of detection reduces the effective detection time in several orders of magnitude. This short detection time in turn reduces the background continuous-wave (CW) level below the electronic system noise as mentioned by Anderssons-Engels et al.[2] The integrator was set to average over 30 pulses in order to reduce the shot to shot intensity variation inherent to the pulsed laser sources.

The synchronization of the laser pulse and the fluorescence signal was achieved with an oscilloscope. The boxcar integrator was triggered with the laser Pockels cell. The electronic gate was delayed and its width appropriately set in order to integrate over the region where the fluorescence signal was present. The analog to digital conversion was made with a high velocity multi-function data acquisition card (Advantech Co., Ltd. Model PCL-812) coupled to a personal computer. The layout is shown in Figure (13.1)

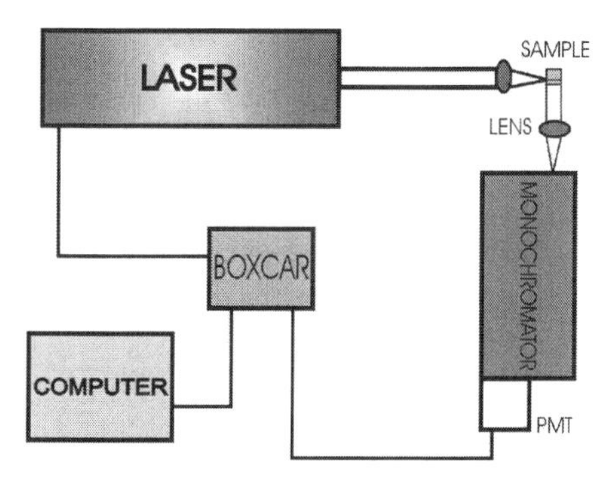

Fig. 13.1. System layout employed for the acquisition of fluorescence spectra.

13.2.6. *Curve-fitting*

The Marquardt-Levenberg algorithm for an iterative nonlinear least squares curve-fitting procedure was used. The initial parameters of the Gaussian fits (amplitude, width and maximum frequency) were varied to insure convergence to the same final results. Curves from different runs were fitted independently in order to estimate the dispersion in the convergence values of the parameters. The curves were smoothed in Origin, V. 7 (Microcal software, Inc.).

13.3. Results and Discussion

13.3.1. *Calibration experiments*

The initial experiments were performed with a Nitrogen laser in order to establish the system operating conditions. This type of low budget laser has been amply used in LIF systems. A strong dependence between the cellular concentration and dispersion was observed. The fluorescence could be adequately detected in the range of $1.6 \times 10^4 cells/mL$ to $7.5 \times 10^4 cells/mL$. The fluorescence signal becomes too weak below the before mentioned lower bound due to the low density of fluorophores. Above the maximum concentration, dispersion does not allow the fluorescence signal to be observed. In subsequent experiments, unless otherwise stated, the cells placed in the quartz cell were suspended in PBS solution at a concentration of 7.5E4 cells/mL. All experiments were performed at room temperature.

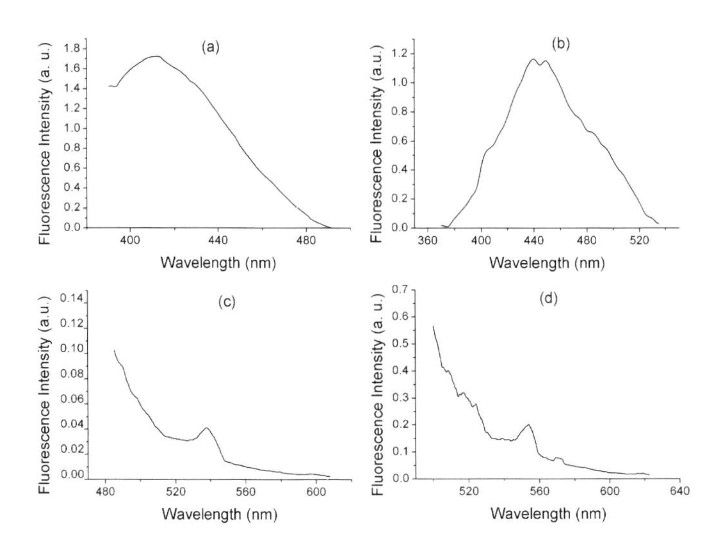

Fig. 13.2. Fluorescent spectra of mononuclear cells in PBS solution obtained at room temperature with different laser sources. (a) Molecular nitrogen laser $\lambda = 337.1$ nm; (b) Nd:YAG laser third harmonic $\lambda = 355$ nm; Argon ion laser lines (c) $\lambda = 457.9$ nm and (d) $\lambda = 472.7$ nm.

13.3.2. *Cell excitation with different laser sources*

The fluorescence spectra of the cells under excitation with different laser sources are shown in Figure (13.2). The minimum laser wavelength employed in this work is 337.1 nm and for this reason the amino acid tryptophan is not excited since its absorption band lies at the 280 - 290 nm range. Shorter wavelength excitation was avoided (using for example, the fourth harmonic Nd:YAG 266 nm emission) so as to avoid DNA damage through the induction of thymine dimmers with UV radiation.[1]

According to Wagnirés et al.,[29] under 330 nm radiation or above only three fluorophores are then possibly excited in our samples, namely, i) pyridoxine, ii) NADH and phosphate NAD(P)H and iii) the flavins (mainly FAD). Furthermore, at 350 nm the pyridoxine is very weakly excited since its absorption band is almost negligible above this wavelength. When the samples are excited with the 337.1 nm Nitrogen laser line, a band peaking at 412 nm with 55 nm bandwidth (FWHM) is obtained. According with the literature,[24,29] this band is associated with pyridoxine and possibly with a contribution from the NADH molecule. Under the 355 nm Nd:YAG third harmonic laser excitation line, a well structured band was obtained peaking around 445 nm with 80 nm bandwidth (FWHM).

In contrast, under incoherent lamp excitation at 350 nm with 1.7 nm bandwidth, the fluorescence band is centered at 450 nm with 120 nm bandwidth.[11] The coherent laser source narrow bandwidth, typically around 0.05 nm, excites very efficiently an extremely narrow absorption band. This highly selective excitation in turn reduces the emission bandwidth. This spectrum will be analyzed in detail in the following subsections. The FAD fluorescence spectrum was apparently not present while exciting with the 457.9 and 472.7 nm Argon ion laser lines as may be seen from the graphs shown in the lower part of Figure (13.2) This spectrum was neither present under 440 nm dye laser excitation (spectrum not shown in figures) probably due to the low cellular concentration and the small laser energy per pulse. At these excitation wavelengths, the NADH molecule no longer absorbs the incoming radiation and its fluorescence spectrum is not present as expected. Under 355 nm excitation, FAD peak emission at 520 nm is not observed. FAD is most efficiently excited at 436 nm and less efficiently at the 365 nm secondary absorption peak.[26,27] Excitation at 355 nm, being 10 nm further away from this secondary peak, is rather inefficient.

The two small features observed in Figures (13.2c) and (13.2d) are certainly due to the Raman scattering of water since they are at the same place relative to the two different excitation wavelengths 457.9 nm and 472.7 nm. The corresponding position of the bumps are 3147 cm^{-1} and 3137 cm^{-1}.[16]

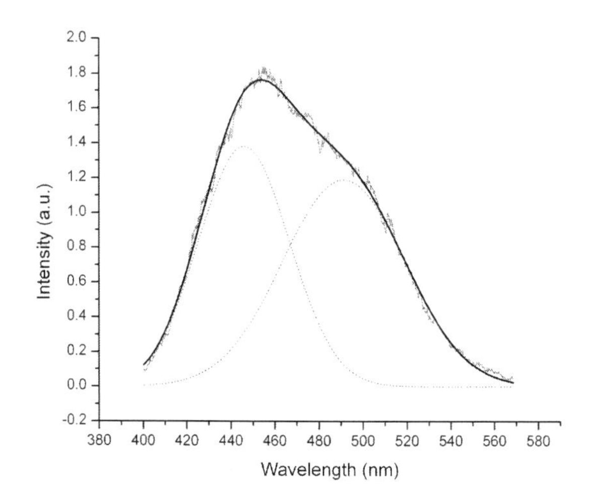

Fig. 13.3. Fluorescence spectrum of NADH molecules in PBS solution excited with the 355 nm laser line at room temperature. The sum (thick solid line) of two Gaussian curves (dotted lines) were used to fit the experimental data.

13.3.3. *NADH fluorescence in PBS solution*

The fluorescent emission of the NADH molecules in solution is very sensitive to chemical as well as physical factors such as pH, polarity, temperature and storage.[20] In fact, only to mention some cases, the emission maximum reported by Andersson-Engels et al.[2] is located at 470 nm whereas Schomacker et al.[28] report it at 445 nm (in a pH=7 solution) and Lakowicz et al.[12] locate it close to 470 nm (in a pH=7 Mops buffer and 10 μM concentration). Excited fluorophores and therefore their fluorescence signals are sensitive to several dynamic processes. In addition the response of the detector can also modify the spectral curves shapes.

These processes include: a) redistribution of electrons in the surrounding solvent molecules induced by the altered (generally increased) dipole moment of the excited fluorophore; b) reorientation of the solvent molecules around the excited state dipole; and c) specific interactions between the fluorophore and the solvent or solutes.[15] All these processes can produce changes in the fluorescence intensity and spectral shifts. The fluorescence spectrum of the free dinucleotide NADH (Sigma, 98% purity) reported in Figure (13.3), was made in PBS solution (pH=7.2) with 10 μM concentration. The light source was a Nd:YAG laser at the third harmonic 355 nm, under the same experimental conditions used to obtain the fluorescence spectra of the cells.

It should be recalled that under pulsed laser excitation there is a shot to shot laser intensity variation that introduces the noise observed in the experimental curve. The maximum emission intensity is located around 456 nm and lies in between the maxima previously reported. This spectrum shows an asymmetry that is not present in previous reports when the molecule is excited with incoherent sources. However, in the recent publication of Yicong Wu et al.[30] with a 349 nm solid state laser excitation, an asymmetric feature is also present. This asymmetry then seems to become apparent due to the highly selective excitation together with the narrow bandwidth detection system. This experimental curve was then fitted with two Gaussians yielding a remarkably good fit. The two Gaussian curves are centered at 446.0 $\pm$0.3 and 491.3 $\pm$0.6 nm with 41 and 55 nm widths respectively. Their relative height is 1.17. The chi squared of the fit is 0.0012.

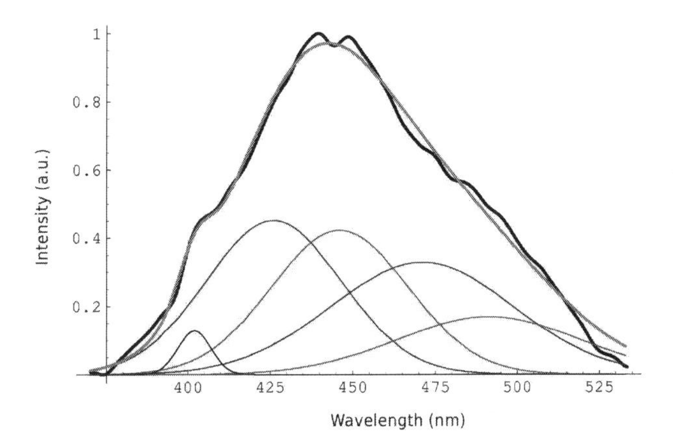

Fig. 13.4. Fluorescence spectrum of mononuclear cells in PBS solution (black solid line). Samples were excited with the 355 nm laser line. The nonlinear fit drawn with a green line arises from the free (red) and bound (blue) components associated with NADH that were previously fitted with two pairs of Gaussians. A residual curve (black thin line) accounts for the residual pyridoxine emission.

13.3.4. *NADH fluorescence in cells*

The fluorescence obtained from mononuclear cells under the same excitation and experimental conditions is shown in Figure 13.4 The un-weighted arithmetic average was performed over four spectra and a rescaling of the curves was carried out in order to obtain unit emission at the maximum intensity. The maximum difference between emission spectra of a given sample at a specific concentration was 2% of the total area under the curve. The maximum difference for different samples under the same spectroscopic conditions was around 4% previously normalized. Reproducible spectra were thus obtained.

We should mention that the peak intensity obtained directly from the detection system arbitrary units is 1.8 for NADH in solution and 1.18 for the cells fluorescence. In order to understand the origin of the fluorescence curve obtained from mononuclear cells, we need to establish to begin with, which fluorophores responsible for this emission spectrum. To this end, we should recall several issues.

13.4. Discarded Fluorophores

On the one hand, the sample was appropriately chosen and carefully prepared so as to avoid the presence of elastin, collagen, porphyrin and lipopigments. On the other hand, a highly selective excitation has been achieved with an adequate laser source, namely Nd:YAG third harmonic at 355 nm. At this excitation wavelength pyridoxine absorption is very weak. The FAD emission peak is not seen at 520 nm because the excitation is inefficient at 355 nm.

13.4.1. *NADH as the dominant fluorophore*

The 355 nm laser line lies in the region where the NADH molecule has a strong absorption peak.[20] Furthermore, Monici et al.[18] reported the largest fluorescence signal from lymphocytes and monocytes samples in between 350 and 360 nm. The spectrum of NADH in PBS solution (13.3) and mononuclear cells excited with 355 nm (13.4) have a similar broad shape and width. We should recall that Lakowicz et al.[14] have reported an approximately 20 nm shift towards the blue of NADH in solution bound to the enzyme MDH (malate dehydrogenase) with respect to unbound NADH in the same solution. Salmon et al.[26] have also reported a blue shift of the order of 25 nm in the fluorescence spectra when NAD(P)H is bound to different dehydrogenases.

The shift that we observe is therefore likely to originate from bound NADH states within the cell. The fact that the observed shift is not so large is an indication that free (unshifted) NADH is also presfent in the mononuclear cells. Therefore, in accordance with the before mentioned arguments, bound and free NADH will be considered as the main fluorophore responsible of the observed emission. This working hypothesis is strengthened by the deconvolution presented hereafter.

13.4.2. *Cell fluorescence deconvolution*

The graph of the natural fluorescence of mononuclear leukocytes under 355 nm excitation is shown in Figure (13.4) Superimposed is the deconvolution of five Gaussian components using the nonlinear least squares fit mentioned in the materials and methods section. Let us consider the contribution of bound and free NADH to this fluorescence spectrum. This structure arises from vibrational progression of about 2000 cm^{-1}. The free NADH is estimated with the NADH spectrum in PBS solution described in the previous section.

Two Gaussians centered at 446.0 nm and 491 nm with 41 nm and 55 nm widths have been used in the nonlinear fit in order to account for this contribution (red lines). The amplitude of these Gaussians have been left as free parameters of the fit. A similar procedure using two half Gaussians in order to account for the asymmetric shape of NADH has been recently reported.[7] However we cannot at this point give the physical origin of each one of the two Gaussians considered in the deconvolution process. On the other hand, according with the previous results of Lakowicz et al.[14] bound NADH is shifted by 20 nm towards the blue. This contribution of bound NADH has been introduced in the fit with another two Gaussians shifted by this amount.

These Gaussians are then centered at 426.0 nm and 471.0 nm with 41 nm and 55 nm widths respectively (blue). The shoulder that the experimental curve shows just above 400 nm is reproducible in the different runs. Consequently, this feature cannot be attributed to noise. A Gaussian curve with free parameters was introduced in order to account for this shoulder. The fit centered the Gaussian at 402 nm with 10 nm width and 0.11 amplitude (thin solid line). This peak is coincident with the pyridoxine emission and it is possibly a contribution arising from the remnants of this molecule. This is peak is most probably due to Raman of water which presents a broad band centered at 3250 cm^{-1}.[16] This fit of five Gaussian curves, two pairs arising from bound and free NADH and a fifth remnant contribution yields a chi squared of 0.0007 ($R^2 = 0.993$). The good quality of the fit gives further support to the proposal that bound and free NADH are the dominant contributions to the observed spectrum. The height of each Gaussian, as mentioned before, was left as a free parameter in the fit.

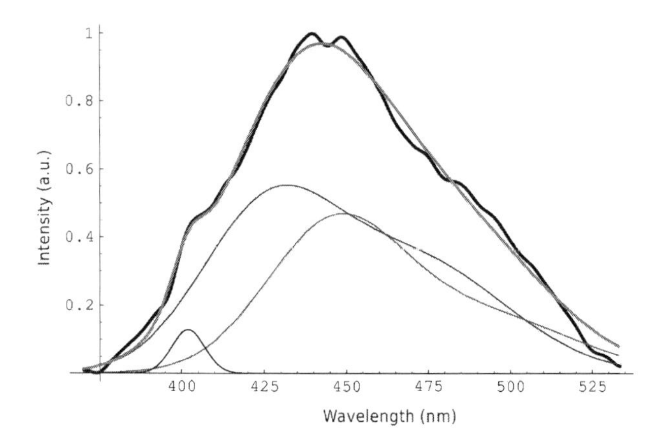

Fig. 13.5. Fluorescence spectrum of mononuclear cells in PBS solution (black solid line). The red line consists of the sum of the two Gaussians depicted in figure 4 that make up the free NADH fluorescence spectrum centered at 450 nm (61 nm FWHM). The blue line includes the sum of the two Gaussians arising from bound NADH, centered at 430 nm (82 nm FWHM).

The ratio of the amplitudes between the Gaussians centered at the bound and free wavelengths gives the ratio of bound to free NADH. This quotient may be obtained from the peaks centered at 426 and 446 nm or the complementary pair at 471 and 491 nm. With the present data we obtain a bound/free NADH ratio within the cell of 1.35 ± 0.45 for mononuclear living cells in a normal physiological state. This fit was made to four spectra coming from cells of different rats as well as to their average spectrum. The bound to free ratio varied within $\pm$ 33% for different samples and thus sets an upper limit to the repeatability of this measure. In Figure (13.5), the two Gaussians that make up either bound or free NADH have been added in order to exhibit the contributions of the two NADH states. The maximum of the free NADH is located at 450 nm with 61 nm width (FWHM). On the other hand the bound state peak is maximum at 430 nm and has a

FWHM of 82 nm. Notice that the widths of the individual Gaussians that make up the free and bound NADH contribution were kept constant.

Nevertheless, since the heights of each Gaussian were left as a free parameter the FWHM of the bound component is significantly broadened. This broadening could reflect the stronger interaction of NADH with different nucleotides within the cell.

The main problem in order to quantify the NAD(P)H bound/free ratio, has been that the fluorescence spectrum obtained form living cells usually involves the sum of emissions due to various fluorophores their differentiated states. Some researchers have attempted to reduce the number of contributions to the fluorescence signal through metabolic perturbations of the cells. In this way, the participation of certain fluorophores been dismissed. Using this procedure, Salmon et al.[27] have reported a slightly modified spectrum of EL2MA cells when subjected from aerobiosis to partial anaerobiosis. The NADH bound/free ratio calculated with their approach was 0.17 for aerobiosis and 0.07 in partial oxygen deficiency. Another approach has been to consider the adjustment coming from the flavins and lipo-pigments in order to optimize the curve fit.[7]

In contrast, the deconvolution process performed here has been attributed to the coenzyme NADH. The bound/free NADH ratio is then directly obtained from the amplitudes of the Gaussian curves associated with each of these states.

Conclusions: The use of various laser sources together with an adequately prepared cells sample has allowed the excitation of the NADH coenzyme with negligible contributions from other fluorophores. The procedure implemented here permitted the acquisition of reproducible spectra as well as improved spectral resolution. These assets permitted the quantification of the fluorescence signal and the direct evaluation of the NADH bound/free ratio.

The NADH molecule is very efficiently excited within mononuclear cells using the 355 nm Nd:YAG third harmonic line. The maximum emission intensity of this molecule is located at 430 nm and has a 82 nm width (FWHM). This fluorescence curve is shifted towards shorter wavelengths by 20 nm and broadened 25% with respect to the spectra of the NADH molecules in PBS solution. The spectrum of NADH in solution has been accurately fitted with two Gaussian curves. This reference spectrum was used to deconvolute the cell fluorescence spectrum. The NADH bound/free ratio estimated with this procedure is 1.35 ± 0.45. Qualitatively, the fluorescent spectrum excited with 355 nm reported here has a particular substructure that is slightly different from the curve shape of the NADH molecules in PBS solution (as seen in Figures 13.3 and 13.4).

These results seem promising for future investigations in two mainstreams. On the one hand, several studies have been reported that discriminate between normal and neoplasic tissue in vivo and in vitro. Rigacci et al.[29] performed an autofluorescence microscopy experiment of lymphoproliferative disorders for the analysis of lymph-node tissue. It is likely that Leukemias involving mononuclear cells may be analyzed using the procedure described in this communication. Leukaemic cell lines, obtained from a patient affected by acute myeloid leukaemia (HL 60) and patients suffering from lymphoid neoplasms (Daudi), are already being studied by epifluorescence microscopy techniques using a high pressure mercury lamp a 366 nm as excitation source.[20] On the other hand, the diagnosis of tissue is based mostly on microscopic analysis of histological samples through standard procedures.

One drawback of these techniques is that the cells may experience chemical modifications along the procedures. The present fluorescence scheme may be used as a complementary technique in order to obtain a metabolic indicator of the cell physiological state. This information could help to diagnose pathologic states in this type of cells.

Acknowledgments

We are grateful to Dr. Alonso Fernández Guasti for the revision of the manuscript and his enlightening advice on several key issues.

A.1. Abbreviation

Abbreviations: CO_2, Carbon Dioxide (laser); FWHM, full width at half maximum; LIF, laser induced fluorescence; NB-UVB, narrow-band ultraviolet bulb; Nd:YAG, Neodymium-Yttrium Aluminum Garnet (laser).

References

1. Ahmed F, Setlow R B, Grist E and Setlow N (1993) DNA damage, photorepair and survival in fish and human cells exposed to UV radiation. Environ Mol Mutagen **22**: 18-25.
2. Andersson-Engels S, Johansson J, Svanberg K and Svanberg S (1991) Fluorescence imaging and point measurements of tissue: applications to the demarcations of malignant tumors and atherosclerotic lesions from normal tissue. Photochem Photobiol **53**: 807-814.

3. Anidjar M., Cussenot O, Avrillier S, Ettori D, Teillac P and Le Duc A (1996) Assessment of different excitation wavelengths for photodetecting neoplastic urothelial lesions by laser-induced autofluorescence spectroscopy. Proc SPIE **2679**: 204-214.
4. Born M and Wolf E (1999) Principles of Optics, pp. 142-223 Cambridge University Press.
5. Chwirot B W, Kowalska M, Sypniewska N, Michniewicz Z and Gradziel M (1999) Spectrally resolved fluorescence imaging of human colonic adenomas. J Photochem Photobiol B:Biol **50**: 174-183.
6. Croce A C, Ferrigno A, Vairetti M, Bertone R, Freitas I and Bottirolli G (2004) Autofluorescence properties of isolated rat hepatocytes under different metabolic conditions Photochem. Photobiol. Sci **3**: 920-927.
7. Croce A C, Spano A, Locatelli D, Barni S, Sciola L and Bottiroli G (1999) Dependence of fibroblast autofluorescence properties on normal and transformed conditions. Role of the metabolic activity. Photochem Photobiol **69**: 364-374.
8. Deng H, Zhadin N and Callender R (2001) Dynamics of protein ligand binding on multiple time scales: NADH binding to lactate dehydrogenase. Biochemistry **40**: 3767-3773.
9. Fernández-Guasti M., Silva-Pérez A, Iturbe-Castillo D, Haro-Poniatowski E, Escobar-Alarcón L and Habichayn-Polloni P (1992) Diseño y construcción de láseres de nitrógeno molecular. Revista Mexicana de Física. **38**: 588-610.
10. Gardner C M, Jacques S L and Welch A J (1996) Fluorescence spectroscopy of tissue: Recovery of intrinsic fluorescence from measured fluorescence. Appl Opt **35**: 1780-1792.
11. Heintzelman D L, Lotan R and Richards-Kortum R (2000) Characterization of the autofluorescence of polymorphonuclear leukocytes, mononuclear leukocytes and cervical epithelial cancer cells for improved spectroscopic discrimination of inflammation from dysplasia. Photochem Photobiol **71**: 327-332.
12. Keijzer M, Richards-Kortum R, Jacques S and Feld M (1989) Fluorescence spectroscopy of turbid media: autofluorescence of the human aorta. Appl Opt **28**: 4286-4292.
13. Koenig F., Mcgovern F J, Althausen A F, Deutsch T F and Schomacker K T (1996) Laser induced autofluorescence diagnosis of bladder cancer. The Journal of Urology **156**: 1597-1601.
14. Lakowicz J R, Szmacinski H, Nowaczyk K and Johnson M L (1992) Fluorescence lifetime imaging of free and protein-bound NADH. Proc Natl Acad Sci USA **89**: 1271-1275.
15. Lakowicz J R (1983) Principles of fluorescence spectroscopy, Plenum Press, New York.
16. Lefevre T, Toscani S, Picquart M, Dugun J (2002) Crystallization of water in multilamellar vesicles. Eur Biophys J **31**: 126-135.
17. Letokhov V S (1986) Laser Analytical Spectrochemistry, English edition, IOP Publishing Ltd, Adam Hilger, Bristol and Boston.
18. Lippert B M, Teymoortash A, Folz B J and Werner J A (2003) Wound healing after laser treatment of oral and oropharyngeal cancer. Lasers Med Sci **18**: 36-42.

19. Monici M, Pratesi R, Bernabei P A, Caporale R, Ferrini P R, Croce A C, Balzarini P and Bottiroli G (1995) Natural fluorescence of white blood cells: spectroscopic and imaging study. J Photochem Photobiol B: Biol. **30**: 29-37.
20. Monici M, Agati G, Mazzinghi P, Fusi F, Bernabei P A, Landini I, Rossi Ferrini P and Pratesi R (1996) Image analysis of cell natural fluorescence. Diagnostic applications in haematology. In Biomedical Systems and Technologies (Edited by N. I. Croitoru, M. Frenz, T. A. King, R. Pratesi, A. M. Verga Scheggi, S. Seeger and O. S. Wolfbeis), pp. 180-187. Proceedings of SPIE, The International Society for Optical Engineering, Bellingham, WA.
21. Novák Z, Bónis B, Baltás E, Ocsovszki I, Ignácz F, Dobozy A and Kemény L (2002) Xenon chloride ultraviolet B laser is more effective in treating psoriasis and in inducing T cell apoptosis than narrow-band ultraviolet B. J Photochem Photobiol B Biol **67**: 32-38.
22. Papazoglou T G (1995) Malignancies and atherosclerotic plaque diagnosis-is laser induced fluorescence spectroscopy the ultimate solution?. J Photochem Photobiol B: Biol. **28**: 3-11.
23. Passonneau J V and Lowry O H (1993) Enzymatic Analysis, pp. 34-56. The Humana Press.
24. Pavlova I, Sokolov K, Drezek R, Malpica A, Follen M, Richards-Kortum R (2003) Microanatomical and biochemical origins of normal and precancerous cervical autofluorescence using laser-scanning fluorescence confocal microscopy. Photoche. Photobiol **77**: 550-555.
25. Ramanujam N, Follen-Mitchell M, Mahadevan-Jansen A, Thomsen S, Malpica A, Wright T, Atkinson N, and Richards-Kortum R (1996) Spectroscopic diagnosis of cervical intraepithelial neoplasia (CIN) in vivo using laser-induced fluorescence spectra at multiple excitation wavelengths. Lasers Surg Med **19**: 63-74.
26. Ramanujam, N., Mitchell M F, Mahadevan A, Warren S, Thomsen S, Silva E and Richards-Kortum R (1994) In vivo diagnosis of cervical intraepithelial neoplasia using 337-nm-excited laser-induced fluorescence. Proc Natl Acad Sci USA **91**: 10193-10197.
27. Rattan S I S, Keeler K D, Buchanan J H and Holliday R (1982) Autofluorescence as an index of ageing in human fibroblast in culture. Biosci Rep **2**: 561-567.
28. Richards-Kortum R and Sevick-Muraca E (1996) Quantitative optical spectroscopy for tissue diagnosis. Annu Rev Phys Chem **47**: 555-606.
29. Rigacci L, Alterini R, Bernabei P A, Ferrini P R, Agati G, Fusi F and Monici M (2000) Multispectral imaging autofluorescence microscopy for the analysis of lymph-node tissues. Photochem Photobiol **71**: 737-742.
30. Salmon J M and Viallet P (1977) *L′*utilisation des spectres électroniques dans *l′*étude des interactions enzymes-piridine nucleotides. J Chim Phys **74**, 239-245.
31. Salmon J M, Kohen E, Viallet P, Hirschberg J G, Wouters A W, Kohen C and Thorell B (1982) Microspectrofluorometric approach to the study of free/bound NAD(P)H ratio as metabolic indicator in various cell types. Photochem Photobiol **36**: 585-593.

32. Schomacker K T, Frisoli J K, Compton C C, Flotte T J, Richter J M, Nishioka N S and Deutsch T F (1992) Ultraviolet laser-induced fluorescence of colonic tissue. Lasers Surg Med **12**: 63-78.
33. Schomacker K T, Frisoli J K, Compton C C, Flotte T J, Richter J M, Deutsch T F and Nishioka N S (1992) Ultraviolet laser-induced fluorescence of colonic polyps. Gastroenterology. **102**: 1155-1160.
34. Svanberg S (1989) Medical applications of laser spectroscopy. Physica Scripta **T26**: 90-98.
35. Wagnieres G A, Star W M and Wilson B C (1998) In vivo fluorescence spectroscopy and imaging for oncological applications. Photochem Photobiol **68**: 603-632.
36. Yicong Wu, Peng Xi, Jianan Y. Qu (2004) Depth-resolved fluorescence spectroscopy reveals layered structure of tissue. Optics Express **12**: 3218-3223.

PART 3

Statistical Physics and Solid State Physics

Chapter 14

Superconductivity in Graphene and Nanotubes

Shigeji Fujita

Department of Physics, University at Buffalo, SUNY, Buffalo, NY 14260, USA

Salvador Godoy

Departamento de Física, Facultad de Ciencias, Universidad Nacional Autónoma de México, México 04510, D.F., México

A new Bloch electron dynamics in which the "electron" ("hole") wave packet, called simply "electron" ("hole"), has a size besides a charge, is introduced to describe the electrical conduction in graphene based on a rectangular unit cell for the graphene's honeycomb lattice. The conduction of a single-wall nanontube (SWNT) is semiconducting or metallic depending on whether the pitch contains an integral number of carbon hexagons along the tube axis or not. The low-temperature residual conductivity of the semiconducting SWNT is shown to arise from the Cooper pairs formed by the phonon exchange attraction. A quantum statistical theory is presented supporting a superconducting state with an ultrahigh critical temperature (1275 K) in the multi-walled nanotubes reported by Zhao and Beeli [Phys. Rev. B **77**, 245433 (2008)].

Contents

14.1. Introduction

In 1991 Iijima[1] discovered carbon nanotubes in the soot created in an electric discharge between two carbon electrodes. A nanotube can be considered as a single sheet of graphite, called a *graphene*, that is rolled up into a tube. A single-wall nanotube (SWNT) has a diameter of one nanometer (nm) while a multiwalled nanotube (MWNT) rolled like a wallpaper stored has a radius exceeding 10 nm in radius and 100 microns in length. Nanotubes have remarkable mechanical properties that can be exploited to strengthen materials. And since they are composed entirely of carbon, nanotubes are light and also have a low specific heat. Ebbesen *et al.*[2] measured the electrical conductivity of individual nanotubes. In the majority cases, the resistance R decreases with increasing temperature T. In contrast the resistance R for a normal metal like copper (Cu) increases with T. This temperature behavior in nanotubes indicates a thermally activated process. Schönenberger and Forro[3] reviewed many aspects of carbon nanotubes; this review[3] and Ref.,[2] and Saito-Dresselhaus-Dresselhaus' book[4] contain many important references. The current band theory based on the Wigner-Seitz (WS) cell model[5] predicts a gapless semiconductor for a nanotube and cannot explain the observed T-behavior. A new theory is required. The WS model[5] is suited for the study of the ground-state energy of a crystal. To treat the electron motion for a non-cubic lattice such as a honeycomb lattice, we may introduce a rectangular unit cell and use a mass tensor. We present a new theoretical model for a Bloch electron dynamics for graphene in which the conduction electron has a size besides a charge ($\pm e$), see Sec.(14.2).

Zhao and his group[6] found an experimental evidence of superconductivity in MWNT with the critical temperature $T_c = 1275$ K after examining the paramagnetic Meissner effect.[7] This is a new record of high T_c, about eight times higher than $T_c = 164$ K found in mercury-based cuprates.[8] Fujita and Godoy developed a quantum statistical theory of superconductivity,[9] extending the Bardeen-Cooper-Shrieffer (BCS) theory[10] by incorporating lattice and band structures of electrons and phonons *and* considering moving Cooper pairs. The Cooper pairs,[11] also called the *pairons*, are formed by the phonon-exchange attraction. These pairons move with linear dispersion relations.[9] The centers of mass (CM) of pairons move as bosons.[9] These pairons undergo a Bose-Einstein condensation (BEC) in two dimensions (2D).[9] We use the quantum statistical theory to support a superconductivity in MWNT at high temperatures ($\sim$ 1275).

Zhang *et al.*[12] showed that graphene under a gate voltage is superconducting at 1.7 K. We discuss this in Section (14.2). The graphene sheet can be wrapped into a singled-wall nanotube. A "hole" ("electrons") is defined as a quasiparticle that has energy lower (higher) than the Fermi energy ε_F *and* that circulates counterclockwise (clockwise) viewed from the tip of the applied magnetic field vector. The "electron" ("hole") is generated on the positive (negative) side of the Fermi surface with the convention that the positive normal vector at the surface points in the energy increasing direction. The "holes" (and not the "electrons") can go through inside the positively charged graphene wall in SWNT. Due to this extra channel, the nanotube's majority carriers are "holes" although the graphene's majority carriers are "electrons", see Section (14.3). The electrical conduction in SWNT is either semiconducting or metallic depending on whether each pitch of the helical line connecting the nearest-neighbor C-hexagon contains an integral number of hexagons or not, which is shown. In MWNT the "holes" are generated more abundantly in shells between the graphene walls. We discuss the superconductivity at ultrahigh critical temperatures in Section 4. Summary and discussion are given in Section (14.5).

14.2. Graphene

In graphene carbon ions (C^+) occupy the two dimensional (2D) honeycomb crystal lattice. See Fig.(14.1).

The conduction electron ("electron", "hole") moves as a wave packet as pointed out by Ashcroft and Mermin in their book.[13] We assume that the "electron" ("hole") wave packet has a negative (positive) charge $-e$ ($+e$) and a size extending over the unit C-hexagon. The positively charged "hole" tends to stay away from the positive ions C^+ and hence its charge is concentrated at the center of the hexagon. The negatively charged "electron" tends to stay close to the C^+ ions and its charge is more concentrated near the C^+ hexagon. Because of the different internal charge distributions, the "electron" and "hole" should have different effective masses m_e and m_h. For the description of the electron motion in terms of the mass tensor,[13] it is convenient to introduce Cartesian coordinates, which do not match with the crystal's natural (triangular) axes. The "electron" may move easily with a smaller effective mass in the direction [110 c-axis] $\equiv$ [110] than perpendicular to it as we see presently. Here we use the conventional Miller indices[13] for the hexagonal lattice with omission of the c-axis index. We

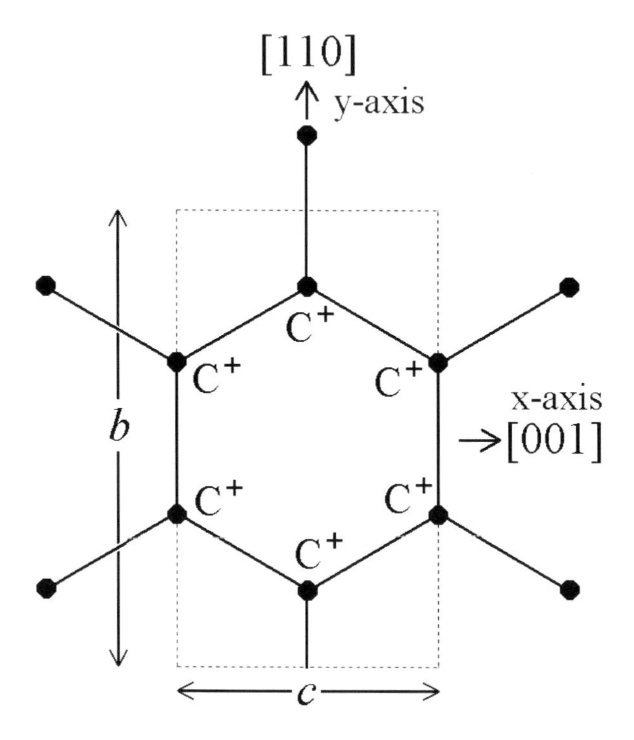

Fig. 14.1. A rectangular unit cell for graphene (dotted line).

may choose the rectangular unit cell (dotted line) with side-lengths (b, c) as shown in Fig.(14.1). Note that the honeycomb lattice has inversion (mirror) symmetry with respect to the x- and the y-axis. Then, the Brillouin zone in the k-space is unique: a rectangle with side lengths $(2\pi/b, 2\pi/c)$. The "electron" may move up or down in [110] to the neighboring hexagon sites passing over one C^+. The positively charged C^+ acts as a welcoming (favorable) potential valley for the negatively charged "electron". The same C^+ acts as a hindering potential hill for the positively charged "hole". The "hole" can however move easily over on a series of vacant sites in [001], each surrounded by six C^+, without meeting the hindering potential hills. Thus, the easy channel direction for the "electrons" and "holes" are [110] and [001], respectively.

Let us consider the system (graphene) at 0 K. If we put an electron in the crystal, then the electron should occupy the center O of the Brillouin zone, where the lowest energy lies. Additional electrons occupy points neighboring O in consideration of Pauli's exclusion principle. The electron distribution is lattice-periodic over the entire crystal in accordance with the Bloch theorem.[13] The graphene is a quadrivalent metal. The first few low-lying bands in the k-space are completely filled. The upper-most partially filled bands are important for the transport properties discussion. We consider such a band. The Fermi surface which defines the boundary between the filled and unfilled k-spaces (areas) is not a circle since the x-y symmetry is broken (Note that $c \neq b$). The effective mass m_e is less in the direction [110] than perpendicular to it. If the electron number is raised by the gate voltage applied perpendicular to the graphene plane, then the Fermi surface more quickly grows in the easy-axis (y-) direction than in the x-direction. The Fermi surface must approach the Brillouin boundary at right angles because of the inversion symmetry possessed by the honeycomb lattice.[14] Hence, the Fermi surface must touch the Brillouin boundary at a certain gate voltage and a "neck" Fermi surface must be developed.

The same easy channels in which the "electron" runs with a small mass, may be assumed for other hexagonal directions [001] and [101]. Hence, the system does not show anisotropy in the charge transport. Zhang *et al.*[12] measured the mobility μ and the resistance R in graphene at 1.7 K at varying gate voltages.

The summary of their data is reproduced in Fig.(14.2) from Ref.[12] Fig.(1). The gate voltage controls the charge density of the conduction electrons ("electrons", "holes"). In fact the charge density and the gate voltage are linearly related as indicated in Fig.(14.2). Experiments showed that (a) both "electrons" and "hole" can be excited in graphene, (b) at zero gate voltage the "electrons" are dominant, (c) the resistance exhibits a sharp maximum at the "electrons" density $n_e \sim 2 \times 10^{11}$ cm^{-2}, and (d) the mobility μ proportional to the conductivity σ shoot up at the "hole" density $n_h \sim 3 \times 10^{11}$ cm^{-2}. The feature (b) should arise from the existence of the welcoming ion C$^+$ for the "electrons" (and not for the holes). The feature (c) is due to the fact that the conductivity, given by the Drude-formula,

$$\sigma = e^2 n_e \tau / m^*, \tag{14.1}$$

where τ is the relaxation time, must decrease since the effective mass m^* shoots up to ∞ in the small-"neck" limit. The feature (d) means that the graphene is superconducting at 1.7 K under the gate voltage. No other explanation exists for the infinite conductivity. In our quantum statistical theory the pairons are generated through the phonon exchange attraction. Since the phonon carries no charge, the system before and after

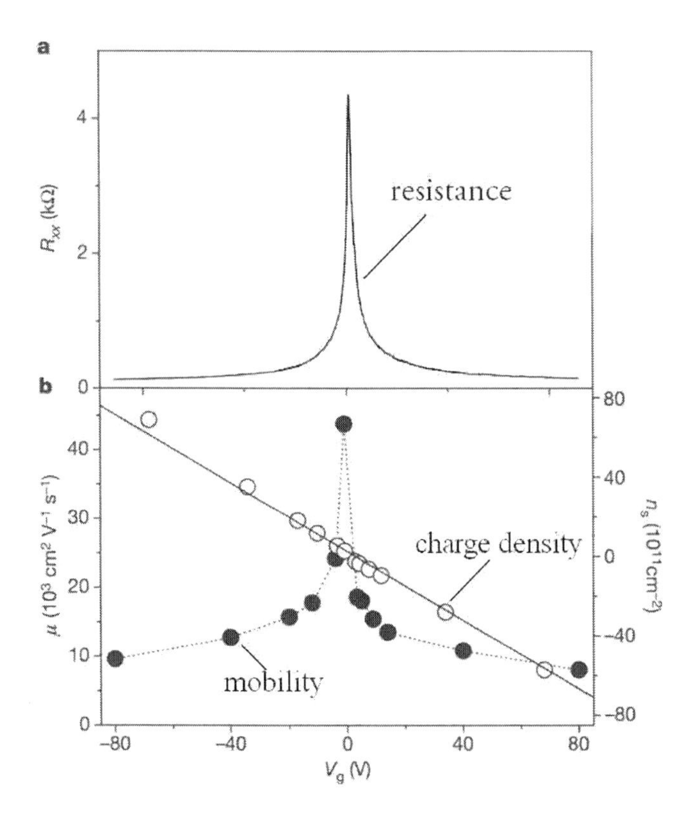

Fig. 14.2. Experimental data (a) for the resistence R_{xx} and (b) the mobility μ versus gate voltage in graphene, reproduced after Zhang *et al.*[12]

the phonon exchange has the same charge state, requiring that positively and negatively charge pairons are created in equal numbers simultaneously. The graphene's majority carriers are "electrons" at zero gate voltage. The mobility-maximum occurs on the "hole" gate voltage side, where the "hole" density is high near the "neck" Fermi surface. The "necks" can occur for both "electron" and "hole" sides. There is a ρ-maximum for the "electron" side and a μ-maximum for the "hole" side in Fig.(14.2). Why is this asymmetry? There are not many "holes" at zero gate voltage and hence the critical temperature T_c, is very low at the "electron" density $n_e = 2 \times 10^{11}$ cm^{-2}. If the system is observed at lower temperatures, then superconducting states should show up on both sides.

14.3. Single Wall Nanotube

Let us now consider a long SWNT rolled with the graphene sheet. The graphene which forms a honeycomb lattice is intrinsically anisotropic as we saw in Sec.(14.2). Moriyama *et al.*[15] fabricated twelve (12) SWNT devices from one chip and observed that two of them are semiconducting and the other ten (10) are metallic, the difference in the room temperature resistance being of two to three orders of magnitude. The semiconducting SWNT samples show an activated-state temperature behavior. Why are there two sets of samples showing very different behaviors? We believe the answer to this question arises as follows.

The line passing the centers of the nearest-neighbor carbon hexagons forms a helical line around the nanotube with a pitch p, see Fig.(14.3), where a section of the circular tube with radius r and pitch p is drawn.

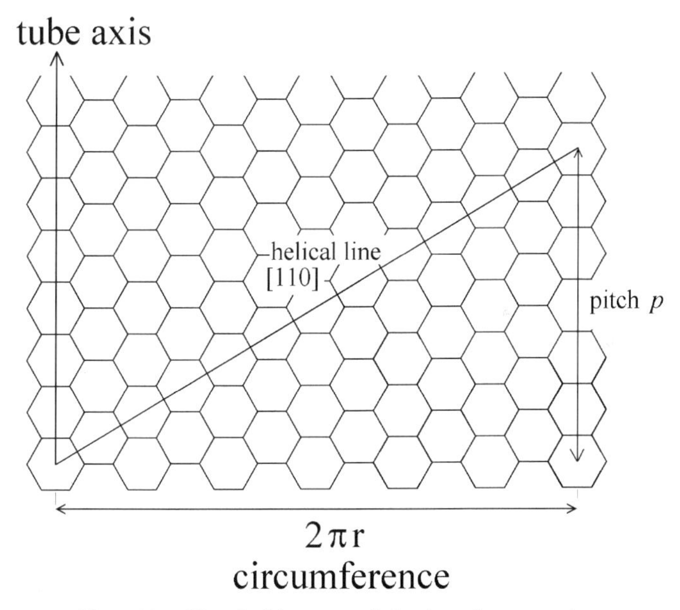

Fig. 14.3. Unrolled honeycomb lattice of a nanotube.

Its unrolled plane is shown and the circumference $2\pi r$, where r is the tube radius, likely contains an integral number of carbon hexagons. The pitch p, however, may or may not contain an integral number of hexagons. In the first alternative, the nanotube is periodic with a period p along the tube axis. Then, there is a one-dimensional (1D) k-vector along the tube. A "hole" which has a positive charge $+e$ and a size of a unit ring of height p and radius r can go through inside the positively charged carbon wall. In contrast, an "electron" having a negative charge $-e$ is attracted by the positively charged carbon wall, and hence it cannot go straight inside the wall. Thus, there is an extra "hole" channel current in the semiconducting SWNT. Moriyama *et al.*[15] observed a "hole" current after examining the gate voltage effect. The system should have the lowest energy if the unit ring contains an integer set (m, n) of carbon hexagons, which may be attained after annealing at high temperatures. This should happen if the tube length is comparable with the circumference. The experimental tube length is much greater than the circumference $2\pi r$. In the fabrication process the pitch is not controlled. Then, the first case in which the unit cell contains an integer set (m, n) of hexagons. must be the minorities. This case then generates a semiconducting transport behavior as shown below.

If the unit ring contains an irrational number of carbon hexagons, which happens more often, then the system does not allow a conduction along the tube axis. The set of irrational numbers is greater in cardinality than the set of rational numbers. The system is still conductive since the conduction electrons ("electrons", "holes") can go through within the tube wall. This conduction is two-dimensional (2D) as can be seen in the unrolled configuration, which is precisely the graphene honeycomb lattice. This means that the conduction behavior in the carbon wall should be the same as that in graphene if the finiteness of the circumference is neglected.

In a four-valence-electron material (carbon) all electrons are bound to ions, and there is no conduction at 0 K. If a "hole" having the charge $-e$ and the size of a unit ring is excited, then this "hole" can move along the tube. The activation energy ε_3 and the mass m_3 should depend on the radius and the pitch.

We are now ready to discuss the electrical conductivity of the semiconducting SWNT. There are four currents carried by:

(a) "Electrons" moving in the graphene wall with the mass m_1 and the density $n_1 \exp(-\varepsilon_1/k_B T)$, running in the channels $\langle 110 \rangle \equiv [110]$, $[011]$, and $[101]$ with n_1 being the "electron" density in the high temperature limit.

(b) "Holes" moving in the graphene wall with the mass m_2 and the density $n_2 \exp(-\varepsilon_2/k_B T)$, running in the channels $\langle 100 \rangle \equiv [100]$, $[010]$, and $[001]$.

(c) "Holes" moving inside the graphene wall with the mass m_3 and the density $n_3 \exp(-\varepsilon_3/k_B T)$, running in the tube-axis direction. The activation energy ε_3 and the effective mass m_3 may vary with the radius and the pitch.

(d) Cooper pairs (pairons), which are formed by the phonon-exchange attraction moving in the graphene wall.

In actuality, one of the currents may be dominant, and be observed.

In the normal Ohmic conduction due to the conduction electrons the resistance R is proportional to the tube length. Then, the conductivity σ is given by the Drude formula:

$$\sigma = \frac{q^2}{m^*} n \tau \equiv \frac{q^2}{m^*} n \frac{1}{\Gamma}, \tag{14.2}$$

where q is the carrier charge ($\pm e$), m^* the effective mass, n the carrier density, and τ the relaxation (scattering) time. The relaxation rate $\Gamma \equiv \tau^{-1}$ is the inverse of the relaxation time. If the impurities and phonons are scatterers , then the rate Γ is the sum of the impurity scattering rate Γ_{imp} and the phonon scattering rate $\Gamma_{\mathrm{ph}}(T)$:

$$\Gamma = \Gamma_{\mathrm{imp}} + \Gamma_{\mathrm{ph}}(T). \tag{14.3}$$

The impurity scattering rate Γ_{imp} is temperature-independent and the phonon scattering rate $\Gamma_{\mathrm{ph}}(T)$ is temperature (T)-dependent. The phonon scattering rate $\Gamma_{\mathrm{ph}}(T)$ is linear in T above around 2 K:

$$\Gamma_{\mathrm{ph}}(T) = aT, \qquad a = \text{constant, (above 2 K)} \tag{14.4}$$

The temperature dependence should arise from the carrier density $n(T)$ and the phonon scattering rate $\Gamma_{\mathrm{ph}}(T)$. Writing the T-dependence explicitly, we obtain from Eqs. (14.2)-(14.4)

$$\sigma = \sum_{j=1,2} \frac{e^2}{m_j^*} n_j \, e^{-\varepsilon_j/k_B T} \frac{1}{\Gamma_{\mathrm{imp}} + aT}. \tag{14.5}$$

where $j = 1$ for "electron" and $j = 2$ for "holes".

Moriyama *et al.*[15] measured the currents in semiconducting SWNT between 2.6 and 200 K, see his Fig.(3) in Ref.[15] which is reproduced in our Fig.(14.4).

By using the Arrhenius plot for the data above 20 K, the activation energy

$$E_3 \sim 3 \text{ meV} \tag{14.6}$$

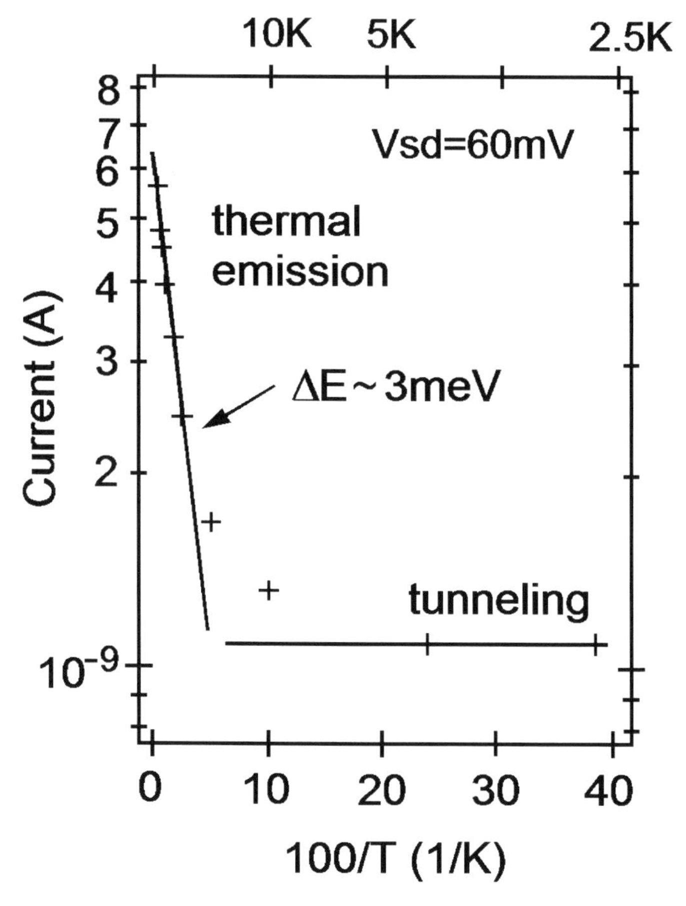

Fig. 14.4. Log-scale plot of the current as a function of the temperature (Arrhenius plot) at a fixed $V_{sd} = 60mV$. Reproduced after Moriyama *et al.*[15]

was obtained. By studying the field (gate voltage) effect, the carriers were found to be "hole"-like. Thus, the major currents observed can be interpreted in terms of the "holes" moving inside the tube wall. This "hole" axial transport depends on the unit ring containing $m \times n$ hexagons, m and n integers. Since both pitch and circumference have a distribution, the activation energy ε_3 should also have a distribution. Hence the obtained value in Eq. (14.6) must be regarded as the average value.

We now go back to the data by Moriyama *et. al.* shown in Fig.(14.2). Below 20 K the currents observed are very small and they appear to approach a constant in the low temperature limit (large T^{-1} limit). These currents, we believe, are due to the Cooper pairs formed by the phonon exchange attraction.

The Cooper pairs (pairons) move in 2D with the linear dispersion relations[9]

$$\varepsilon = c^{(j)} p \tag{14.7}$$

$$c^{(j)} = \frac{2}{\pi} v_F^{(j)}, \tag{14.8}$$

where $v_F^{(j)}$ is the Fermi velocity of the "electron" ($j = 1$) ["hole" ($j = 2$)]. This relation is obtained, starting with a BCS-like Hamiltonian, setting up and solving an energy-eigenvalue problem for the moving pairon, see reference.[9] These pairons move similar to massless particles with the common speed $(2/\pi)v_F^{(j)}$.

Consider first "electron"-pairs. The velocity $\mathbf{v}$ is given by (omitting superscript)

$$\mathbf{v} = \frac{\partial \varepsilon}{\partial \mathbf{p}} \qquad \text{or} \qquad v_x = \frac{\partial \varepsilon}{\partial p}\frac{\partial p}{\partial p_x} = c\,\frac{p_x}{p}$$

$$p \equiv (p_x^2 + p_y^2)^{1/2}, \tag{14.9}$$

where Eq. (14.7) was used. The equation of motion along the E-field (x-) direction is

$$\frac{dp_x}{dE} = q'E, \qquad q' = -2e \tag{14.10}$$

whose solution is given by

$$p_x = q'Et + p_x^{(0)}, \tag{14.11}$$

where $p_x^{(0)}$ is the initial momentum component. The current density j is calculated from (charge q')$\times$(number density n_p)$\times$(average velocity $\overline{v}$). The average velocity $\overline{v}$ is calculated by using Eqs. (14.9) and (14.11) with the assumption that the pair is accelerated only for the collision time τ. We obtain

$$j \equiv q' n_p \overline{v} = q' n_p\, c\,\frac{1}{p}(q' E\,\tau) = q'^2\,\frac{c}{p}\, n_p\, E\,\tau. \tag{14.12}$$

For stationary currents, we may take n_p to be the Bose distribution function $f(\varepsilon_p)$

$$n_p = f(\varepsilon_p) = [\exp(-\alpha + \varepsilon_p/k_B T) - 1]^{-1}, \tag{14.13}$$

where $\exp(\alpha)$ is the fugacity. Integrating over all 2D p-space, and using Ohm's law

$$j = \sigma E, \tag{14.14}$$

we obtain for the conductivity σ:

$$\sigma^{(1)} = \frac{1}{(2\pi\hbar)^2} q'^2 c \int d^2p \frac{1}{p} f(\varepsilon_p) \tau. \tag{14.15}$$

In the temperature range between 2 K and 20 K we may assume the Boltzmann distribution function for $f(\varepsilon_p)$:

$$f(\varepsilon_p) \simeq \exp(\alpha - \varepsilon/k_B T). \tag{14.16}$$

We assume that the relaxation time arises from the phonon scattering so that $\tau = (aT)^{-1}$, see Eqs. (14.2) - (14.4). After performing the p-integration we obtain

$$\sigma = \frac{2}{\pi} \frac{e^2 k_B}{a\hbar^2} e^{\alpha} \tag{14.17}$$

which is T-independent. If there are "electron" and "hole" pairs, then they contribute additively to the conductivity.

These pairons should undergo a Bose-Einstein Condensation (BEC) if the measured temperature is lowered. The currents running in metallic SWNT do not show the activated-state behavior. We suspect that these currents are carried by not by fermionic conduction electrons but by bosonic Cooper pairs. We discuss this point further in the following section.

Bachtold *et. al.*[16] used electrostatic force microscopy (FEM) to study the potential drop of metallic SWNT between two electrodes. They found that the ac-FEM signals at room temperature is flat along the tube length. We may interpret this as supercurrents running without the potential drop. An alternative explanation is by the ballistic charge transport. It is difficult, however, to explain why ballistic electrons are not scattered by impurities and phonons? It is also difficult to answer why the ballistic electrons (fermions) are generated without any activation energy. There are no ballistic electrons at 0 K.

In our interpretation metallic SWNT show a superconducting state at room temperature. This means that the superconducting temperature T_c is higher than 300 K.

The picture of a superconducting (metallic) SWNT is supported by other experiments as follows. Moriyama *et al.* observed in his Fig.(1)

of Ref.[15] that the resistance of metallic SWNT is unchanged (while that of semiconducting SWNT decreases) with increasing gate voltage. The stability of the metallic (superconducting) sample arises from the neutral supercurrents and the superconducting energy gap. Note that this stability cannot be explained based on the ballistic electron model.

14.4. Multiwalled Nanotube

A graphene sheet can be rolled like a wallpaper stored to produce a MWNT. The tube diameter may reach 10 nm or more. The measured currents are most likely to come from the currents in the outermost shells between carbon walls and the outermost wall, since the circumference is greatest there. Because of the open-ended structure in circumference, the quasi-unit-ring containing an integral number of carbon hexagons, in each pitch along the tube axis similar to the unit-ring for the semiconducting MWNT may be generated. The experimental works by Ebbessen *et al.*[2] indicate that the conductivities in MWNT are varied greatly, from diffusive (semiconducting) to superconducting, which was discussed fully by Fujita and Suzuki[17].

The charge carriers are "electrons", "holes" and Cooper pairs as in SWNT. Significant differences will be (a) the carrier densities are higher in MWNT since the circumference is greater and (b) the "holes" run in shells between walls.

In graphene the "electron" pairons, having the greater speed, dominate the transport and the BEC. The critical (superconducting) temperature T_c is given by

$$k_B T_c = 1.24\ \hbar v_F^{(1)} n_0^{1/2}, \tag{14.18}$$

where n_0 is the pairon density. Briefly, the BEC occurs when the chemical potential vanishes at a finite T. The critical temperature T_c can be determined from

$$n_0 = \frac{1}{(2\pi\hbar)^2} \int d^2p\, [e^{\beta_c \varepsilon} - 1]^{-1}, \qquad \beta_c \equiv (k_B T_c)^{-1}. \tag{14.19}$$

After expanding the integrand in powers of $e^{-\beta_c \varepsilon}$ and using $\varepsilon = cp$, we obtain

$$n_0 = 1.654\ (2\pi)^{-1} (k_B T_c/\hbar c)^2, \tag{14.20}$$

yielding formula (14.18) with $c = (2/\pi) v_F$. Experiments[18] indicate that the linear coefficient c equals 1.02×10^6 m/s, implying that the Fermi velocity v_F is approximately equal to 1.57×10^6 m/s. Using Eq. (14.18), we obtain the critical temperature $T_c = 1275$ K, for $n_0 = 7.33 \times 10^{11}$ cm^{-2}. The measurements of the saturation magnetization of MWNT bundle[6] yield the 3D electron density 2.12 $\times 10^{19}$ cm^{-3}, which correspond to the 2D density $(21.2)^{2/3} \times 10^{12}$ cm^{-2}. The pairon density must be smaller than the electron density. Hence, the quoted pairon density $n_0 = 7.33 \times 10^{11}$ cm^{-2} is reasonable.

In our quantum statistical theory the critical (superconducting) temperature T_c is identified as the BEC temperature of the pairons. In the original BCS theory the critical temperature is identified as the temperature at which the quasi-electron energy $E_k = (\varepsilon_k^2 + \varepsilon_0^2)^{1/2}$ becomes ε_k, where ε_k is the electron energy and ε_0 is a temperature-dependent energy gap. Our definition of T_c is more fundamental since we can prove that the electrons in the presence of a supercondensate below the critical temperature have the energy expression E_k. Besides, we can discuss the pairon density dependence of the critical temperature as given by Eq. (14.18).

14.5. Summary and Discussion

We have shown, based on the quantum statistical theory, that a superconducting state exists in MWNT with a ultrahigh critical temperature ($\sim$ 1275 K). The linear dispersion relation in Eq. (14.7) may be probed by the angle-resolved photo-emission spectroscopy (ARPES). Lanzara *et al.*[19] applied ARPES to demonstrate the linear dispersion relation in the quantum Hall effect in graphene, which supports our theoretical formula (14.7) for the linear dispersion relation.

Acknowledgments

The authors wish to thank Professors Guo-meng Zhao, Manuel de Llano and Murthy Ganapathy for enlightening discussions. One of the authors (S.F.) thank Leo and Judith for ever-lasting friendship since our student days at the University of Maryland.

References

1. S. Iijima, *Nature* (London) **354**, 56 (1991); *Mater. Sci. Eng.* ***B19***, 172 (1993).
2. T. W. Ebbesen *et al.*, *Nature*, **382**, 54 (1996).
3. C. Schönenberger and L. Forro, physicsworld.com, *Multiwall carbon nanotubes* (2000).

4. R. Saito, G. Dresselhaus and M. S. Dresselhaus, *Physical Properties of Carbon Nanotubes* (Imperial College Press, London 1998).
5. E. Wigner and F. Seitz, *Phys. Rev.* **43**, 804 (1933).
6. G-M. Zhao and P. Beeli, Phys. Rev. B **77**, 245433 (2008).
7. B. I. Spivak and S. A. Kivelson, Phys. Rev. B **43**, 3740 (1991).
8. Y. Maeno *et al.*, Nature **372**, 532 (1994).
9. S. Fujita and S. Godoy, *Quantum Statistical Theory of Superconductivity* (Plenum, New York, 1996), pp. 167, 181-183, 184-185, 202-204.
10. J. Bardeen, L. N. Cooper, and J. R. Schrieffer, Phys. Rev. **108**, 1175 (1957).
11. L. N. Cooper, Phys. Rev. **104**, 1189 (1956).
12. Y. Zhang *et al.*, Nature Physics **438**, 201 (2005).
13. N. W. Ashcroft and N. D. Mermin, *Solid State Physics* (Saunders, Philadelphia, 1974). p. 217, pp. 91-92, pp. 133-135.
14. S. Fujita and K. Ito, Quantum Theory of Conducting matter, (Springer, New York, 2007), pp. 106-107.
15. S. Moriyama, K. Toratani, D. Tsuya, M. Suzuki, Y. Aoyagi and K. Ishibashi, Physica E **24**, 46 (2004).
16. A. Bachtold *et al.* Phys. Rev. Lett. **84**, 6082 (2000).
17. S. Fujita and A. Sizuki, J. Appl. Phys.,(2010), accepted for publication.
18. M. Orlita *et al.*, J. Phys: Condens. Matter **20**, 45423 (2008).
19. S. Y. Zhou *et al.*, Nature Physics **2**, 595 (2006).

Chapter 15

Electron Correlations in Strongly Interacting Systems

J.J. Quinn and G.E. Simion

Department of Physics and Astronomy, University of Tennessee, Knoxville, TN, 37996, USA

Fractional quantum Hall systems are the prime example of strongly interacting electron systems. At very high values of the applied magnetic field, there is only one relevant energy scale, the Coulomb interaction $V_c \approx e^2/\lambda$, where λ is the magnetic length. Conventional many-body perturbation theory is inapplicable to such systems. Insight into the nature of the correlations was first given by Laughlin. This important contribution together with contributions by Haldane, Halperin, and Jain are reviewed in light of some rigorous mathematical theorems in an attempt to gain further insight into the correlations.

Contents

15.1. Introduction

The study of the electronic properties of quasi-two-dimensional systems has led to several discoveries of major importance. The integer quantum

Hall (IQH) effect discovered by von Klitzing and collaborators[1] has given us a more accurate value for an important fundamental constant and a new resistance standard. The observed behavior of magnetoconductivity resulted from the energy gap $\hbar\omega_c$ between the single particle Landau levels (LLs) and the existence of both localized and extended states within the LL. The observation of similar behavior at fractional filling of the lowest LL by Tsui et al.[2] was a surprise. It has to be associated with the interactions among the N electrons at particular fractional values of the filling factor $\nu = N/N_\Phi$ (where N_Φ is the number of degenerate single particle states in the LL). It was unexpected because it involved a quasi-two-dimensional system of spin polarized electrons with repulsive interactions. The energy gap causing the incompressible quantum liquid (IQL) ground state was not forseen before the observation of the fractional quantum Hall effect. In this paper we review the theoretical contributions of Laughlin,[3,4] Haldane,[5,6] Halperin,[7,8] Jain[9–12] and others[13–22] that resulted in an understanding of how IQL states at fractional factors ν arise. We address the novel IQL states observed by Pan et al.[23] and show that they can't be understood without introducing correlations among quasiparticles that are different from the Laughlin correlations among electrons in the lowest Landau level (LL0). We emphasize that the Laughlin correlations (avoidance of pair states with the largest values of the pair angular momentum L_2) depend strongly on the pair pseudopotential $V(L_2)$ (the interaction energy of a pair with pair angular momentum L_2).

15.2. Quantum Hall Systems

In the presence of an applied magnetic field $\vec{B}$ perpendicular to the $x-y$ direction, the kinetic energy of an electron confined to the plane $z=0$ becomes quantized into Landau levels. In the symmetric gauge, where the vector potential producing the field $\vec{B}$ is taken as $\vec{A}(\vec{r}) = \frac{1}{2}B(y,-x,0)$, the eigenstates of the electrons can be written $\Psi_{nm}(r,\theta) = e^{-im\theta}u_{nm(r)}$, and $E_{nm} = \frac{1}{2}\hbar\omega_c(2n+1+|m|-m)$. Here $n=0,1,2,\ldots$, and m is an integer. The lowest energy level has $n=0$ and $m=0,1,2\ldots$. It is convenient to introduce a complex coordinate $z = re^{-i\theta} = x - iy$, and to write the wave function for the lowest Landau level as[3] $\Psi_{0m} \propto z^m \exp\left(-|z|^2/(4\lambda^2)\right)$, where $\lambda = (\hbar c/(eB))^{1/2}$ is the magnetic length. The function $|\Psi_{0m}|^2$ has a maximum at a radius r_m proportional to $\sqrt{m}$. For a sample of finite radius the allowed values of m will go from $m=0$ to $m=N_\Phi - 1$, where $N_\Phi = BA/\Phi_0$ ($\Phi_0 = hc/e$ is the quantum of flux) is the number of flux quanta of the applied magnetic field threading the sample of area A. If the filling factor $\nu = N/N_\Phi$ is an integer, there is a gap $\hbar\omega_c$ between the filled LLs and the first empty one.

Tsui et al.[2] investigated the magnetoconductance of a very high mobility GaAs sample at low temperature and high applied magnetic field. They found behavior for σ_{xx} and σ_{xy} very similar to that observed by von Klitzing et al.[1] at $\nu = 1, 2, \cdots$, but at values of the filling factor $\nu = 1/3, 2/5, \cdots$. Since each LL consists of N_Φ degenerate states, the gap causing this behavior could only result from the interactions among the electrons. Laughlin correctly surmised that the fractional quantum Hall (FQH) states at filling factor ν equal to the reciprocal of an odd integer m resulted when the electrons were able to avoid pair states with largest repulsion (i.e. with relative angular momentum smaller then m). He proposed that the N electron ground state at $\nu = m^{-1}$ could be $\Psi_m(1, 2, \cdots N) = \prod_{i<j} z_{ij}^m \exp\left[-\sum_k |z_k|^2 /(4\lambda^2)\right]$

For $m = 1$ it corresponds to a completely filled LL0. For $m = 3$, it has $\nu = N/N_\Phi = 1/3$. It's antisymmetric, and it avoids pair states with relative pair angular momentum $m = 1$ (or total pair angular momentum in Haldane's spherical geometry with $L_2 = 2\ell - 1$).

15.3. Haldane's Spherical Geometry and Hierarchy

The first explanation of IQL states not belonging to the Laughlin set $\nu = m^{-1}$, where m is an odd integer was proposed by Haldane.[5] Haldane realized that Laughlin's fractionally charged quasiparticles (QPs) would have Landau levels of their own. He suggested that putting N_{QP} QPs into N'_Φ states of a QP Landau level was essentially the same problem as the original one of putting N electrons in N_Φ states of the electron LL. Because the number of QP states in a QP LL was equal to N, the number of electrons in the electron LL, Haldane said IQL daughter states should occur when the number of available QP states $N'_\Phi (= N)$ in the QP Landau level was equal to an even integer $2p$ times the number of QPs. He chose an even integer $2p$ in the relation $\nu_{\mathrm{QP}} = N_{\mathrm{QP}}/N = (2p)^{-1}$ in place of the odd integer $m = 2p + 1$ used by Laughlin (in $\nu = N/N_\Phi = (2p + 1)^{-1}$) because he considered the QPs as Bosons instead of as electron-like Fermions. The Haldane hierarchy contained all odd denominator fractions as IQL states. Slightly different versions of the hierarchy were later independently proposed by Halperin[8] and by Laughlin.[4] Haldane made it quite clear that the hierarchy scheme depended on the QP interactions (about which very little was known) being sufficiently similar to the interactions among the electrons in LL0.

In addition to proposing the hierarchy of IQL states, Haldane introduced the idea of performing exact numerical diagonalization (within lowest Landau level subspace of the full Hilbert space) in a spherical geometry. He put N electrons on the surface of a sphere of radius R with magnetic monopole of strength $2Q\Phi_0$ (where $2Q$ is an integer) at its center. The magnetic monopole gives rise to a magnetic field $B = 2Q\Phi_0/(4\pi R^2)$ which is perpendicular to the spherical surface. This geometry has the advantage of having a finite surface area without the imposition of boundary conditions, and rotational invariance in place of full translational invariance of an infinite plane. The single particle eigenstates are called monopole harmonics and denoted by $|Q, \ell, m>$.[24,25] They are eigenfunctions of $\hat{\ell}^2$ and $\hat{\ell}_z$ with eigenvalues $\ell(\ell+1)$ and m respectively, and $|m| \le \ell$. The energy of the state $|Q, \ell, m>$ is given by $E_\ell = (\hbar\omega_c)/(2Q)[\ell(\ell+1) - Q^2]$. An N electron eigenstates can be written $|L, L_z, \alpha>$, where L and L_z are the total angular momentum and its z-component. The multiplet index α distinguishes different multiplets that have the same value of total angular momentum L. Because the system is rotationally invariant, the matrix elements of $H' = \sum_{i<j} e^2|\vec{r}_i - \vec{r}_j|^{-1}$ satisfy the relation $\langle L', L'_z, \alpha'|H'|L, L_z, \alpha\rangle = \delta(L', L)\delta(L'_z, L_z)V_{\alpha\alpha'}(L)$, and $V_{\alpha\alpha'}(L)$ is independent of L_z (via the Wigner-Eckart theorem). This property greatly reduces the size of the H' matrix that has to be diagonalized.

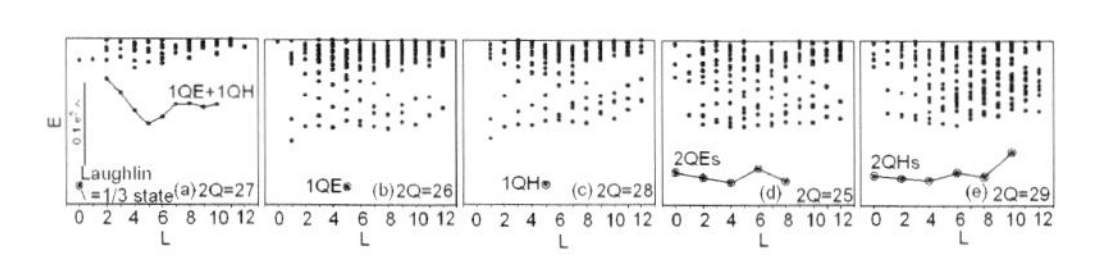

Fig. 15.1. The spectra of 10 electrons in the lowest Landau level calculated on a Haldane sphere with $2Q$ between 25 and 29. The open circles and solid lines mark the lowest energy bands with the fewest composite Fermion quasiparticles.[26]

In Fig.(15.1) we display the energy spectra[26] of a $N = 10$ electron system for values of $2Q$ between 25 and 29. It is clear that at $2Q = 27 = 3(N-1)$ an $L = 0$ Laughlin IQL ground state occurs. As the value of $2Q$ is increased (or decreased) by unity a state containing one quasihole (QH) (or one quasielectron (QE)) of angular momentum $\ell_{\rm QP} = 5$ is found. For an increase of $2Q$ by two, a band of states with two QHs each with

$\ell_{\rm QH} = 11/2$ and L values equal to $L_{\rm 2QH} = 2\ell_{\rm QH} - j$ (with j an odd integer) is found. For a decrease $2Q$ by two QE states each having $\ell_{\rm QE} = 9/2$ and $L_{\rm 2QE} = 2\ell_{\rm QE} - j$ are found. These bands of 2QP states give us, up to an overall constant $V_{\rm QP}(L')$, the QP interaction.

The energy of the multiplet $|L\alpha>$ can be written

$$E_\alpha = \frac{1}{2}N(N-1)\sum_{L_2} P_{L\alpha}(L_2)V(L_2) \;, \tag{15.1}$$

where $P_{L\alpha}(L_2)$ is the probability that the multiplet $|L\alpha>$ contains pairs with pair angular momentum L_2, and $V(L_2)$ is the energy of interaction of a pair with total pair angular momentum $L_2 = 2\ell - \mathcal{R}_2$. We use $\mathcal{R}_2 = 1, 3, 5 \cdots$ to denote "relative pair angular momentum," and sometimes write $V(\mathcal{R}_2)$, understanding this to mean $V(L_2)$ with L_2 replaced by $2\ell - \mathcal{R}_2$. $V(L_2)$ is simply the expectation value of $e^2/|\vec{r}_1 - \vec{r}_2|$ in the antisymmetric pair state $|\ell^2; L_2>$.

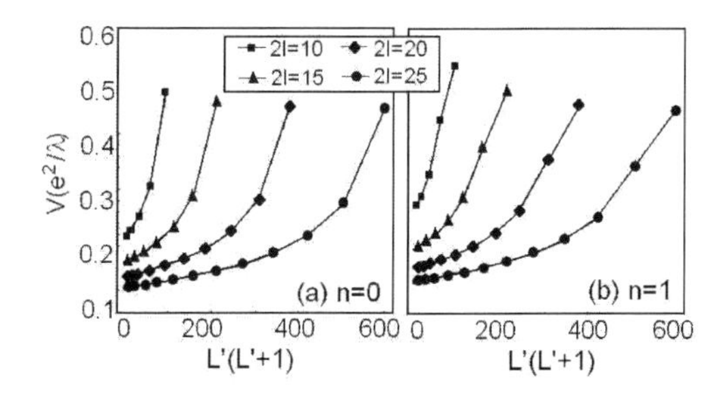

Fig. 15.2. Pseudopotential $V(L')$ of the Coulomb interaction in the lowest (a) and the first excited Landau level (b) as a function of squared pair angular momentum $L'(L'+1)$. Squares ($\ell = 5$), triangles ($\ell = 15/2$), diamonds ($\ell = 10$), and circles ($\ell = 25/2$) indicate data for different values of $Q = \ell + n$.[27]

In Fig.(15.2) we show plots $V(L_2)$ vs. $L_2(L_2+1)$, the eigenvalues of $\hat{L}_2^2$, for electrons in LL0 and LL1. Different symbols indicate different values of 2ℓ, twice the angular momentum of the shell in which the electrons reside. The two pseudopotentials are quite similar, except for LL1 at the largest values of L_2. $V(L_2)$ appears to increase linearly with $L_2(L_2+1)$ from $L_2 = 2\ell - 5$ to $2\ell - 1$. For all other values of L_2 in LL1 and for all values in LL0, the slope of $V(L_2)$ as a function of $L_2(L_2+1)$ increases with increasing L_2. We refer to a potential $V(L_2)$ which increases linearly with $L_2(L_2+1)$ as a "harmonic" pseudopotential. One with slope vs. $L_2(L_2+1)$ that decreases (or increases) with increasing L_2 is referred or as "subharmonic" (or "superharmonic").

15.4. Chern-Simons Gauge Field and Jain's Composite Fermion Picture

Jain[9,10,12] made a remarkable observation that the most robust fractional quantum Hall states observed experimentally could be understood in a very simple way in terms of a composite Fermion picture. This CF picture made use of a Chern-Simons (CS) transformation (see, e.g.[28]) and a CS gauge field familiar to field theorists. The CS transformation can be described as attaching to the jth electron $(1 \leq j \leq N)$ a flux tube carrying a magnetic field $\vec{b} = \alpha\Phi_0\delta(\vec{r} - \vec{r}_j)\hat{z}$. Here α is a constant, and $\hat{z}$ a unit vector normal to the 2D layer. Simplification results only when the mean field (MF) approximation is made. This is accomplished by replacing the density operator in the CS vector potential and in the Coulomb interaction by its MF value n_s, the uniform 2D electron density. The resulting Hamiltonian is the sum of single particle Hamiltonians in which an "effective" magnetic field $B^* = B - \alpha\phi_0 n_s$ appears.

Jain introduced the idea of a CF to represent an electron with an attached flux tube which carried an even number $\alpha(= 2p)$ of flux quanta.[11] In the MF approximation the CF filling factor ν^* is given by $\nu^{*-1} = \nu^{-1} - \alpha$, i.e. the number of flux quanta per electron of the dc field less the CS flux per electron. When ν^* is equal to an integer $n = \pm 1, \pm 2, \cdots$, then $\nu = n(1 + \alpha n)^{-1}$ generates (for α=2) quantum Hall states at $\nu = 1/3, 2/5, 3/7, \cdots$, and $\nu = 1, 2/3, 3/5, \cdots$. These are the most pronounced FQH states observed.

In the spherical geometry one can introduce an effective monopole strength seen by one CF.[13] It is given by $2Q^* = 2Q - 2p(N - 1)$ since the $2p$ flux quanta attached to every other CF must be subtracted from the original monopole strength $2Q$. Then $|Q^*| = \ell_0^*$ plays the role of the angular momentum of the lowest CF shell just as $Q = \ell_0$ is the angular momentum of the lowest electron shell. When $2Q$ is equal to an odd integer $(1 + 2p)$ times $(N - 1)$, the CF shell ℓ_0^* is completely filled, and an $L = 0$ incompressible Laughlin state at filling factor $\nu = (1 + 2p)^{-1}$ results. When $2|Q^*| + 1$ is smaller (larger) than N, QEs (QHs) appear in the shell $\ell_{\text{QE}} = \ell_0^* + 1$ $(\ell_{\text{QH}} = \ell_0^*)$. The low energy sector of the energy spectrum consists of the states with the minimum number of QP excitations required by the value of $2Q^*$ and N. By comparing with numerical results presented in Fig.(15.1), we readily observe that the total angular momentum multiplets appearing in the lowest energy sector are always correctly predicted by this simple MF CS picture.[26,27,29]

15.5. Beyond Mean Field: Two Energy Scales

Despite the satisfactory description of the allowed angular momentum multiplets, the magnitude of the MFCF energies is completely wrong. The magnetoroton energy does not occur at the effective cyclotron frequency $\hbar\omega_C^* = eB^\star/mc$. This MF energy is irrelevant at large values of B (if we keep $m_{\mathrm{CF}} = m_e$), so it is a puzzle why the CF picture does so well at predicting the structure of the energy spectrum.

For large values of B the MF energy $\hbar\omega_C^*$ is much larger than the Coulomb scale e^2/λ. Therefore the low lying multiplets of interacting electrons will be contained in a band of width proportional to e^2/λ about the lowest electron LL. The noninteracting CF spectrum contains a number of bands separated by $\hbar\omega_C^*$. Interactions (Coulomb and CS gauge interactions) among fluctuations beyond the MF essentially have to restore the original noninteracting electron spectrum when $B \to \infty$. Halperin et al.[30] and Lopez and Fradkin[20,31,32] have used conventional many-body perturbation theory to treat fluctuations. However, there is no small parameter to guarantee convergence or to justify simple approximations like the random phase approximation (RPA).

15.6. The Composite Fermion Hierarchy

Sitko et al.[14] introduced a very simple CF hierarchy picture in an attempt to understand the connection between Haldane's hierarchy of Laughlin correlated QP daughter states and Jain's sequence of IQL states with integrally filled CF Landau levels. Jain's CF picture neglected interactions between QPs. The gaps causing incompressibility were energy separations between the single particle CFLLs. Not all odd denominator fractions occurred in Jain's sequence $\nu = n(2pn \pm 1)^{-1}$ where n and p are integers. The missing IQL states, which occurred for partially filled QP shells (or CFQP Landau levels), had to depend on "residual interactions" between QPs, neglected in Jain's mean field CF picture.

In the CF hierarchy picture[14,33–36] an initial electron filling factor ν_0 was related to an effective CF filling factor ν_0^* by the relation ${\nu_0^*}^{-1} = \nu_0^{-1} - 2p_0$. This says that the total number of flux quanta (of both the dc magnetic field and CS gauge field) was equal to the dc flux per electron minus the CS flux per electron subtracted in the CF transformation. If ν_0^* were an integer n, then the IQL state of the CFs would occur at $\nu_0 = n(2p_0 n \pm 1)^{-1}$. This is the Jain sequence of integrally filled CF LLs.

What happens if ν_0^* is not equal to an integer? Sitko et al.[14,33] suggested that one should write ν_0^* as $\nu_0^* = n_1 + \nu_1$, where n_1 was an integer and ν_1 represented the filling factor of the partially filled CFQP level. If, as Haldane[5] suggested, the residual interactions between QPs were sufficiently similar to the Coulomb interaction between electrons in the lowest LL, one could assume Laughlin correlations among QPs. By reapplying the CF transformation to them and writing $\nu_1^{*-1} = \nu_1^{-1} - 2p_1$, ν_1^* could be an integer n_2 resulting in $\nu_1 = n_2(2p_1 n_2 \pm 1)^{-1}$ and an IQL daughter state at $\nu_0^{-1} = 2p_1 + \left[n_1 + n_2(2p_1 n_2 + 1)^{-1}\right]^{-1}$. This is a new odd denominator fraction not belonging to the Jain sequence. If ν_1^* is not an integer, simply set $\nu_1^* = n_2 + \nu_2$ and reapply the CF transformation to the new QPs in the shell of filling factor ν_2. In general one finds $\nu_l^{-1} = 2p_l + (n_{l+1} + \nu_{l+1})^{-1}$ at the lth level of the hierarchy. When $\nu_{l+1} = 0$, there is a filled shell of CFs at the lth level of the hierarchy. The procedure generates Haldane's continued fraction leading to IQL states at all odd denominator fractions. It gives the Jain sequence as a special case in which integral CF filling $\nu_0^* = n$ of the CFQP shell is found at the first level of the CF hierarchy.

It is not difficult to show by numerical diagonalization that hierarchy picture can't be correct in general. The reason, as suggested by Sitko et al.,[33] has to do with the residual QP interactions. Consider, for example, the electron system with $(N, 2\ell)$ given by $(8, 18)$. Applying the CF transformation with $2p_0 = 2$ gives $2\ell_1^* = 4$. Thus, the lowest CF shell has $\ell_1^* = 2$; it can accommodate five CFs. The remaining three CFs must go into the first excited CF shell with $\ell_{\text{QE}} = 3$. The five CFs in the lowest shell would give an IQL state if three CFQEs were not present. Only the CFs in the partially filled CF shell are considered to be QPs. Three Fermions each with $\ell_{\text{QE}} = 3$ give the multiplets $L = 0 \oplus 2 \oplus 3 \oplus 4 \oplus 6$. If the CF hierarchy were correct, applying a second CF transformation with $2p_1 = 2$ to the three CF QEs would give $2\ell_{\text{QE}}^* = 2$ and an $L = 0$ IQL ground state. Numerical diagonalization of the $(N, 2\ell) = (8, 18)$ system gives the spectrum shown in Fig.(15.3). The low lying multiplets are exactly as predicted at the first CF level. However, the $L = 0$ multiplet is clearly not the ground state as predicted by reapplying the CF transformation. It should be emphasized that the numerical results are obtained for a spin polarized system (with total spin $S = N/2 = 4$). The reason for this failure [the "subharmonic" behavior of the CFQE pseudopotential[36]] will be explained later.

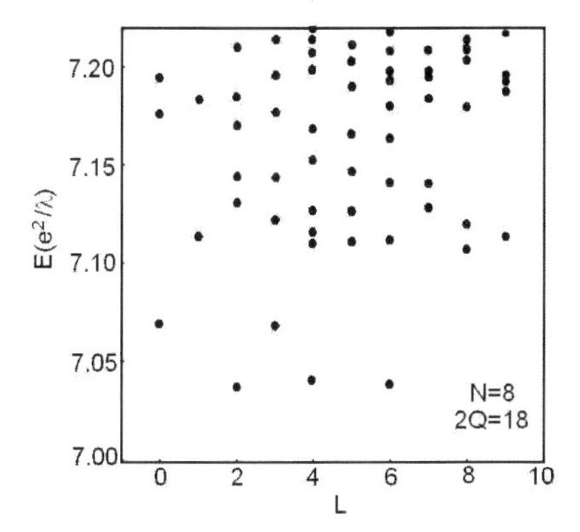

Fig. 15.3. Low energy spectrum of 8 electrons at $2\ell = 18$. The lowest band contains 3 QEs each with $\ell_{QE} = 3$. Reapplying the CS mean field approximation to these QEs would predict an $L = 0$ daughter state corresponding to $\nu = 4/11$. The data makes it clear that this is not valid.

15.7. Residual Interactions

The QEs and QHs have residual interactions that are more complicated than simple Coulomb interactions. They are difficult to calculate analytically, but if we look at an N electron system at a value of $2\ell = 3(N-1) \pm 2$, we know that the lowest band of states in the spectrum will correspond to 2 QEs or 2 QHs of the Laughlin $\nu = 1/3$ FQH state (for the minus and plus signs respectively). Fig.(15.1) gives the spectrum for $N = 10$ electrons at $2\ell = 25$ (2 QE case) and $2\ell = 29$ (2 QH case). It is clear that the low energy bands are not degenerate, but that the energy E depends on L, which (as we have seen) can be understood as the total angular momentum of the QP pair. For QEs, $E(L)$ has a maximum at $L = 2\ell_{QE} - 3$ and a minima at $L = 2\ell_{QE} - 1$ and $2\ell_{QE} - 5$. For QHs, $E(L)$ has a maximum at $L = 2\ell_{QH} - 1$ and $L = 2\ell_{QH} - 5$, and a minimum at $L = 2\ell_{QH} - 3$. This is quite different from the pseudopotentials for electrons, and it is undoubtedly the reason why the CF picture fails when it is reapplied to QEs.

In Fig.(15.4) we display the pseudopotentials for electrons in LL0 and LL1 with that for QEs of the Laughlin $\nu = 1/3$ IQL state in CF LL1. The electron pseudopotentials are the same ones presented in Fig.(15.2) but are presented here as a function of $\mathcal{R} = 2\ell - L'$, the relative angular momentum of a pair. The QE pseudopotentials in frame (c) were taken from the calculations of Lee et al.[21,37] and from the diagonalization of small electron systems done by Wójs et al.,[38,39] and are known up to a constant. The magnitude of interaction of CFQEs is much smaller than the interaction between electron, and has a sharp maximum at $\mathcal{R} = 3$ and minima at $\mathcal{R} = 1$ and 5.

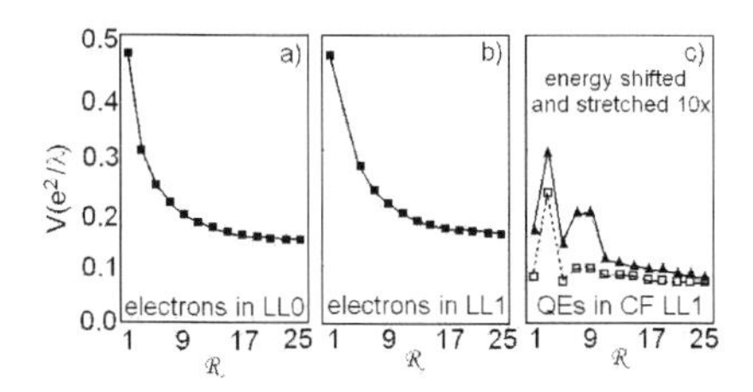

Fig. 15.4. Pair interaction pseudopotentials as a function of relative angular momentum $\mathcal{R}$ for electrons in LL0 (a), LL1 (b) and for the QEs of the Laughlin $\nu = 1/3$ state calculated by Lee et al.[21,37] (squares) and by Wójs et al.[38] (triangles), displaced from each other for clarity.

These pseudopotentials have been obtained for 2D electron layers of zero width. It is well known[40–42] that the finite extent of the subband wavefunction in the direction perpendicular to the layer introduces a correction to the electron pseudopotentials. The QP pseudopotentials are also sensitive to the layer width since they are obtained from the energy of the two QP band obtained by exact diagonalization of the appropriate electron system including the specific form of the (lowest) subband wave function.

Laughlin correlated states belonging to the Laughlin-Jain sequence $\nu = n(2pn \pm 1)^{-1}$ occur for LL0 for $p = 1$ and 2, and for $n = 1, 2, 3 \cdots$. For electrons in LL1, robust FQH states occur $\nu = 5/2, 7/3$, and $11/5$ (corresponding to $\nu_1 = \nu - 2 = 1/2, 1/3$, and $1/5$), and their $e - h$ conjugate at $\nu_1' = 1 - \nu_1$. However, the Jain states at $\nu_1 = 2/5, 3/7, \cdots$, and their $e-h$ conjugates are either not observed at all, or appear as very weak minima in ρ_{xx}. FQH daughter states arising from interacting QPs in CFLL1 occur at $\nu_{\rm QE}$ or $\nu_{\rm QH} = 1/3$ (corresponding to $\nu = 4/11$ and $4/13$) and $\nu_{\rm QE} = 1/2$ ($\nu = 3/8$). These states are thought to be fully spin polarized, but that is not absolutely certain. In addition, our numerical studies of the interactions between CFQPs suggest that the spin polarized states are not Laughlin correlated.

Because electrons in LL0 and LL1, and QPs in CFLL1 are interacting Fermions in a degenerate Landau level, the differences in their properties can only be attributed to the differences in the pseudopotentials describing their interactions. For QPs in CFLL1, $V_{\rm QE}(L_2)$ and $V_{\rm QH}(L_2)$ are clearly not monotonic functions. They have maxima and minima at values of $\mathcal{R} \leq 7$. It seems that Jain's picture is valid for LL0, but not for LL1 when $2/3 > \nu_1 > 1/3$ or for CFQPs of Laughlin $\nu = 1/3$ state.

In the following sections we explore the conditions for which the CF picture is valid. We give examples of situations in which it is not valid, and we suggest new kinds of correlations that might occur in such cases.

15.8. Validity of the CF Hierarchy Picture

From the experimental results of Pan et al.[23] it is clear that there are IQL states which do not belong to the Jain sequence of integrally filled CF levels (e.g. the totally spin polarized $\nu = 4/11$ state). This state should occur in the CF (or equivalent Haldane) hierarchy if the interaction between CFQEs results in Laughlin correlations among them. Numerical diagonalization[14] of small systems did not find a Laughlin correlated $L = 0$ ground state of the CFQEs at $\nu_{\rm QE} = 1/3$. Furthermore, Pan et al. observed strong minima in ρ_{xx} at even denominator filling factors ($\nu = 3/8$ and $\nu = 3/10$), suggesting IQL states which do not belong to the CF hierarchy.

Our research group has made an important contribution to this field, by rigourously proving[43] under which conditions Jain's elegant CF picture is applicable. Because there is no small parameter to guarantee the convergence of many body perturbation theory, our proof does not involve treating the interactions between fluctuations by a perturbative expansion. It involves proving several rigorous mathematical theorems and applying them, together with well-known concepts used frequently in atomic and nuclear physics. We ouline these theorems, referring to earlier publications for proofs:

Theorem 15.1. *If $\left|\ell^N; L\alpha\right\rangle$ is the αth multiplet of total angular momentum L formed from N Fermions in a shell of angular momentum ℓ, then*

$$\left\langle \ell^N; L\alpha \left| \sum_{\langle i,j \rangle} \hat{L}_{ij}^2 \right| \ell^N; L\alpha \right\rangle = L\left(L+1\right) + N\left(N-2\right)\ell\left(\ell+1\right) \ . \quad (15.2)$$

Here $\hat{L}_{ij}^2$ is the square of the angular momentum operator $\hat{\ell}_i + \hat{\ell}_j$ for electrons i and j, and the sum is over all pairs $\langle i,j \rangle$.[43]

Theorem 15.2. *It is well-known in atomic and nuclear shell model studies that $\left|\ell^N; L\alpha\right\rangle$ can be written as a sum of product functions*[43]

$$\left|\ell^N; L\alpha\right\rangle = \sum_{L_2} \sum_{L'\alpha'} G_{L\alpha,L'\alpha'}\left(L_2\right)\left|\ell^2, L_2; \ell^{N-2}, L'\alpha'; L\right\rangle . \qquad (15.3)$$

Here $G_{L\alpha,L'\alpha'}\left(L_2\right)$ is called a coefficient of fractional parentage. The ket vector $\left|\ell^2, L_2; \ell^{N-2}, L'\alpha'; L\right\rangle$ is a product of $\left|\ell^2, L_2\right\rangle$ and $\left|\ell^{N-2}, L'\alpha'\right\rangle$ selected to give a state of total angular momentum L. $|\ell^N; L\alpha>$ is totally antisymmetric even through $|\ell^2, L_2; \ell^{N-2}, L'\alpha'; L>$ is not.

Theorem 15.3. *Because $\left|\ell^N; L\alpha\right\rangle$ is totally antisymmetric*

$$\left\langle \ell^N; L\alpha \left| \sum_{\langle i,j\rangle} \hat{L}_{ij}^2 \right| \ell^N; L\alpha \right\rangle = \frac{N\left(N-1\right)}{2} \sum_{L_2} L_2\left(L_2+1\right) P_{L\alpha}\left(L_2\right). \quad (15.4)$$

This is simply a statement that the sum over all pairs can be replaced by a sum over all allowed values of the pair angular momentum L_2 of one pair, multiplied by the total number of pairs $N\left(N-1\right)/2$. $P_{L\alpha}(L_2)$ is defined in Eq. 15.1.

Theorem 15.4. *If the pseudopotential $V\left(L_2\right)$ is harmonic by which we mean $V\left(L_2\right) = V_H(L_2) = A + BL_2\left(L_2+1\right)$ where A and B are constants, then every multiplet α with the same total angular momentum L has the same energy, given by*

$$E_\alpha\left(L\right) = N\left[\frac{1}{2}N\left(N-1\right)A + B\left(N-2\right)\ell\left(\ell+1\right)\right] + BL\left(L+1\right). \quad (15.5)$$

This means that the degeneracy of the angular momentum multiplets of non-interacting Fermions is not removed by a harmonic pseudopotential for different multiplets having the same L.

Theorem 15.5. *If $G_{N\ell}\left(L\right)$ is the number of independent multiplets of total angular momentum L that can be formed from N Fermions in a shell of angular momentum ℓ, then $G_{N\ell^*}\left(L\right) \le G_{N\ell}\left(L\right)$ for every L, if $\ell^* = \ell - \left(N-1\right)$.*[44]

Theorem 15.6. *The subset $G_{N\ell^*}\left(L\right)$ of angular momentum multiplets of* *the set $G_{N\ell}(L)$ avoids the largest allowed pair angular momentum $L_2 = 2\ell - 1$, which for LL0, corresponds to the largest pair repulsion.*

This is obvious for $N = 2$ where $L_2^{MAX} = 2\ell - 1$ and $L_2^{*MAX} = 2\ell^* - 1 = 2\ell - 3$, but it is true for arbitrary N. This theorem means that the set of states selected by Jain's mean field CF picture (where ℓ^* plays the role of the effective CF angular momentum) is subset of $G_{N\ell}(L)$. This subset avoids pair states with $L_2 = 2\ell - 1$, and contains multiplets with low angular momentum and low energy.

Theorem 15.7. *By adding an integral number, α, of Chern-Simons flux quanta (oriented opposite to the applied magnetic field) to the Hamiltonian for N electrons, not via a gauge transformation, but adiabatically, the pair eigenstate (in the planar geometry) $\Psi_{nm} = e^{im\phi}u_{nm}(r)$, where $u_{nm}(r)$ is the radial wavefunction, transforms to $\tilde{\Psi}_{nm} = e^{im\phi}u_{n,m+\alpha}(r)$.*[45]

These theorems (or conjectures for cases where we have not worried too much about mathematical rigor) justify Jain's CF picture when applied to LL0. Is this important? In our opinion, Jain's mean field CF picture has been a brilliant success. It is used very often to interpret experimental data. However, because Coulomb and CS gauge interactions beyond the mean field involve two entirely different energy scales ($\hbar\omega_c^* = \nu B$ and $e^2/\lambda \propto \sqrt{B}$), these two interactions between fluctuations beyond the mean field can't possibly cancel for all values of B.

Because correlations (i.e. the lifting of the degeneracy of the angular momentum multiplets $|\ell^N; L\alpha\rangle$ of non-interacting electrons in partially filled LL0) depend on the deviation of the actual pseudopotential from harmonic behavior (i.e. on $\Delta V(L_2) = V(L_2) - V_H(L_2)$), it is interesting to explore the simplest possible anharmonicity. If we assume that $\Delta V(L_2) = k\delta(L_2, 2\ell - 1)$ with $k > 0$, then it is obvious that the lowest energy multiplet $|L\alpha\rangle$ for every value of L is the one that has that has the smallest value of $P_{L\alpha}(L_2 = 2\ell - 1)$. This is exactly what is meant by Laughlin correlations, and is the reason why the Laughlin wavefunction is the exact solution to the short range pseudopotential $V(L_2) = \delta(L_2, 2\ell - 1)$. It should be noted that if $k < 0$, the opposite is true. For this case, the lowest multiplet for each value of L has a maximum value of $P_{L\alpha}(L_2 = 2\ell - 1)$, and the particles have a tendency to form pairs with $L_2 = 2\ell - 1$. Laughlin correlations at a given value of L_2 occur only is the pseudopotential is superharmonic at that value of L_2.

15.9. Spin Polarized Quasiparticles in a Partially Filled Composite Fermion Shell

15.9.1. *Heuristic picture*

We have demonstrated that the simplest repulsive anharmonic pseudopotential $V(\mathcal{R}_2) = V_H(\mathcal{R}_2)+k\delta(\mathcal{R}_2, 1)$ caused the lowest energy state for each value of the total angular momentum L to be Laughlin correlated. For a spin polarized LL0 with $1/3 \leq \nu \leq 2/3$ such a potential (superharmonic at $\mathcal{R} = 1$) gives rise to the Laughlin-Jain sequence of integrally filled CF levels with $\nu_{\pm} = n(2n \pm 1)^{-1}$, where n is an integer. Haldane[5] suggested that if the highest occupied CF level is only partially filled, a gap could result from the residual interactions between the QPs, in the same way that the original gap resulted from the electron interactions. However, this would require $V_{\mathrm{QP}}(\mathcal{R})$ to be "superharmonic" at $\mathcal{R} = 1$ to give rise to Laughlin correlations. We have already shown that in a Laughlin $\nu = 1/3$ or $1/5$ state $V_{\mathrm{QE}}(\mathcal{R})$ was not superharmonic at $\mathcal{R} = 1$ and $\mathcal{R} = 5$, and that V_{QH} was not at $\mathcal{R} = 3$. This means that many of the novel IQL states observed by Pan et al.[23] have to result from correlations among the QPs that are quite different from the Laughlin correlations.

Just as electrons in LL1 tend to form clusters,[46] we expect QPs in CF LL1 to tend to form pairs or larger clusters. The major differences between electrons in LL1 and QPs in CF LL1 are: (i) the pseudopotential $V_1(L')$ for electrons in LL1 (shown in Fig.(15.4)) is an increasing function of L', but it is not superharmonic at $\mathcal{R} = 1$, while $V_{\mathrm{QE}}(L')$ is strongly subharmonic having a maximum at $\mathcal{R} = 2\ell - L' = 3$ and minima at $\mathcal{R} = 1$ and 5 and (ii) the e-h symmetry of LL1 is not applicable to QEs and QHs in CF LL1.[47] The QEs are quasiparticles of the Laughlin $\nu = 1/3$ IQL state, while QHs in CF LL1 are actually quasiholes of the Jain $\nu = 2/5$ state. The QE and QH pseudopotentials in frames a) and c) are similar, but not identical as shown in Fig.(15.5). The QHs of the $\nu = 1/3$ state reside in CF LL0 and have a different pseudopotential (frame b).

The experimental results of Pan et al.[23] suggest that the novel $\nu = 4/11$ IQL ground state is fully spin polarized. Because $V_{\mathrm{QE}}(L')$ is not superharmonic at $R' \equiv 2\ell - L' = 1$, the CF picture could not be reapplied to interacting QEs in the partially filled CF shell.[36] This led to the suggestion[48] that the QEs forming the daughter state had to be spin reversed and reside in CF LL0 as quasielectrons with reverse spin (QERs). Szlufarska et al.[49] evaluated $V_{\mathrm{QER}}(L')$, the pseudopotential of QERs. They showed that

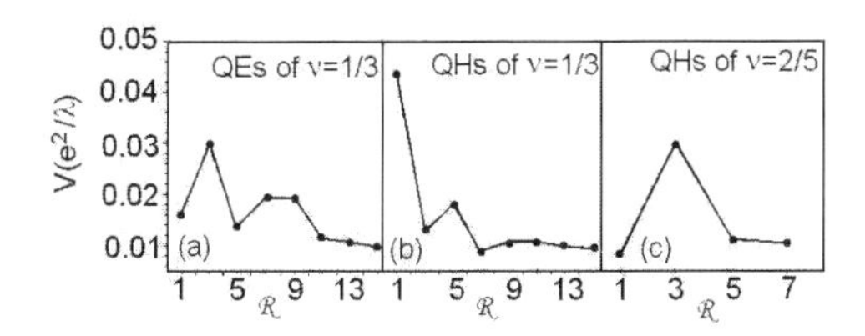

Fig. 15.5. $V_{\text{QE}}(\mathcal{R})$ and $V_{\text{QH}}(\mathcal{R})$ for (a) QEs of $\nu = 1/3$ state (b) QHs of $\nu = 1/3$ state, and (c) QHs of $\nu = 2/5$ state.

$V_{\text{QER}}(L')$ was superharmonic at $\mathcal{R} = 1$, so that unlike majority spin QEs, they could support Laughlin correlations at $\mathcal{R} = 1$.

This leaves at least two possible explanations of the $\nu = 4/11$ IQL state. It could be a Laughlin correlated daughter state of spin reversed QEs (i.e. QERs), or it could be a spin polarized state in which the QEs form pairs or larger clusters. Here we investigate only the completely polarized case.

The simplest idea is exactly that used for electrons in LL1,[46] namely the formation of pairs with $\ell_{\text{P}} = 2\ell - 1$, where ℓ is the angular momentum of the shell occupied by the QEs. If we assume that the QEs form pairs and treat them as Fermions,[36] the effective angular momentum of Laughlin correlated Fermion pairs (FPs) is given by $2\ell^* = 2\ell_{\text{FP}} - 2p(N_{\text{P}} - 1)$ where $2\ell_{\text{FP}} = 2(2\ell - 1) - 3(N_{\text{P}} - 1)$. The term $-3(N_{\text{P}} - 1)$ keeps the CF pair separation large enough to avoid violation of the Pauli principle. The FP filling factor satisfies $\nu_{\text{FP}}^{-1} = 4\nu^{-1} - 3$. The factor of four results from the uncorrelated pairs N_{P} having charge $-2e$, and the number of pairs N_{P} being equal to $N/2$. Correlations between FPs are introduced in the standard way, by attaching $2p$ CS flux quanta to each FP to obtain the effective angular momentum $2\ell_{\text{FP}}^*$ for correlated FPs. For $\nu_{\text{FP}} = m^{-1}$, where m is an odd integer, we can obtain the value of ν_{QE} corresponding to the Laughlin correlated state of FPs (pairs of quasielectrons with $\ell_{\text{P}} = 2\ell - 1$). Exactly the same procedure can be applied to QHs in CF LL1 since $V_{\text{QE}}(\mathcal{R})$ and $V_{\text{QH}}(\mathcal{R})$ are qualitatively similar at small values of $\mathcal{R}$. Here we are assuming that $V_{\text{QE}}(\mathcal{R})$ and $V_{\text{QH}}(\mathcal{R})$ are dominated by their short range behavior $\mathcal{R} \leq 5$. The QH pseudopotential is not as well determined for $\mathcal{R} > 5$ because it requires larger N electron systems then we can treat numerically. The electron filling factor is given by $\nu^{-1} = 2 + (1 + \nu_{\text{QE}})^{-1}$ or by $\nu^{-1} = 2 + (2 - \nu_{\text{QH}})^{-1}$. For QHs in CFLL0 $\ell_{\text{P}} = 2\ell - 3$ and the term that prevent violation of the Pauli principle is $-7(N_{\text{P}} - 1)$.

The value of ν_{FP}^{-1}, ν_{QE} for CFLL1 and ν_{QH} for CFLL0 together with the

Table 15.1. Values of $\nu_{\rm FP} = m^{-1}$ the resulting values of $\nu_{\rm QE}$, $\nu_{\rm QH}$ and the electronfilling factors that they generate (only for the observed FQH states).

QEs in CFLL1			QHs in CFLL0		
$\nu_{\rm FP}^{-1}$	5	9	$\nu_{\rm FP}^{-1}$	9	13
$\nu_{\rm QE}$	1/2	1/3	$\nu_{\rm QH}$	1/4	1/5
ν	3/8	4/11	ν	3/10	4/13

resulting values of the electron filling factor ν for novel IQL states observed experimentally by Pan et al.[23] are given in Table 15.9.1.

QHs in CFLL1 with $\nu_{\rm QH} = 2/3$ and $1/2$ produce the same $\nu = 4/11$ and $\nu = 3/8$ states as the QEs if we assume QE-QH symmetry. IQL states at $\nu_{\rm FP}^{-1} = 7$ in CFLL1, and $\nu_{\rm FP}^{-1} = 11$ in CFLL0 could possibly occur, but they have not been observed.

15.9.2. *Numerical studies of spin polarized QP states*

Standard numerical calculations for N_e electrons are not useful for studying such new states as $\nu = 4/11$, because convincing results require large values of N_e. Therefore we take advantage of the knowledge[21,33,36,37] of the dominant features of the pseudopotential $V_{\rm QE}(\mathcal{R})$ of the QE-QE interaction, and diagonalize the (much smaller) interaction Hamiltonian of $N_{\rm QE}$ systems. This procedure was earlier[33] shown to reproduce accurately the low energy N_e-electron spectra at filling factors ν between 1/3 and 2/5.

One might question whether using the pair pseudopotential for QPs obtained by diagonalization of a finite system of N electrons (containing two QEs or two QHs) gives a reasonable accurate description of systems containing more than a few QPs. We have attempted to account for finite size effects[36,38,47,49] by plotting the values of $V_{\rm QP}(\mathcal{R})$ for each value of $\mathcal{R}$ as a function of N^{-1}, where N is the number of electrons in the system that produced the two CFQPs.[50] We extrapolate $V_{\rm QP}(\mathcal{R})$ to the macroscopic limit. In addition, the low energy spectra of an N electron system (obtained by diagonalization of $V_0(\mathcal{R})$) that contains $N_{\rm QP}$ quasiparticles is compared with the spectrum of $N_{\rm QP}$ quasiparticles [obtained by diagonalization of $V_{\rm QP}(\mathcal{R})$]. The results for the system with $(N, 2\ell) = (12, 29)$ and the one obtained after applying a CF transformation, $(N_{\rm QE}, 2\ell_{\rm QE}) = (4, 9)$ are shown in Fig.(15.6).

We realize that using the values of $V_{\rm QE}(L')$ obtained by extrapolation to the macroscopic limit ($N_e \to \infty$) for systems containing $N_{\rm QE} \leq 16$

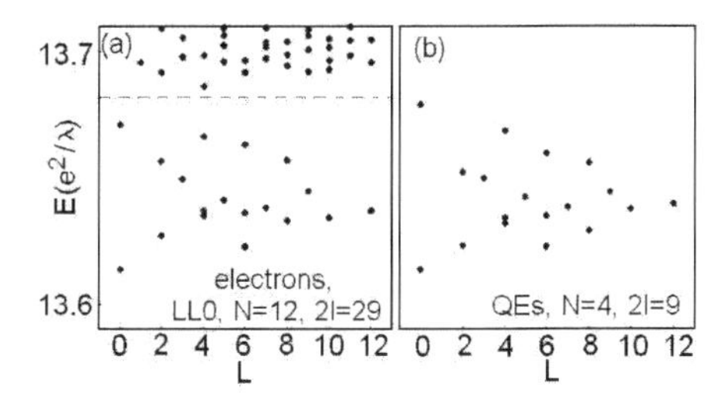

Fig. 15.6. Energy spectra for $N = 12$ electrons in the lowest LL with $2\ell = 29$ and for $N = 4$ QEs in the CF LL1 with $2l = 9$. The energy scales are the same, but the QE spectrum obtained using $V_{\rm QE}(\mathcal{R})$ is determined only up to an arbitrary constant.

QEs is an inconsistency. We believe it introduces only small errors since $N_{\rm QE}$ systems result from a much larger N electron systems. However, this assumption should be checked carefully.

The fact that $(N_{\rm QE}, 2\ell_{\rm QE})$ system has an $L = 0$ ground state at $2\ell_{\rm QE} = 3N_{\rm QE} - 3$ led a number of researchers[51–55] to suggest that it represented a second generation of CFs giving rise to a daughter state and resulting $\nu = 4/11$ spin polarized IQL state observed by Pan et al.[23] This idea can't be correct because $V_{\rm QE}(L')$ is not superharmonic at $\mathcal{R} = 1$ and cannot cause a Laughlin correlated CF daughter state of spin polarized QEs.

The fact that the magnitude of $V_{\rm QE}(\mathcal{R})$ is only about one fifth as large as the energy necessary to create an additional QE-QH pair in a Laughlin correlated state permits diagonalization in the subspace of the partially filled QE Landau level with reasonably accurate results [see e.g, Fig.(15.1) (d) and (e)]. For situations in which the width of the band of two QP states is closer to the energy needed to create a QE-QH pair, higher bands (or higher QP LL) cannot be neglected.

The value of 2ℓ at which the IQL state at filling factor ν occurs in the spherical geometry is given by $2\ell = \nu^{-1}N + \gamma(\nu)$, where N is the number of particles and $\gamma(\nu)$ is a finite size effect shift.[5] For Laughlin correlated electrons in LL0 at filling factor ν equal to the inverse of an odd integer, $\gamma(\nu) = -\nu^{-1}$, so that the $\nu = 1/3$ IQL states occur at $2\ell = 3N - 3$. For quasielectrons of Laughlin $\nu = 1/3$ state, an IQL state occurs at $(N, 2\ell) = (4, 9)$.

As mentioned earlier, we believe that because QEs will not support Laughlin correlations at $\nu = 1/3$. We believe that it is an "aliased" state[56,57] at $2\ell = 2N + 1$ (conjugate to $2\ell = 2N - 3$) that supports pairing correlations. By "aliased" states we mean two states with the same values of N and 2ℓ that belong to different sequences $2\ell = \nu^{-1}N + \gamma(\nu)$. Different values of $\gamma(\nu)$ for IQL states of electrons in LL0 and QEs in CFLL1 suggest that the QE correlations are different from the Laughlin correlations for electrons in LL0. It also gives emphasis to how important is to select a value of N and then to diagonalize the N particle system for many different values of 2ℓ. One cannot assume that $\gamma(\nu)$ is known. For example, when $\nu = 1/3$, we assume $2\ell = 3N - j$, where j is an integer, and we diagonalize for many different values of j. $L = 0$ IQL ground states with a substantial gap separating them from the lowest excited state are found to fall into families with the values of j (or of $\gamma(\nu)$) depending on the kind of correlations. Elaborate calculations for N particles systems only at $2\ell_{\text{QE}} = 3N_{\text{QE}} - 3$ totally miss most of the IQL states.

In Fig.(15.7) a) we present the energy spectrum of a system of ten QEs in a shell of angular momentum $\ell = 17/2$. It is obtained by numerical diagonalization of the QP interaction presented in Fig.(15.5) a). The spectrum contains an $L = 0$ ground state separated from the lowest excited state by a substantial gap. Frame b) shows the probability $P(\mathcal{R})$ that the ground state contains pairs with total pair angular momentum $L' = 2\ell - \mathcal{R} = 1, 3, 5, \cdots$. The solid dots represent the results for the ten QE system; open circles show $P(\mathcal{R})$ for ten Laughlin correlated electrons in LL0 for contrast. The maxima in $P(\mathcal{R})$ at $\mathcal{R} = 1$ and 5 and the minimum at $\mathcal{R} = 3$ for the QE system are in sharp contrast to the Laughlin correlated $P(\mathcal{R})$ of the ten electron system in LL0. The QE maximum at $\mathcal{R} = 1$ and minimum at $\mathcal{R} = 3$ suggests formation of QE pairs with $\ell_{\text{P}} = 2\ell - 1$ and the avoidance of pairs with $\mathcal{R} = 3$, the pair state with the largest repulsion. This IQL ground state occurs at $2\ell = 2N - 3$ and corresponds to $\nu_{\text{QE}} = 1/2$ and $\nu = 3/8$.

The $\nu_{\text{QP}} = 1/2$ state should occur at the conjugate values of 2ℓ given by $2\ell = 2N - 3$ and $2N + 1$. Thus Fig.(15.7) can be thought of as $N_{\text{QP}} = 10$ or $N_{\text{QP}} = 8$, the former corresponding to $2\ell = 2N - 3$ and the latter to $2\ell = 2N + 1$. We have already mentioned that QEs in the CFLL1 are Laughlin QEs of the $\nu = 1/3$ IQL, while QHs in the CFLL1 are QHs of the Jain $\nu = 2/5$ state. It seems reasonably to diagonalize $V_{\text{QP}}(\mathcal{R})$ for QHs when CFLL1 is more than half filled and for QEs when it is less than half filled. If only $V_{\text{QP}}(\mathcal{R})$ for $\mathcal{R} \leq 5$ is important, $V_{\text{QE}}(\mathcal{R})$ and $V_{\text{QH}}(\mathcal{R})$ are qualitatively similar (but not identical). We should then expect the same correlations independent of which $V_{\text{QP}}(\mathcal{R})$ is used in the numerical diagonalization. This would suggest that Fig.(15.7) be interpreted as containing $N_{\text{QH}} = 8$ and $2\ell = 2N_{\text{QH}} + 1 = 17$ instead of as $N_{\text{QE}} = 10$ and $2\ell = 2N - 3 = 17$.

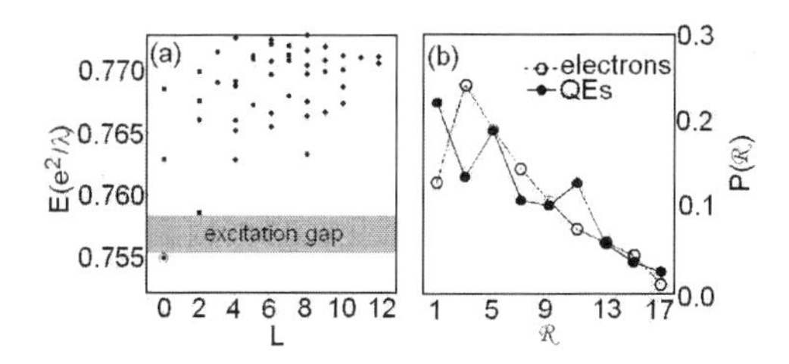

Fig. 15.7. (a) Energy spectra as a function of total angular momentum L of 10 QEs at $2\ell = 2N - 3 = 17$ corresponding to $\nu_{\text{QE}} = 1/2$ and $\nu = 3/8$. It is obtained in exact diagonalization in terms of individual QEs interacting through the pseudopotential shown in Fig. 15.4 (c) (triangles) (b) Coefficient of $P(\mathcal{R})$, the probability associated with pair states of relative angular momentum $\mathcal{R}$, for the lowest $L = 0$ state. The solid dots are for 10 QEs of the $\nu_{\text{QE}} = 1/2$ state in a shell of angular momentum $\ell = 17/2$. The open circles are for 10 electrons in the LL0 at $\ell_0 = 17/2$.[27]

We have evaluated spectra for many values of $(2\ell, N)$. The FQH states with the largest gaps are found to fall into families. The $\nu_{\text{QP}} = 1/2$ state occurs at $2\ell = 2N - 3$ (and its conjugate $2N + 1$). The $\nu_{\text{QE}} = 1/3$ state is found at $2\ell = 3N - 7$.

In our numerical studies the $\nu_{\text{QP}} = 1/2$ state occurs only when the number of QPs is even, suggesting that QP pairs are formed. However, IQL states are formed only when the number of minority QPs in CFLL1 is 8 or 12, but not when it is 10 or 14. This could indicate that the CF pairs form quartets (i.e. pairs of CF pairs) in the IQL state. This is completely speculative since we have very little knowledge of the pseudopotential describing the interaction between CF pairs.

The "shift" describing the $2\ell = 3N - 7$ sequence identified here ($\gamma = 7$) is different not only from $\gamma = 3$ describing a Laughlin state, but also from $\gamma = 5$ that results for a Laughlin correlated state of Fermion pairs (FPs). This precludes the interpretation of these finite-size $\nu_{\text{QE}} = 1/3$ ground states found numerically (and also of the experimentally observed $\nu = 4/11$ FQH state) as a state of Laughlin correlated pairs of QEs (i.e., particles in the partially filled CF LL1). However, it is far more surprising that paired state of QEs turns out as an invalid description for these states as well. Clearly, the correlations between the pairs of QEs at $\nu_{\text{QE}} = 1/3$ must be of a different, non-Laughlin type, and we do not have a simple model to describe this state.

While we do not completely understand the correlations between QEs at $\nu_{\text{QE}} = 1/3$, it may be noteworthy that the value of $\gamma = 7$ appropriate for the series of incompressible states found here can be obtained for the Laughlin state of QE triplets (QE$_3$s), each with the maximum allowed angular momentum, $\ell_{\rm T} = 3\ell - 3$, or of quartets (made up of pairs of pairs) with maximum allowed angular momentum of the quartet $\ell_{\rm Q} = 4\ell - 10$. The quartet state can be thought of as consisting of four filled states ($\ell, \ell - 1, \ell - 4, \ell - 5$) separated by two empty states ($\ell - 2, \ell - 3$). Both of these heuristic pictures give $2\ell = 3N - 7$ for the $\nu = 1/3$ state.

15.10. Summary and Conclusions

In this paper we have reviewed exact numerical diagonalization of small systems within the Hilbert subspace of a single partially occupied LL. The numerical results are thought as "numerical experiment", and simple intuitive models fitting the numerical data are sought, to better understand the underlying correlations. We describe calculations for N electrons confined to a Haldane spherical surface, and present simple results at different values of the LL degeneracy $g = 2\ell + 1$. We demonstrate that Jain's remarkable CF picture predicts not only the values of 2ℓ at which IQL ground states occur for different values of N, but also predicts the angular momenta L of the lowest band of multiplets for any value of 2ℓ in a very simple way. We emphasize that Jain's CF picture is valid, not because of some magical cancelations of Coulomb and Chern-Simons gauge interactions beyond mean-field, but because it introduces Laughlin correlations by avoiding pair states with the lowest allowed relative angular momentum $\mathcal{R} = 2\ell - L'$. The allowed angular momentum multiplets which avoid pair states with $\mathcal{R} = 1$ form a subset of the set of multiplets $G_{N\ell}(L)$ that can be formed from N Fermions in a shell of angular momentum ℓ. This subset avoids the largest repulsion and has the lowest energy. Adiabatic addition of Chern-Simons flux introduces Laughlin correlations without the necessity of introducing an irrelevant mean field energy scale $\hbar\omega_c^* = \nu\hbar\omega_c$. The experiment of Pan et al. showed that neither Jain's CF picture nor Haldane's hierarchy was the whole story.

The energy of a multiplet $|\ell^N; L\alpha>$ formed from N electrons in a shell of angular momentum ℓ is given by Eq. 15.1. We have proved a simple theorem, that led to the conclusion that a pseudopotential of the form $V_H(\hat{L}') = A + B\hat{L}'^2$ (referred to as a " harmonic" pseudopotential) failed to lift the degeneracy of the multiplets α that had the same total angular momentum L. Correlations (removal of this degeneracy) were caused only by the anharmonic part of $V(L')$, i.e. by $\Delta V(L') = V(L') - V_H(L')$. For

$\Delta V(L') = k\delta(\mathcal{R}, 1)$, where $k > 0$, the lowest energy state for each value of L is the multiplet for which $P_{L\alpha}(\mathcal{R} = 1)$is a minimum. Here $P_{L\alpha}(\mathcal{R} = 1)$ is the probability that $|\ell^N; L\alpha>$ has pairs with pair angular momentum $L' = 2\ell - 1$. This is exactly the condition for Laughlin correlations at $\nu = (2\mathcal{R} + 1)^{-1} = 1/3$. If the anharmonic part of $V(L')$ is negative (i.e. $k < 0$), then the lowest energy for each angular momentum L occurs for the multiplet with $P_{L\alpha}(\mathcal{R} = 1)$ equal to a maximum, indicating a tendency to form pairs with $\mathcal{R} = 1$.

An important point to emphasize is the relation between the pseudopotential $V(L')$ describing the interaction between a pair of identical Fermion in a degenerate Landau level of angular momentum ℓ and correlations. For charged Fermions in a strong perpendicular magnetic field $V(L')$ is Coulomb like at small values of L' (i.e. large separation), being proportional to the square of the total charge of Fermion. At large values of L' the Fermions are close enough that the spatial distribution of charge (not just the total charge) can be important. If $V(L')$ has strong maxima within the range of allowed values of L', these maxima will be avoided in the lowest energy multiplets $|L, \alpha>$. For minima, just the opposite is true. This has been checked by numerical diagonalization from which $P(\mathcal{R})$ can easily be evaluated to reveal the nature of the correlations.

Abbreviations

IQL-incompressible quantum liquid
FQH -fractional quantum Hall
LL-Landau level
QP -quasiparticle
QE -quasielectron
QH-quasihole
CF-composite Fermion
MF- mean-field
CS-Chern-Simons

References

1. K. von Klitzing, G. Dorda, and M. Pepper, *Phys. Rev. Lett.* **45**(6), 494–497 (Aug, 1980).
2. D. C. Tsui, H. L. Stormer, and A. C. Gossard, *Phys. Rev. Lett.* **48**(22), 1559–1562 (May, 1982).
3. R. B. Laughlin, *Phys. Rev. Lett.* **50**(18), 1395–1398 (May, 1983).

4. R. B. Laughlin, *Surf. Sci.* **142**(1-3), 163–172 (July, 1984).
5. F. D. M. Haldane, *Phys. Rev. Lett.* **51**(7), 605–608 (Aug, 1983).
6. F. D. M. Haldane and E. H. Rezayi, *Phys. Rev. Lett.* **60**(10), 956–959 (Mar, 1988).
7. B. I. Halperin, *Phys. Rev. Lett.* **52**(18), 1583–1586 (Apr, 1984).
8. B. I. Halperin, *Helv. Phys. Acta.* **56**(1-3), 75–102, (1983).
9. J. K. Jain, *Phys. Rev. Lett.* **63**(2), 199–202 (Jul, 1989).
10. J. K. Jain, *Phys. Rev. B.* **40**(11), 8079–8082 (Oct, 1989).
11. J. K. Jain, *Phys. Rev. B.* **41**(11), 7653–7665 (Apr, 1990).
12. J. K. Jain, *Science.* **266**(5188), 1199–1203 (November, 1994).
13. X. M. Chen and J. J. Quinn, *Solid State Commun.* **92**(11), 865–497, (1994).
14. P. Sitko, K.-S. Yi, and J. J. Quinn, *Phys. Rev. B.* **56**(19), 12417–12421 (Nov, 1997).
15. A. Wójs and J. J. Quinn, Hund's rule for monopole harmonics, or why the composite fermion picture works, *Solid State Commun.* **110**(1), 45–49, (1999).
16. G. Fano, F. Ortolani, and E. Colombo, *Phys. Rev. B.* **34**(4), 2670–2680 (Aug, 1986).
17. G. Moore and N. Read, *Nucl. Phys. B.* **360**(2-3), 362–396 (Apr, 1991).
18. M. Greiter, X.-G. Wen, and F. Wilczek, *Phys. Rev. Lett.* **66**(24), 3205–3208 (Jun, 1991).
19. M. Greiter, X.-G. Wen, and F. Wilczek, *Nucl. Phys. B.* **374**(3), 567–614, (1992).
20. A. Lopez and E. Fradkin, *Phys. Rev. B.* **44**(10), 5246–5262 (Sep, 1991).
21. S.-Y. Lee, V. W. Scarola, and J. K. Jain, *Phys. Rev. Lett.* **87**(25), 256803 (Nov, 2001).
22. S. S. Mandal and J. K. Jain, *Phys. Rev. B.* **66**(15), 155302 (Oct, 2002).
23. W. Pan, H. L. Stormer, D. C. Tsui, L. N. Pfeiffer, K. W. Baldwin, and K. W. West, *Phys. Rev. Lett.* **90**(1), 016801 (Jan, 2003).
24. T. T. Wu and C. N. Yang, *Nucl. Phys. B.* **107**(3), 365–380, (1976).
25. T. T. Wu and C. N. Yang, *Phys. Rev. D.* **16**(4), 1018–1021 (Aug, 1977).
26. J. J. Quinn and A. Wójs, *J. Phys.: Condens. Matter.* **12**, R265–R298, (2000).
27. J. J. Quinn, A. Wójs, K.-S. Yi, and J. J. Quinn. Composite fermions in quantum hall systems. In *The electron liquid paradigm in condensed matter physics*, pp. 469–497. IOS Press, Amsterdam, (2004).
28. F. Wilczek, *Fractional Statistics and Anyon Supercondcutivity.* (World Scientific, Singapore, 1990).
29. J. J. Quinn and J. J. Quinn, *Solid State Commun.* **140**(2), 52–60, (2006).
30. B. I. Halperin, P. A. Lee, and N. Read, *Phys. Rev. B.* **47**(12), 7312–7343 (Mar, 1993).
31. A. Lopez and E. Fradkin, *Phys. Rev. Lett.* **69**(14), 2126–2129 (Oct, 1992).
32. A. Lopez and E. Fradkin, *Phys. Rev. B.* **47**(12), 7080–7094 (Mar, 1993).
33. P. Sitko, S. N. Yi, K. S. Yi, and J. J. Quinn, *Phys. Rev. Lett.* **76**(18), 3396–3399 (Apr, 1996).
34. K. S. Yi, P. Sitko, A. Khurana, and J. J. Quinn, *Phys. Rev. B.* **54**(23), 16432–16435 (Dec, 1996).

35. K. Yi and J. J. Quinn, *Solid State Commun.* **102**(11), 775, (1997).
36. A. Wójs and J. J. Quinn, *Phys. Rev. B.* **61**(4), 2846–2854 (Jan, 2000).
37. S.-Y. Lee, V. W. Scarola, and J. Jain, *Phys. Rev. B.* **66**(8), 085336 (Aug, 2002).
38. A. Wójs, D. Wodziński, and J. J. Quinn,*Phys. Rev. B.* **74**(3), 035315 (Jul, 2006).
39. A. Wójs, G. Simion, and J. J. Quinn, *Phys. Rev. B.* 75(15):155318, (2007).
40. A. Wójs and J. J. Quinn, *Phys. Rev. B.* **75**(8), 085318 (Feb, 2007).
41. M. R. Peterson and S. Das Sarma, (2008). arXiv:0801.4819v1.
42. S. He, F. C. Zhang, X. C. Xie, and S. Das Sarma, *Phys. Rev. B.* **42**(17), 11376–11379 (Dec, 1990).
43. A. Wójs and J. J. Quinn, *Solid State Commun.* **108**(7), 493–497, (1998).
44. A. T. Benjamin, J. J. Quinn, J. J. Quinn, and A. Wójs, *Journal of Combinatorial Theory, Series A.* **95**(2), 390–397 (august, 2001).
45. J. J. Quinn and J. J. Quinn, *Phys. Rev. B.* **68**(15), 153310 (Oct, 2003).
46. G. E. Simion and J. J. J. Quinn, *Physica E.* **41**, 1–5, (2008).
47. A. Wójs, *Phys. Rev. B.* **63**(23), 235322 (May, 2001).
48. K. Park and J. K. Jain, *Phys. Rev. B.* **62**(20), R13274–R13277 (Nov, 2000).
49. I. Szlufarska, A. Wójs, and J. J. Quinn, *Phys. Rev. B.* **64**(16), 165318 (Oct, 2001).
50. X. C. Xie, S. Das Sarma, and S. He, *Phys. Rev. B.* **47**(23), 15942–15945 (Jun, 1993).
51. J. H. Smet, *Nature.* **422**, 391 (march, 2003).
52. M. Goerbig, P. Lederer, and C. M. Smith, *Physica E.* **34**(1–2), (2006).
53. M. O. Goerbig, P. Lederer, and C. M. Smith, *Phys. Rev. B.* **69**(15), 155324, (2004).
54. A. López and E. Fradkin,*Phys. Rev. B.* **69**(15), 155322 (Apr, 2004).
55. C.-C. Chang and J. K. Jain, *Phys. Rev. Lett.* **92**(19), 196806 (May, 2004).
56. R. H. Morf, *Phys. Rev. Lett.* **80**(7), 1505–1508 (Feb, 1998).
57. R. H. Morf, N. d'Ambrumenil, and S. Das Sarma, *Phys. Rev. B.* **66**(7), 075408 (Aug, 2002).

Chapter 16

Kinetic and Calorimetric Behavior of Polymers in the Glass Transition Region

A. Andraca and P. Goldstein
Departamento de Física, Facultad de Ciencias UNAM, 04510, México, D.F.
patricia.goldstein@ciencias.unam.mx

L.F. del Castillo
Instituto de Investigaciones en Materiales, UNAM, 04510, México, D.F.
lfelipe@servidor.unam.mx

We revisit the Gibbs-di Marzio and di Marzio-Dowell models for the calorimetric properties of polymers in the glass transition region. We find an expression similar to the one given by di Marzio-Dowell for the specific heat that fits very well with the experimental data reported in the ATHAS database, and using these results we find the logaritmic shift factor that fits well with the newest experimental data available for the relaxation times.

Contents

16.1. Introduction

The study of the relaxation processes that take place in supercooled liquids in the vicinity of the glass transition temperature T_g has been extensively presented in the literature in the last three decades. An important behavior of a liquid approaching the glass transition is the rapid increase of its viscosity. Theoretically, many efforts have been undertaken to study the temperature dependence of the viscosity and other thermodynamic properties.

One of the most important empirical equations that deals with the behavior of the relaxation time as the system approaches T_g is the Vogel-Fulcher-Tammann (VFT) equation, namely (16.1)-(16.3),

$$\log \tau = A - \frac{B}{T - T_0} \tag{16.1}$$

where A and B are independent parameters and T_0 may be interpreted as the isoentropic temperature, namely, the temperature where the configurational entropy vanishes.

In the case of polymers, Williams, Landel and Ferry proposed another empirical equation that may be written in terms of the VFT equation, the Williams-Landel-Ferry (WLF) Eq. (16.4),

$$\log \tau = \frac{C_1(T - T_s)}{C_2 + T - T_s} \tag{16.2}$$

where T_s is a reference temperature and C_1 and C_2 are constants.

The relaxation phenomena described by the VFT or WLF equations correspond to the very slow α-relaxation processes. One may find, however, that fast relaxation processes occur in the vicinity of T_g, namely the β-relaxation processes.[5,15] Relaxation and diffusion mechanisms present drastic changes around a cross-over temperature T_c which lies within the interval $[1.15T_g, 1.28T_g]$.[16,35] There are two important aspects that characterize this cross-over region. Both the VFT and the WLF empirical equations do no longer describe the experimental results for the viscosity below T_c, and, furthermore, the diffusion mechanisms undergo changes.[36,37]

We shall consider in this work the behavior of glass forming polymers in the region above T_c, thus, the WLF equation holds to describe the relaxation phenomena that take place under this thermal circumstances. For the last two decades we have been working on the phenomenology of the glass transition in fragile and polymeric glasses[36,39]]. Back in 1988, Dr. L. S. García-Colín proposed two of us (P.G. and L.F.C.) the project of studying the phenomenology of the glass transition. The first part of that work was revisiting the paper by Gibbs and di Marzio[40] where they propose a form for the configurational entropy of a glass former polymer. Gibbs and di Marzio supposed in 1958 that the glass transition might be considered as a second order phase transition. By that time the VFT and the

WLF empirical equations had been already predicted. In 1965, Adam and Gibbs[41] studied the temperature dependence of the relaxation processes in glass formers and obtained a relation between the logarithmic shift factor and the configurational entropy of the system, the well known Adam-Gibbs equation,

$$\log a_T = -K\left[\frac{1}{T_s S_c(T_s)} - \frac{1}{T S_c(T)}\right] \tag{16.3}$$

This equation establishes an important link between the kinetic relaxation processes and the calorimetric properties of the glass former.

More than 50 years after it was introduced, the WLF equation is still an important issue as one can find in very recent literature.[42,47] We may find its applications in food chemistry, such as the loss of vitamins in vegetables as temperature is lowered,[48] the advantages of different ways of food storage[49,50]], and the properties of different kinds of sugars and honeys;[51,52] its application in the study of natural materials composed of viscoelastic systems such as wood;[53] and several applications in the pharmaceutical industry.[54]

In this work, we go back to an earlier article we published with Dr. García-Colín. We shall consider the behavior of the glass formers for temperatures above T_c. In the first section we present the results obtained in[39]]. The second section consists on the study of the heat capacity of polymers obtained experimentally and the comparison with the di Marzio-Dowell expression.[55] Finally we evaluate the logarithmic shift factor, compare the values predicted by the WLF equation we obtained earlier and the ones obtained in this work, and show how the latter fit the experimental data reported in 2008.

16.2. Summary of Previous Results

In our paper published in 1993,[39] we arrived to an expression for the logarithmic shift factor for five polymers using the Adam-Gibbs equation and the form for the specific heat evaluated by di Marzio-Dowell.[55] This expression is of the form

$$\Delta C_P = \frac{A}{T^2} + BT - CT^2 \tag{16.4}$$

where the coefficients A, B and C were calculated using the data they reported in their work. Using the expression given in Eq. (16.4), we evaluated the configurational entropy and using the Adam-Gibbs equation, Eq. (16.3), we arrived to the following form for the logarithmic shift factor,

$$\log a_T = -K\left[\frac{1}{F(T_s)(T_s - T_0)} - \frac{1}{G(T)}\frac{1}{T^2}\right] \tag{16.5}$$

with,

$$F(T) = \left(1 + \frac{T_0}{T}\right)\left(\frac{A}{2T_0^2} - \frac{CT^2}{2}\right) + BT \tag{16.6}$$

and,

$$G(T) = \left(1 - \frac{T_0^2}{T^2}\right)\left(\frac{A}{2T_0^2 T} - \frac{CT}{2}\right) + B\left(1 - \frac{T_0}{T}\right) \tag{16.7}$$

where K is a constant that is different for every polymer. We proved that the values for the logarithmic shift factor evaluated with the Williams-Landel-Ferry, Eq. (16.2) and those calculated with our form, Eq. (16.5) agreed very well.

16.3. The Specific Heat. The Athas Database

At present, data for thermodynamic properties of polymers are available and we are able to check the experimental values with those proposed by different theories. One of the most important databases is the ATHAS (Advanced Thermal Analysis System) [56]. We may find in this database different thermodynamic properties of polymers, particularly specific heats for the glass forming region. We took the experimental values for ΔC_p [6] for three polymers, PIB (Polyisobytylene), PS (Polystyrene) and PVAc (Polyvinyl acetate) and fitted them to a form such as the one given by di Marzio-Dowell, Eq. (16.4),

$$C_{pliq} - C_{pcryst} = \Delta C_P = \frac{A'}{T^2} + B'T - C'T^2 \tag{16.8}$$

The fit was quite good as we can see in Figures (16.1), (16.2) and (16.3), but the values of the coefficients A', B' and C', given in Table 1, are different from those given by di Marzio Dowell. It is important to point out that in this work we are using very recent experimental data, and the ones reported by Di Marzio- Dowell are taken from experimental data thirty years old.

Even though the coefficients A, B, C in Eq. (16.4) are quite different from the ones we found in Eq. (16.8), we obtain values for ΔC_P at T_g very similar to the ones reported by Di Marzio-Dowell as it can be seen in Table (16.3).

Table 16.1. Data for the coefficients A', B' and C' for each polymer.

POLYMER	T_g	A' (JK/g)	B' (J/gK^2)	C' (J/gK^3)
PIB	200	3655.976	0.00314	8.284×10^{-6}
PS	373	32695.22	0.00071	1.369×10^{-6}
PVAc	304	40090.74	0.00108	2.849×10^{-6}

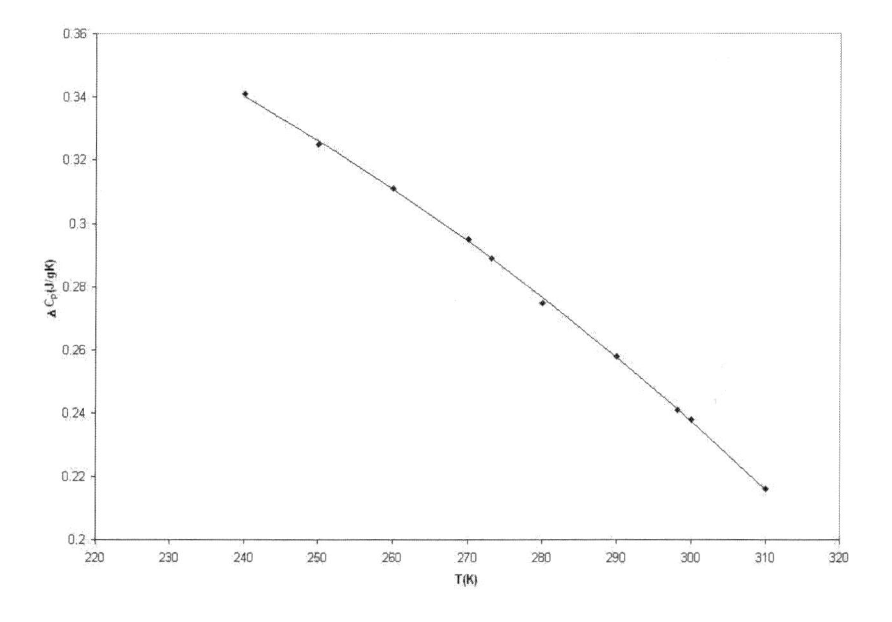

Fig. 16.1. ΔC_p for PIB, T>T_c. The points represent the experimental data (ATHAS) and the solid line Eq. (16.8).

Here, we are before an important question. Though the analytical form for the fit of the experimental data and diMarzio-Dowell model are the same, the coefficients are different.

In di Marzio-Dowell paper, the exact form for ΔC_p is.

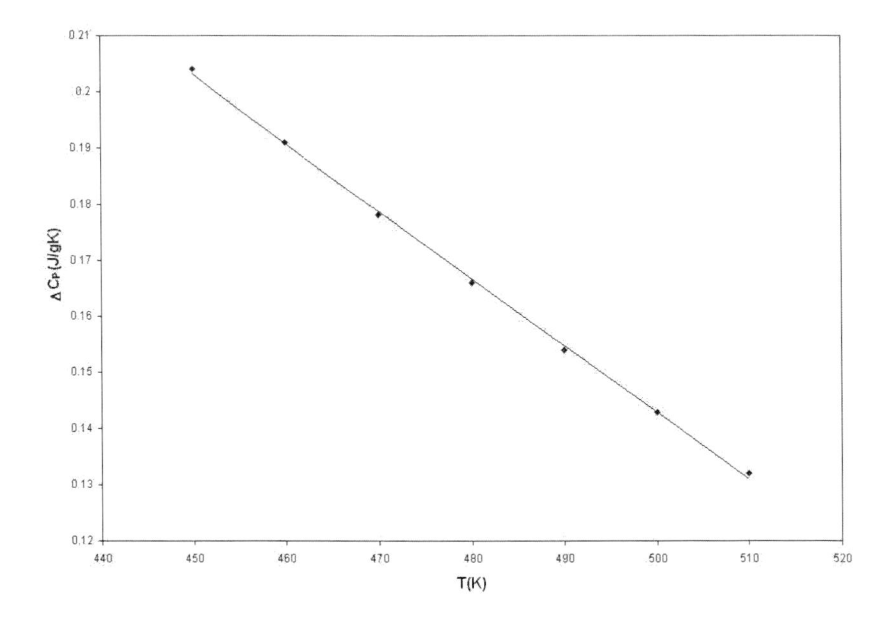

Fig. 16.2. ΔC_p for PS T>T_c. The points represent the experimental data (ATHAS) and the solid line Eq. (16.8).

$$\Delta C_p = R\left(\frac{\Delta\varepsilon}{k_BT}\right)^2 f(1-f) + 4RT\Delta\alpha(1-4.17T\Delta\alpha) + 0.5T\Delta\alpha c_p(T_g-)$$

The first term is related to the change of the specific heat at constant volume at T_g (Δc_v). This is due mainly to the change in the conformational equilibrium (number of states "trans" and "gauche") in the polymeric chain in terms of the temperature. One may evaluate Δc_v,[57–60]

$$\Delta c_v = \frac{R(\Delta\varepsilon/k_BT)^2 2\exp(-\Delta\varepsilon/k_BT)}{1+2\exp(-\Delta\varepsilon/k_BT)^2}$$

where,

$$\Delta\varepsilon = \varepsilon_{gauche} - \varepsilon_{trans}$$

the difference in the conformational energy, so that

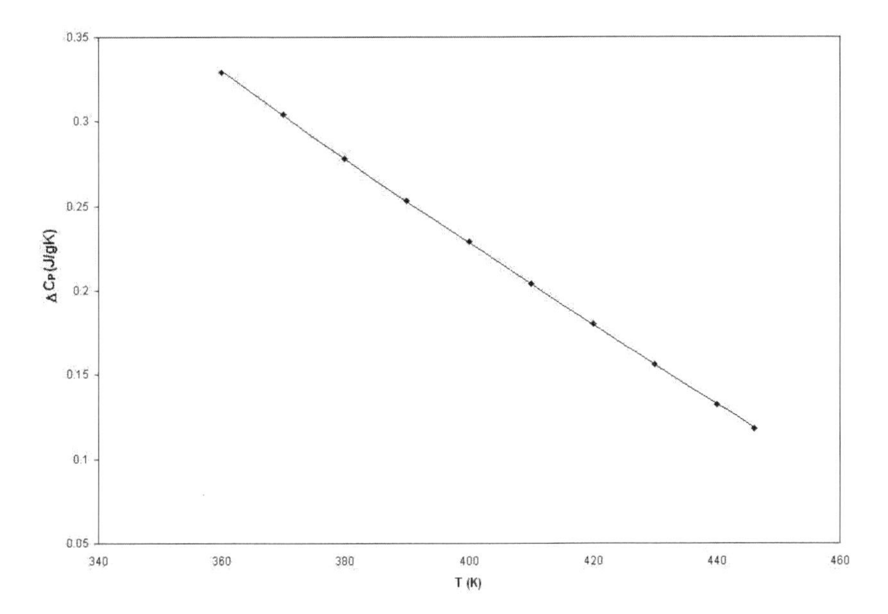

Fig. 16.3. ΔC_p for PVAc T>T_c. The points represent the experimental data (ATHAS) and the solid line Eq. (16.8).

$$f = \frac{2\exp(-\Delta\varepsilon/k_B T)}{1 + 2\exp(-\Delta\varepsilon/k_B T)}$$

and f is the fraction of chain bonds in the gauche state. The second term is related to the number of holes, that is, the free volume, temperature dependent, and α is the thermal expansivity. The last contribution is of a vibrational kind and we have not considered it for our temperatures are high. In Ref. 55, some of the coefficients are considered the same for all polymers and thirty years ago the data they considered may now be measured in a better way. We have checked that some of the properties of the polymers we are working with, such as the number of flexes are different in[55] an in other sources. We are planning to recalculate all these terms using all the values that are here considered using the data from ATHAS, so that we may find why our coefficients are not the same.

In the next section we show that the entropy evaluated using our espression for ΔC_P, fits quite well with the experimental logaritmic shift factor for the polymers we have studied.

Table 16.2. Comparison of ΔC_Pdata evaluated in T_g. First column data of reference.[55] Second one data obtained with Eq. (16.8).

POLYMER	ΔC_P(J/gK)	ΔC_P'(J/gK)
PIB	0.43	0.38
PS	0.31	0.31
PVAc	0.47	0.49

16.4. The Logarithmic Shift Factor

To prove that our form for ΔC_p Eq. (16.8) is correct, we have recalculated the logaritmic shift factor for our polymers, using the same values of K and T_s, Table 2, reported in Ref. 39 and evaluate Eq. (16.5). We then plot the experimental values for $\log a_T$ reported recently[61] we can see that the agreement is rather good. If we compare the WLF equation reported in[39] with the experimental data, there is no way to fit both. We must reproduce the actual experimental data for the logarithmic shift factor, so it is clear that the coefficients of the di Marzio-Dowell expression do not fit with the real $\log a_T$. The good news is that our expression for the specific heat does.

Table 16.3. Values of the isoentropic and reference temperatures, the constant K, and the molecular weight of each polymer.

POLYMER	T_0 (K)	T_s (K)	K	M. W. (g/mole)
PIB	135	243	462.625	56
PS	341	408	66.875	104
PVAc	270	349	159.32	86

For example we present in Figure (16.4 both the plots of Eq. (16.5) and the experimental values reported in Ref. 57 for PIB.

16.5. Discussion

We have revisited the di Marzio-Dowell model for the specific heat in the α relaxation region, that is for temperatures above T_c in order to be sure that the relaxation time may be written in terms of the WLF equation. We have now, sixteen years after our previous work new data both for the relaxation times, the calorimetric data, and thermodynamic properties in order to work on a new comparison. We still are sure that the previous work is completely physically well done. We are now revisiting it to include

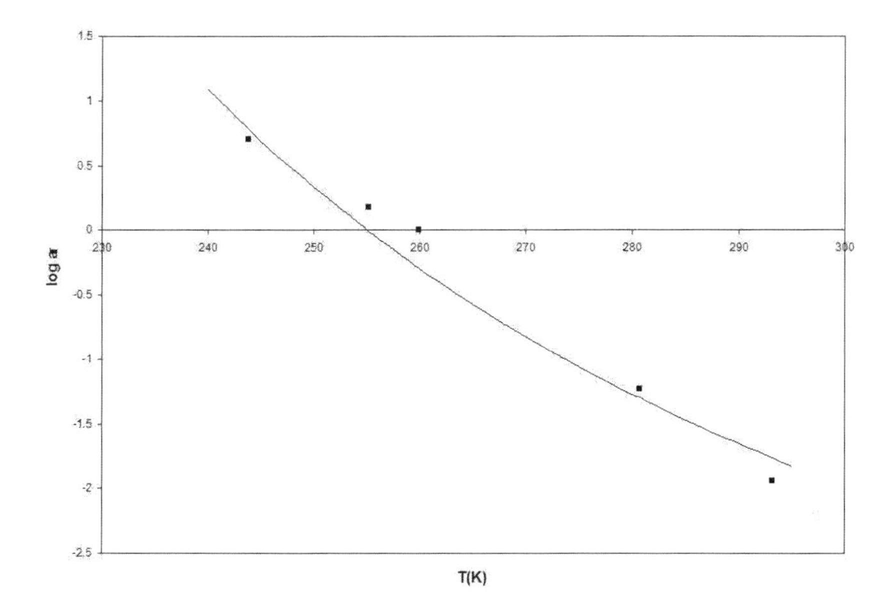

Fig. 16.4. Log a_T for PIB, $T>T_c$. The points represent the experimental data given in Ref. 57 and the solid line the one obtained using the form for ΔC_p given in Eq. (16.8).

the new ideas that have risen in the phenomenology of the glass transition and the newest experimental data. In the near future we must also analyze if the same model for the specific heat may be used for temperatures below T_c, where we have proven that neither VFT nor WLF equation hold anymore .[36,37]

References

1. H. Vogel, Z. Phys. **22** (1921) 645.
2. G. S. Fulcher, J. Am. Ceram. Soc. **8** (1925) 339.
3. G. Tamann, G. Hesse, Z. Anorg. Allg. Chem. **156** (1926) 245.
4. M. L. Williams, R. F. Landel, J. D. Ferry, J. Am. Chem. Soc. **77** (1955) 3701.
5. J. Gerardin, S. Mohanti, U. Mohanti, J. Chem. Phys. **119** (2003) 4473.
6. E. Donth, *The glass transition. Relaxation dynamics in liquids and disordered materials*, Springer Verlag, Berlin Heidelberg, 2001.
7. M. Beiner, J. Korus, E. Donth, Macromolecules **30** (1997) 8420.
8. M. Beiner, S. Kahle, E. Hempel, K. Schöter, E. Donth, Macromolecules **31** (1998) 8973.
9. M. Beiner, S. Kahle, E. Hempel, K. Schröter, E. Donth, Europhysics Lett. **44** (1998) 321.

10. S. Kahle, E. Hempel, M. Beiner, R. Unger, K. Schöter, E, Donth, J. Molec. Struct. **749** (1999) 149.
11. D. Gómez, A. Alegría, A. Arbe, J. Colmenero, Macromolecules **34** (2001) 503.
12. K. L Ngai, Phys. Rev. E **57** (1998) 7346.
13. C. León, K. L. Ngai, J. Chem. Phys. **110** (1999) 11585.
14. C. León, K. Ngai, J. Phys. Chem. B **103** (1999) 4045.
15. L. Andreozzi, M. Faetti, M. Giordano, D. Leporini, J. Phys.: Condensed Matter **11** (1999) A131.
16. C. A. Angell, J. Non-Cryst. Solids **131-133** (1991) 13.
17. F. H. Stillinger, Science **267** (1995) 1935.
18. B. Frick, D. Richter, Science **267** (1995) 1939.
19. R. Böhmer, K. L. Ngai, C. A. Angell, D. J. Plazek, J. Chem. Phys. **99** (1993) 4201.
20. J. Colmenero, A. Alegría, J. M. Alberdi, F. Álvarez, B. Frick, Phys. Rev. B **44** (1991) 7321.
21. M. D. Ediger, C. A. Angell, S. R. Nagel, J. Phys. Chem. **100** (1996) 13200.
22. C. A. Angell, P. H. Poole, J. Shao, Il Nuovo Cimento **16D** (1994) 993.
23. B. Frick, D. Richter, Phys. Rev B **47** (1993) 14795.
24. E. Rössler, U. Warschewski, P. Eiermann, A. P. Sokolov, D. Quitmann, J. Non-Cryst. Solids **172-174** (1994) 113.
25. A. Sokolov, A. Kisliuk, D. Quitmann, A. Kudlik, E. Rössler, J. Non-Cryst. Solids **172-174** (1994) 133.
26. A. Sokolov, E. Rössler, A. Kisliuk,, D. Quitmann, Phys. Rev. Lett. **71** (1993) 262.
27. F. Stickel, E. W. Fischer, R. Richert, J. Chem. Phys. **102** (1995) 6251.
28. F. Stickel, E. W. Fischer, R. Richert, J. Chem. Phys. **104** (1996) 2043.
29. C. Hansen, F. Stickel, T. Berger, R. Richert, E. W. Fischer, J. Chem. Phys. **107** (1997) 1086.
30. C. Hansen, F. Stickel, R. Richert, E. W. Fischer, J. Chem. Phys. **108** (1998) 6408.
31. R. Richert, C. A. Angell, J. Chem. Phys. **108** (1998) 9016.
32. K. Duvvuri, R. Richert, J. Chem. Phys. **117** (2002) 4414.
33. R. Richert, K. Duvvuri, L. T. Duong, J. Chem. Phys. **118** (2003) 1828.
34. L. M. Wang, R. Richert, J. Chem. Phys. **121** (2004) 11170.
35. E. Rössler, Phys. Rev. Lett. **65** (1990) 1595.
36. P. Goldstein, L. S. García-Colín, L. F. del Castillo, Physica A **275** (2000) 325.
37. A. Andraca, P. Goldstein, L. F. Del Castillo, Physica A **387** (2008) 4531.
38. L. S. García-Colín, L. F. del Castillo, P. Goldstein, Phys. Rev. B **40** (1989) 7040.
39. P. Goldstein, L. F. del Castillo, L. S. García-Colín, Macromolecules **26** (1993) 655.
40. J. H. Gibas, E. di Marzio, J. Chem. Phys. **28** (1958) 373.
41. G. Adam, J. H. Gibbs, J. Chem. Phys. **43** (1965) 139.
42. C. A. Angell, Polymer **38** (1997) 6261.

43. L. Dagdug, L. S. García-Colín, Physica A **250** (1998) 133.
44. M. R. Guvich, A. T. Andonian, Journal of Materials Science **35** (2000) 289.
45. K. Hagiwara, T. Ougizawa, T. Inoue, K. Hirata, Y. Kobuyashi, Radiation Physics and Chemistry **58** (2000) 525.
46. S. Sills, R. M. Overney, Phys. Rev. Lett. **91** (2003) 095501-1.
47. C. Y. Liu, J. He, R. Keunings, C. Bailly, Macromolecules **39** (2006) 8867.
48. M. C. Giannakourou, P. D. Taoukis, Food Chemistry **83** (2003) 33.
49. C. Ratti, Journal of Food Engineering **49** (2001) 311.
50. M. E. Yildis, J. L. Hokini, J. Rheol. **45** (2001) 903.
51. A. L. Ollett, R. Parker, Journal of Texture Studies **21** (2007) 355.
52. P. A. Sopade, P. Hally, B. Banhdari, B. D'Arcy, C. Doebler, N. Caffin, Journal of Food Engineering **56** (2003) 67.
53. S. S. Lelley, T. G. Rials, W. G. Glasser, Journal of Materials Science **22** (1987) 617.
54. B. C. Hancock, S. L. Shamblin, G. Zografi, Pharmaceutical Research **12** (1995) 799.
55. E. Di Marzio, F. Dowell, J. Appl. Phys. **50** (1979) 6061.
56. B. Wunderlich, Pure and Appl. Chem. **67** (1995) 1019.
57. J. M. O'Reilly, F. E. Karauz, J. Polym. Sci. **14** (1996) 49.
58. E. Di Marzio, J, H, Gibbs, P. D. Fleming, I. C. Sánchez, Macromolecules **9** (1976) 763.
59. T. M. Birshtein, O. B. Ptitsyn, *Conformations of Macromolecules*(New York, Interscience, 1966).
60. J. M. O'Reilly, J. Appl. Phys. **48** (1977) 10.
61. K. L. Ngai, D. J. Plazek, C, M Roland, Macromolecules **41** (2008) 3925.

PART 4

Statistical Physics, Cosmology and Relativity

Chapter 17

On the Fokker-Planck Equation for the Relativistic Lorentz Gas

Guillermo Chacón-Acosta, Leonardo Dagdug and Hugo A. Morales-Técotl

Depto. de Física, Universidad Autonóma Metropolitana-Iztapalapa, Distrito Federal, 09340, México

gca, dll, hugo@xanum.uam.mx

Incorporating relativity principles with those of kinetic theory is important not only to understand their theoretical foundations but to actually interpret current high energy and astrophysical phenomena. In this work we study the relativistic Lorentz gas: a binary mixture composed of light relativistic particles that diffuse in a background gas made up of heavy non-relativistic particles, so that $m \ll m_G$ and $n \ll n_G$, with suffix $_G$ indicating the non-relativistic component whereas quantities without index indicate the relativistic one; m and n stand for the rest masses and particle densities, respectively. We start by considering the relativistic Boltzmann equation for the relativistic component while component $_G$ is assumed to be in equilibrium. As it is well known the transport coefficients can be investigated by making approximations for the collision term in Boltzmann's equation, in particular by replacing it with a Fokker-Planck differential operator in which diffusion and friction coefficients appear directly. In the case of the relativistic Lorentz gas we consider we get an extra term proportional to the distribution function besides the usual Fokker-Planck contribution. Our analysis can be considered complementary to a previous one in which, where the background gas was relativistic and the diffusing component was the non-relativistic one[1]

Contents

17.1. Introduction

Relativistic kinetic theory was boosted by F. Jüttner's derivation of the equilibrium distribution function for relativistic simple (1911) and quantum (1928) gases through an entropy maximization procedure.[2] The equation to be fulfilled by this distribution was found in 1935 by Walker for the collisionless case,[3] and for the complete relativistic generalization of the Boltzmann equation by Lichnerowicz and Marrot in 1940.[4]

In the sixties several authors including Synge[5] and Israel,[6] adapted Chapmann-Enskog-Hilbert and Grad's methods of the non-relativistic kinetic theory, to the domain of special relativity. The main objective was to calculate the transport coefficients for the relativistic gas. Follow up generalizations were given by Chernikov,[7] Ehlers[8] and Stewart[9] to the general relativistic regime, (for comprehensive relativistic kinetic theory reviews see for instance[10,11]). However, there have been in the literature several different versions of it. For example, there are proposals alternatives to the Jüttner distribution,.[12–15] Furthermore, there are some attempts to establish the foundations of the relativistic irreversible thermodynamics on the basis of a first order relativistic kinetic theory.[16] Indeed, the development of the correct relativistic generalization of kinetic theory has been in constant debate, recently including numerical simulations.[17–19]

The importance of developing a relativistic kinetic theory is not limited to the compatibility of the foundations of relativity and kinetic theories. It actually has applications including cosmology,[20–22] astrophysical processes such as the Sunyaev-Zeldovich effect (which is the distortion of the black body spectrum of the Cosmic Microwave Background (CMB) produced by the Compton scattering of CMB photons by electrons in galaxy clusters[23]), and recently in nuclear physics experiments at the Relativistic Heavy Ion Collider (RHIC) and the Large Hadron Collider (LHC) for describing the quark-gluon plasma.[24] The involved species in this experiments are relativistic and therefore a theoretical basis is required to understand the corresponding transport coefficients of these systems. No doubt it is an important and active research area.[17–19,25–28]

One way to study the kinetic equations is through approximations in the collision term. This is the case of the models of Marle[29] and Anderson-Witting[30] that are the relativistic generalizations of the well known Bhatnagar, Gross and Krook model (BGK),[31] where the collision integral is replaced by the collision frequency. Another possibility is to approximate the collision integral by a differential operator that models the diffusion in the system. In the relativistic case diffusion equations have been proposed to study the evolution of dark matter in the universe,[32] as well as diffusion in RHIC and LHC,[33] and relativistic corrections to the Sunyaev-Zeldovich effect.[34] So far there have been several works which proposed different versions of the relativistic Brownian motion,[35] however, even the physical interpretation of these processes is yet unclear and still there seems to be no general agreement.[36] Amusingly relativistic Brownian motion has been used in quantum gravity proposals, where a minimum length scale, possibly Planck's length, plays key role. This has led to phenomenological approaches that seek to reveal the effects of such minimum length, replacing the paths of free particles by diffusive processes in spacetime.[37] In some others proposals a stochastic equation has been used to model the effects of the micro-structure of spacetime by assuming a fluctuating metric.[38]

This paper is devoted to the study of the diffusion process driven by collisions rather than by stochastic forces, i.e., through an approximation of the collision term of the relativistic Boltzmann equation. This approximation is possible thanks to the characteristics of the system: a binary mixture of relativistic light particles diffusing in a background gas made up of non-relativistic heavy particles. In the non-relativistic case this system was proposed by Lorentz to describe electrons in a metal, so it is known as *Lorentz gas*.[39,40] Also it was used, e.g. to model a weakly ionized plasma or the thermalization of particles in a heat bath.[41–43] The relativistic counterpart has been studied in[10,44,45] and[46] for the case of a relativistic electron beam in a plasma. The complementary case of a relativistic gas of light particles in which non-relativistic heavy particles diffuse, has been previously studied in.[1] It corresponds to a relativistic Brownian motion (or the Rayleigh gas) and will be reviewed in Section 17.3 after briefly describing the relativistic Boltzmann equation in Section 17.2. In Section 17.4 we show the details of how to approximate the collision integral of the relativistic Boltzmann equation, into a differential Fokker-Planck-type equation for light relativistic diffusing particles in a non-relativistic heavy gas. Finally, in section 17.5 we discuss our results and touch upon some open problems.

17.2. Relativistic Boltzmann Equation

For a simple gas composed of particles of the same species that interact through binary elastic collisions, we can write the relativistic Boltzmann equation, just like in the non relativistic case. One calculates the difference in the number of particles that leave a differential volume element with respect to those that enter in it and which changes due to the collisions of the particles. An essential feature of the relativistic case is that one has to take into account the way in which each of the quantities involved behave under Lorentz transformations.

For a gas with particles of mass m, four-momentum $p^\mu = (p^0, \mathbf{p})$, and spacetime coordinates $x^\mu = (ct, \mathbf{x})$, the one-particle distribution function $f(x^\mu, p^\mu) = f(\mathbf{x}, \mathbf{p}, t)$ characterizes the state of the system. It gives the number of particles $d\mathcal{N} = f(\mathbf{x}, \mathbf{p}, t) d^3x d^3p$ at time t in the volume element d^3x located at $\mathbf{x}$ having momentum $\mathbf{p}$ in the volume element d^3p in momentum space.

The relativistic distribution function f is Lorentz invariant.[47,48] This can be seen because the number of particles in the volume element remain unchanged under change of reference frame. Also one can show,[49] using arguments of simultaneity and the mass shell constrain $p^\mu p_\mu = m^2c^2 = (p^0)^2 - |\mathbf{p}|^2$, that the volume element in phase space is invariant $d^3xd^3p = d^3\tilde{x}d^3\tilde{p}$, where tilde refer to quantities in a boosted reference frame.

If two particles, with momentum p^μ and p_*^μ respectively, only interact through elastic collisions, we have: $p^\mu + p_*^\mu = p'^\mu + p_*'^\mu$, where primed quantities indicate they are evaluated after the collision.

The corresponding Boltzmann equation that should satisfy the relativistic distribution function is the following:

$$p^\mu \frac{\partial f}{\partial x^\mu} + m \frac{\partial f \mathcal{K}^\mu}{\partial p^\mu} = \int (f' f'_* - f f_*) F \sigma \, d\Omega \frac{d^3 p_*}{p_*^0}, \tag{17.1}$$

here we use $f'_* = f(\mathbf{x}, \mathbf{p}'_*, t)$, $\mathcal{K}^\mu$ denotes an external four-force, σ is the differential cross section of the scattering process and $d\Omega$ is the element of solid angle, while $F = \sqrt{(p^\mu p_{*\mu})^2 - m^2 m_*^2 c^4}$ is the so-called invariant flux. In (17.1) we use the invariant measure on the mass shell $\frac{d^3p}{p^0}$. Written in this way, the relativistic Boltzmann equation is manifestly covariant. An equivalent form that is often found in the literature, which however is not manifestly covariant, is

$$\frac{\partial f}{\partial t} + v^i \frac{\partial f}{\partial x^i} + \frac{\partial f \mathrm{F}^i}{\partial p^i} = \int (f' f'_* - f f_*) g_\Omega \sigma d\Omega d^3 p_*, \tag{17.2}$$

where F^i are the spatial components of the external force, and g_Ω the Moller velocity,[10,11] that is related to the invariant flux by $g_\Omega = \frac{cF}{p^0 p_*^0}$. This velocity is reduced to the usual relative velocity in the non relativistic case. This latter form of the Boltzmann equation is the one that we use throughout the work.

17.3. Relativistic Brownian Motion as an Approximation to the Relativistic Boltzmann Equation

The characteristics of the system studied in this work are reversed in the case of Brownian motion, namely, $m_G \ll m$. Then, here we review the relativistic Brownian motion starting from the kinetic theory point of view[a], instead of dealing with relativistic stochastic processes as in.[35,36] This was done recently, in a previous work, by Chacón-Acosta and Kremer.[1] They found a relativistic generalization for the Fokker-Planck equation starting from the relativistic Boltzmann equation (17.2). There, the diffusion of Brownian heavy non-relativistic particles in a gas of light relativistic particles was considered. The authors of[1] adapted to the relativistic case the method by Wang Chang and Ulhenbeck[50] and Green[51] for the non relativistic Brownian motion.

Here we summarize the key results of.[1] Using similar methods to those shown in the next section, it was possible to write a Fokker-Planck operator for the collision integral:

$$\frac{\partial f_B}{\partial t} + v^i \frac{\partial f_B}{\partial x^i} + F^i \frac{\partial f_B}{\partial p^i} = \eta \left(m_B kT \frac{\partial^2 f_B}{\partial p^i \partial p_i} + \frac{\partial (f_B p^i)}{\partial p^i} \right), \qquad (17.3)$$

where f_B is the distribution function for the Brownian particles and η is the viscous-friction coefficient that was expressed as integrals of microscopic quantities.[1] The viscous-friction coefficient η for a constant cross section σ is

$$\eta = \frac{32\pi n_G \sigma kT}{3 m_B c K_2(z_G)} \frac{e^{-z_G}}{z_G^2} (3 + 3z_G + z_G^2), \qquad (17.4)$$

where K_2 is the modified Bessel function of second kind, and $z_G = m_G c^2/kT$ is the ratio between the rest energy and temperature, and the index $_G$ stands for the gas particles. For the ultra relativistic limit corresponding to $z_G \ll 1$ the viscous-friction coefficient is

$$\eta \cong \frac{16\pi n_G \sigma kT}{m_B c} \left(1 + \frac{z_G^2}{12} + \dots \right). \qquad (17.5)$$

[a]This system is also known as the Rayleigh gas in the non relativistic case.

For the non-relativistic limit for which $z_G \gg 1$, the value of η becomes

$$\eta \cong \frac{32 n_G \sigma}{3 m_B} \sqrt{2 m_G \pi k T} \left(1 + \frac{9}{8 z_G} + \ldots\right). \tag{17.6}$$

The leading term of (17.6) has the well known form[50,51] for the non relativistic Brownian motion.

In the next section we address the complementary case, namely relativistic light particles diffusing in a heavy non relativistic gas: the Lorentz gas.

17.4. Fokker-Planck Equation for the Relativistic Lorentz Gas

Let us consider a relativistic Lorentz gas, i.e., a binary mixture where one of its constituents is taken as a background gas of heavy non relativistic particles with rest mass m_G, and $v_G \ll c$. The other particles species are considered relativistic and having mass and particle number density much smaller than those of the gas, i.e. $n \ll n_G$ and $m \ll m_G$.

In general the evolution of the system is governed by a system of coupled Boltzmann equations, one for each component. Nevertheless, due to the characteristics of the mixture, we can make a well known assumption:[1,50,51] The equation for the distribution function of gas particles is not affected by the other component. Indeed we suppose that the gas is in equilibrium. In the equation for the diffusing particles, there are contributions only due to the interactions between particles of different species, therefore the relativistic Boltzmann equation (17.2) can be written as

$$\frac{\partial f}{\partial t} + v^i \frac{\partial f}{\partial x^i} + \frac{\partial (f \mathrm{F}^i)}{\partial p^i} = \int (f_G^{(0)\prime} f' - f_G^{(0)} f) g_\Omega \sigma d\Omega d^3 p_G. \tag{17.7}$$

Here we shall suppose that the distribution function f does not deviate far from equilibrium and hence

$$f = f^{(0)} h(p^i), \tag{17.8}$$

where $h(p^i)$ is a deviation function that tells us how the system deviates from equilibrium and $f^{(0)} \propto e^{-\frac{U_\alpha p^\alpha}{kT}}$ is the relativistic distribution function for the equilibrium state that is known as the Jüttner distribution.[2]

In the right hand side of (3.4) we can factorize $f^{(0)}$, such that the collision term is $f^{(0)} \mathcal{I}$, where $\mathcal{I}$ is the following integral

$$\mathcal{I} = \int (h' - h) f_G^{(0)} g_\Omega \sigma d\Omega d^3 p_G, \tag{17.9}$$

and $h' = h(p')$. Expanding the post-collision deviation in Taylor series around the pre-collision momentum up to second order, the difference in (17.9) takes the form

$$h' - h \approx \Delta p^i \frac{\partial h}{\partial p^i} + \frac{1}{2} \Delta p^i \Delta p^j \frac{\partial^2 h}{\partial p^i \partial p^j}. \tag{17.10}$$

To deal with $\mathcal{I}$ it is necessary to introduce the relative velocity g^i in terms of the momenta of the components of the mixture. The difference between the pre and post collision relative velocities is given by

$$\frac{\Delta \mathrm{g}^i}{c} = \Delta \left(\frac{p^i_{_G}}{p^0_{_G}} \right) - \Delta \left(\frac{p^i}{p^0} \right), \tag{17.11}$$

where Δ indicates the difference of pre- and post-collision quantities. Because heavy gas particles are non-relativistic $p^{0\,\prime}_{_G} \approx p^0_{_G}$, and because of the conservation of energy, (17.11) becomes

$$\Delta \mathrm{g}^i = -c \left(\frac{1}{p^0_{_G}} + \frac{1}{p^0} \right) \Delta p^i. \tag{17.12}$$

As in,[1] we will translate the smallness condition on the ratio between the masses of the constituents into a condition on the ratio of their energies. When one component is relativistic and the other is not, then $p^0_{rel}/p^0_{norel} \ll 1$. In this case the relativistic component is the diffusing one, and the non relativistic particles are the constituents of the gas

$$\frac{p^0}{p^0_{_G}} \ll 1. \tag{17.13}$$

With (17.13) we can approximate (17.12) as follows

$$\Delta \mathrm{g}^i \cong -c \frac{\Delta p^i}{p^0}. \tag{17.14}$$

Notice that this relationship involves only the zero component of the momentum of the particles that diffuse as opposed to the relativistic Brownian motion discussed in the previous section for which instead of p^0 we have $p^0_{_G}$. Next one has to incorporate the above approximations to determine the integral $\mathcal{I}$. This is given in detail in the appendix. From the results of the appendix $\mathcal{I}$ can be written as a differential operator:

$$\mathcal{I} = -C^i \frac{\partial h}{\partial p^i} + D^{ij} \frac{\partial^2 h}{\partial p^i \partial p^j} \tag{17.15}$$

where

$$C^i \equiv 2\pi c \frac{|\mathbf{p}|}{p^0} n_{_G} a_1 \left[1 + 2 \left(\frac{p^0}{|\mathbf{p}|} \right)^2 \frac{kT}{m_{_G} c^2} \right] p^i, \tag{17.16}$$

and

$$D^{ij} \equiv \frac{\pi n_{_G}}{2} c \frac{|\mathbf{p}|}{p^0} \Bigg\{ \left[p^i p^j \left(4a_1 - 3a_2 \right) - a_2 \eta^{ij} |\mathbf{p}|^2 \right] \\ + \frac{1}{2} \left(\frac{p^0}{|\mathbf{p}|} \right)^2 \frac{kT}{m_{_G} c^2} \left[5p^i p^j \left(4a_1 - 3a_2 \right) - \eta^{ij} |\mathbf{p}|^2 \left(8a_1 + 5a_2 \right) \right] \Bigg\}. \tag{17.17}$$

In (17.16) and (17.17) we introduced the following integrals

$$a_1 \left(|\mathbf{p}| \right) = \int \sigma \left(|\mathbf{p}|, \chi \right) \left(1 - \cos \chi \right) \sin \chi d\chi, \tag{17.18}$$

$$a_2 \left(|\mathbf{p}| \right) = \int \sigma \left(|\mathbf{p}|, \chi \right) \sin^3 \chi d\chi. \tag{17.19}$$

To evaluate these integrals it is necessary to specify the differential cross section $\sigma \left(|\mathbf{p}|, \chi \right)$.

Our goal is to approximate the collision integral of the Boltzmann equation (3.4) by a differential operator. Now the right hand side of Boltzmann equation, Eq. (3.4), is given by $f^{(0)}\mathcal{I}$. Then, after a lengthy but otherwise elementary calculation that consists of reshuffling of the functions h, f^0 to express $f^{(0)}\mathcal{I}$ in terms of f and its derivatives one gets

$$\frac{\partial f}{\partial t} + \frac{\partial}{\partial p^i} (F^i f) = D^{ij} \frac{\partial^2 f}{\partial p^i \partial p^j} - A^i \frac{\partial f}{\partial p^i} - \nu f, \tag{17.20}$$

where we have defined

$$A^i \equiv C^i + \frac{2c}{kTp^0} D^{ij} p_j, \tag{17.21}$$

$$\nu \equiv \frac{c}{kTp^0} \left\{ D^{ij} \left[\eta_{ij} + \frac{p_i p_j}{(p^0)^2} \left(1 - \frac{cp^0}{kT} \right) \right] - C^i p_i \right\}. \tag{17.22}$$

Making use of (17.16) and (17.17) we have that A^i and ν can be written as:

$$A^i = \widetilde{\eta}\, p^i \left[1 - 2\left(1 - \frac{a_2}{a_1}\right)\frac{cp^0}{kT}\left(\frac{|\mathbf{p}|}{p^0}\right)^2 - 2\left(\frac{p^0}{|\mathbf{p}|}\right)^2 \frac{kT}{m_{_G} c^2}\right.$$
$$\left. - 7\left(\frac{p^0}{p^0_{_G}}\right)\left(1 - \frac{5}{14}\frac{a_2}{a_1}\right)\right], \tag{17.23}$$

$$\nu = \widetilde{\eta}\,\frac{cp^0}{kT}\left(\frac{|\mathbf{p}|}{p^0}\right)^2 \left\{\left(1 - \frac{a_2}{2a_1}\right)\left(1 - \frac{cp^0}{kT}\right)\left(\frac{|\mathbf{p}|}{p^0}\right)^2\right.$$
$$\left. + \frac{7}{2}\frac{kT}{m_{_G} c^2}\left(1 - \frac{5}{14}\frac{a_1}{a_2} - \left(\frac{p^0}{|\mathbf{p}|}\right)^2\right) - \frac{7}{2}\frac{p^0}{p^0_{_G}}\right\}, \tag{17.24}$$

where we can factorize the viscous-friction coefficient $\widetilde{\eta}$ as a function of the diffusing particle's momentum:

$$\widetilde{\eta}\,(|\mathbf{p}|) \equiv 2\pi n_{_G}\, c\frac{|\mathbf{p}|}{p^0}\, a_1\,(|\mathbf{p}|)\,. \tag{17.25}$$

The equation (17.20) is a generalized Fokker-Planck-type equation, where A^i as the dynamic friction vector, but with an extra term proportional to f whose coefficient can be associated to a characteristic frequency ν, similarly to the BGK model.[31]

From (17.17), (17.23) and (17.24) it is evident that they are functions of $|\mathbf{p}|$. To determine them explicitly we need to specify the differential cross section. We will use the hard sphere model regarding σ as a constant[b]. With this approximation the integration in χ can be done in equations (17.18)-(17.19) yielding

$$a_1 = 2\sigma \quad \text{and} \quad a_2 = \frac{4}{3}\sigma. \tag{17.26}$$

[b]For a physical justification of this approximation see for example the cross-section of the scattering of photons in a non-relativistic electron gas.[10]

Therefore the corresponding coefficients for this case are

$$D^{ij} = \widetilde{\eta}_{HS}\left[\left(p^i p^j - \frac{|\mathbf{p}|^2}{3}\eta^{ij}\right) + \frac{5}{4}\left(\frac{p^0}{|\mathbf{p}|}\right)^2 \frac{kT}{m_G c^2} \times\left(p^i p^j - \frac{58}{15}|\mathbf{p}|^2\eta^{ij}\right)\right], \tag{17.27}$$

$$A^i = \widetilde{\eta}_{HS}\, p^i\left[1 - \frac{2}{3}\frac{cp^0}{kT}\left(\frac{|\mathbf{p}|}{p^0}\right)^2 - \left(\frac{p^0}{|\mathbf{p}|}\right)^2\frac{kT}{m_G c^2} - \frac{16}{3}\frac{p^0}{p^0_G}\right], \tag{17.28}$$

$$\nu = \widetilde{\eta}_{HS}\frac{c|\mathbf{p}|^2}{kTp^0}\left[\frac{2}{3}\left(1 - \frac{cp^0}{kT}\right)\left(\frac{|\mathbf{p}|}{p^0}\right)^2 + \frac{7}{2}\frac{kT}{m_G c^2} \times\left(\frac{16}{3} - \left(\frac{p^0}{|\mathbf{p}|}\right)^2\right) - \frac{7}{2}\frac{p^0}{p^0_G}\right], \tag{17.29}$$

where the viscous-friction coefficient for hard-spheres is

$$\widetilde{\eta}_{HS} = 4\pi n_G c\frac{|\mathbf{p}|}{p^0}\,\sigma. \tag{17.30}$$

Notice that the coefficients (17.27)-(17.29) have several contributions. For instance, the diffusion tensor has two terms, one of those multiplied by $(\beta^2 z_G)^{-1}$. The friction vector and the characteristic frequency also have a term like this, but for these coefficients there is another contribution of order $p^0/p^0_G \ll 1$ that we can neglect due to the Lorentz gas approximation.

From (17.30) we see that the viscous-friction coefficient $\widetilde{\eta}$ does not depend on the temperature, unlike the previous case of relativistic Brownian motion (17.6), where $\eta \propto kT$. In this case, $\widetilde{\eta}$ depends on the particle's velocity of the diffusing component $|\mathbf{p}|/p^0 = v/c$. Physically this difference can be considered reasonable. Indeed, in the case of Brownian, non relativistic particles that diffuse in a relativistic gas, the speed of the Brownian particles is approximately zero, on average, and then it does not contribute to the relevant coefficients which turns out to be proportional to the temperature. In the case of the Lorentz gas the situation is reversed; the relevant coefficients are expected to be proportional to the speed of the diffusing particles that are fully relativistic as opposed to the gas component. Such behavior makes the difference between the two complementary situations. As we can see from equation (17.30) in the ultra relativistic limit the viscous friction coefficient is constant and proportional to the speed of light. In the opposite non-relativistic regime this coefficient is just proportional to the speed.[41]

Now we briefly describe the non-relativistic and ultra relativistic limiting cases of D^{ij}, A^i and ν corresponding to Eqs. (17.27), (17.28) and (17.29) respectively. Let us notice each coefficient include a couple of terms containing, a factor of $(kT/m_G c^2)(p^0/|p|)^2$ and $c|p|^2/kTp^0$. In both limiting cases the first factor is negligible. In the non relativistic case this is due to the fact that the factor takes the form $(kT/mv^2)(m/m_G) \sim m/m_G \ll 1$, where $kT/mv^2 \sim kT/\langle mv^2 \rangle \sim 1$. On the other hand, in the ultra-relativistic $p^0/|\mathbf{p}| \sim 1$ and, since the background gas is non-relativistic, $kT/m_G c^2 \ll 1$, this makes the first factor negligible in this limit too. The second factor in both limiting cases takes the form of the ratio between the corresponding energy and thermal energy, kT. At this point in the non-relativistic case we have

$$D^{ij}_{NR} = \widetilde{\eta}_{NR}\left(p^i p^j - \frac{|\mathbf{p}|^2}{3}\eta^{ij}\right), \tag{17.31}$$

$$A^i_{NR} = \widetilde{\eta}_{NR} p^i \left(1 - \frac{2}{3}\frac{mv^2}{kT}\right), \tag{17.32}$$

$$\nu_{NR} = -\frac{2}{3}\widetilde{\eta}_{NR}\left(\frac{mv^2}{kT}\right)^2, \tag{17.33}$$

where $\widetilde{\eta}_{NR} = 4\pi n_G v\,\sigma$. In the ultra relativistic case they become

$$D^{ij}_{UR} = \widetilde{\eta}_{UR}\left(p^i p^j - \frac{|\mathbf{p}|^2}{3}\eta^{ij}\right), \tag{17.34}$$

$$A^i_{UR} = \widetilde{\eta}_{UR} p^i \left(1 - \frac{2}{3}\frac{cp^0}{kT}\right), \tag{17.35}$$

$$\nu_{UR} = \frac{2}{3}\widetilde{\eta}_{UR}\frac{cp^0}{kT}\left(1 - \frac{cp^0}{kT}\right), \tag{17.36}$$

where $\widetilde{\eta}_{UR} = 4\pi n_G c\,\sigma$. Interestingly, in both limiting cases, the diffusion tensor and the dynamic friction vector share the same form with the only change that in the non-relativistic limit the energy is mv^2, while in the relativistic one it takes the form cp^0. On the other hand, the characteristic frequency ν, behaves quadratically in the ratio between the corresponding kinetic and thermal energies, for both limiting cases Eqs. (17.33) and (17.36).

17.5. Discussion

The need to interpret several current high energy and astrophysical phenomena leads naturally to investigate further the incorporation of the kinetic theory principles with those of relativity. In particular, in dealing with the relativistic Boltzmann equation it can be useful to approximate the so called collision term by a Fokker-Planck operator to study the behavior of the corresponding coefficients accompanying the distribution function and its derivatives.

In this work we have considered a Lorentz gas which is formed by two components one of which is made of light relativistic particles and diluted with respect to the other which in turn is made of heavier particles and described by the non-relativistic Maxwell's distribution. This system has been considered previously in the literature (See for instance[10,44]) as well as the complementary case in which the component in equilibrium is the relativistic one whereas the diluted component is not relativistic.[1] In[44] an analysis was advanced to identify a Fokker-Planck equation for the relativistic Lorentz gas in which the distribution is a function in the 4-momentum space. This allows to split additively the dependence in energy and spatial momentum; each part then obeys a corresponding Fokker-Planck equation. The part depending on energy can be identified in the non-relativistic limit with that obtained by Chapmann and Cowling.[40]

The starting point for our analysis is the relativistic Boltzmann equation in which the distribution is a function of spatial momentum, namely the mass shell condition has been taken into account. We consider the non-relativistic component as described by the Maxwell's distribution whereas the relativistic component is expressed as a small deviation of the relativistic Jüttner's distribution. This allows us to expand the collision term as a series in the transferred spatial momentum which can be reexpressed in terms of a ratio of energies of the light particles and the heavy ones. This is what is traded from the non-relativistic case in which the relevant expression was a ratio of the masses of the particles of the two species. Hereby we arrive to a Fokker-Planck equation (17.20) containing a non-derivative term of the distribution function. It is remarkable that for the present case of the Lorentz gas the very conditions defining it lead to the Fokker-Planck description. The resulting coefficients (17.17), (17.23) and (17.24) in front of the distribution function and its derivatives turn out to be functions of spatial momentum in accord with the literature in the non-relativistic[40,41] as well as in the relativistic case.[44] To proceed further we used a hard-sphere cross section thus obtaining rather simple coefficients in which two contributions can be distinguished easily, each multiplied by the factors $(kT/m_G c^2)(p^0/|p|)^2$ and $c|p|^2/kTp^0$, respectively, and that are qualitatively similar to those of[44] for the energy and spatial momentum terms.

We noticed that the first kind of factor, namely $(z_G \beta^2)^{-1}$, is the same one in the energy contribution as in,[44] and may be neglected in the ultra and non-relativistic limits. The second factor can not be neglected and turns out to be, in both cases, the ratio between a corresponding kinetic energy and thermal energy kT. However these quantities should not be expected to coincide with those in[44] because of the different nature of the assumed approximations in both treatments. Amusingly in our work as well as in[44] a non-derivative term of the distribution function is included. Our characteristic frequency ν in front of such term, (17.29), has a quadratic behavior in the ratio between the corresponding kinetic and thermal energies, for both ultra and non-relativistic limiting cases. For the diffusion tensor D^{ij} and the dynamical friction vector A^i we found that these quantities have the same form, in both, ultra and non-relativistic cases, changing only the corresponding energy as can be seen in (17.31)-(17.32) and (17.34)-(17.35).

It is left open to investigate in detail the behavior of the above coefficients (17.17), (17.23) and in particular, ν, Eq. (17.24). Such kind of terms have been used in literature to model driving forces,[52] creation or annihilation rates, growth rates,[53] chemical reaction rates,[43] and production-like processes. It also can describe delay or memory effects.[55] To best of our knowledge it is not known to date what is the role of such terms in relativistic systems.

Moreover it would be clearly useful to study a manifestly covariant version of the above analysis. It is often rewarding to do so not only to better understand important features of the relativistic system but to clarify open issues associated to the use of particular frames (See for instance[56]).

Acknowledgments

This work was partiality supported by Mexico's National Council of Science and Technology (CONACyT) through: grant 51132F and a PhD-grant 131138 (GCA).

Appendix

In this appendix we provide the details of the calculation of $\mathcal{I}$, Eq. (17.9). We start by substituting (17.14) in (17.9) and using (17.10). Then we integrate with respect to the angle ϵ of the solid angle $d\Omega = \sin\chi d\chi d\epsilon$. For this purpose the following expressions are used

$$\int_0^{2\pi} \Delta \mathrm{g}^i d\epsilon = -2\pi\,(1-\cos\chi)\,\mathrm{g}^i, \tag{17.37}$$

$$\int_0^{2\pi} \Delta \mathrm{g}^i \Delta \mathrm{g}^j d\epsilon = 2\pi \left\{ \left[(1-\cos\chi)^2 - \frac{1}{2}\sin^2\chi \right] \mathrm{g}^i \mathrm{g}^j - \frac{|\mathbf{g}|^2}{2}\eta^{ij}\sin^2\chi \right\}, \tag{17.38}$$

where $\eta^{ij} = \mathrm{diag}\{-1,-1,-1\}$ stands for the spatial part of the Minkowski metric. Then, we can write the integral as

$$\mathcal{I} = 2\pi \int f_G^{(0)}(g_\Omega \sigma)\sin\chi d\chi d^3p_G \left[\frac{p^0}{c}\frac{\partial h}{\partial p^i}\mathrm{g}^i(1-\cos\chi) \right.$$
$$\left. + \frac{(p^0)^2}{4c^2}\frac{\partial^2 h}{\partial p^i \partial p^j}\left\{(1-4\cos\chi+3\cos^2\chi)\mathrm{g}^i\mathrm{g}^j - |\mathbf{g}|^2\eta^{ij}\sin^2\chi\right\} \right]. \tag{17.39}$$

Using spherical coordinates $d^3p_G = \sin\phi d\phi d\theta |\mathbf{p}_G|^2 d|\mathbf{p}_G|$ and integrating over the azimuthal angle ϑ with

$$c\int_0^{2\pi} \left(\frac{p_G^i}{p_G^0} - \frac{p^i}{p^0} \right) d\vartheta = 2\pi c \left(\frac{|\mathbf{p}_G|}{|\mathbf{p}|}\frac{p^0}{p_G^0}\cos\phi - 1 \right)\frac{p^i}{p^0}, \tag{17.40}$$

$$c^2 \int_0^{2\pi} \left(\frac{p_G^i}{p_G^0} - \frac{p^i}{p^0} \right)\left(\frac{p_G^j}{p_G^0} - \frac{p^j}{p^0} \right) d\vartheta = 2\pi c^2 \left\{ \left[\left(\frac{|\mathbf{p}_G|}{|\mathbf{p}|}\frac{p^0}{p_G^0}\cos\phi - 1 \right)^2 \right.\right.$$
$$\left.\left. - \frac{1}{2}\frac{|\mathbf{p}_G|^2}{|\mathbf{p}|^2}\frac{(p^0)^2}{(p_G^0)^2}\sin^2\phi \right] \frac{p^i p^j}{(p^0)^2} - \frac{1}{2}\frac{|\mathbf{p}_G|^2}{(p_G^0)^2}\eta^{ij}\sin^2\phi \right\}, \tag{17.41}$$

the integral $\mathcal{I}$ becomes

$$\mathcal{I} = 4\pi^2 \int f_G^{(0)}(g_\Omega\sigma)|\mathbf{p}_G|^2 d|\mathbf{p}_G| \sin\chi d\chi \sin\phi d\phi \left\{ (1-\cos\chi)p^0 \right.$$
$$\times \frac{\partial h}{\partial p^i}\frac{p^i}{p^0}\left(\frac{|\mathbf{p}_G|}{|\mathbf{p}|}\frac{p^0}{p_G^0}\cos\phi - 1 \right) + \frac{(p^0)^2}{4}\frac{\partial^2 h}{\partial p^i \partial p^j}\left[(4(1-\cos\chi) - 3\sin^2\chi) \right.$$
$$\times \left\{ \frac{p^i p^j}{(p^0)^2}\left(\frac{|\mathbf{p}_G|^2}{|\mathbf{p}|^2}\frac{(p^0)^2}{(p_G^0)^2}\cos^2\phi - 2\frac{|\mathbf{p}_G|}{|\mathbf{p}|}\frac{p^0}{p_G^0}\cos\phi + 1 - \frac{1}{2}\frac{|\mathbf{p}_G|^2}{|\mathbf{p}|^2}\frac{(p^0)^2}{(p_G^0)^2}\sin^2\phi \right) \right.$$
$$\left.\left.\left. - \frac{1}{2}\eta^{ij}\frac{|\mathbf{p}_G|^2}{(p_G^0)^2}\sin^2\phi \right\} - \frac{|\mathbf{g}|^2}{c^2}\eta^{ij}\sin^2\chi \right] \right\}. \tag{17.42}$$

To proceed further we need to approximate the Moller velocity. Due to $p^0/p^0_{G} \ll 1$ and because the gas is non relativistic, the approximation to the invariant flow $F = p^0 p^0_{G} g_{\Omega} c^{-1}$ follows

$$F \cong p^0_{G}|\mathbf{p}| \left(1 - \frac{p^0}{p^0_{G}}\frac{|\mathbf{p}_{G}|}{|\mathbf{p}|}\cos\phi + \frac{1}{2}\frac{(p^0)^2}{(p^0_{G})^2}\frac{|\mathbf{p}_{G}|^2}{|\mathbf{p}|^2}\cos^2\phi\right). \tag{17.43}$$

Because the cross-section σ is a function of F, given (17.43), we can approximate it as $\sigma(F,\chi) \simeq \sigma(|\mathbf{p}|,\chi)$. At this order it does not depend on the $|\mathbf{p}_{G}|$. Thus the product $g_{\Omega}\sigma$, with the use of (17.43), gives[c]

$$g_{\Omega}\sigma(F,\chi) \cong c\frac{|\mathbf{p}|}{p^0}\sigma\left(|\mathbf{p}|,\chi\right)\left(1 - \frac{p^0}{p^0_{G}}\frac{|\mathbf{p}_{G}|}{|\mathbf{p}|}\cos\phi + \frac{1}{2}\frac{(p^0)^2}{(p^0_{G})^2}\frac{|\mathbf{p}_{G}|^2}{|\mathbf{p}|^2}\cos^2\phi\right). \tag{17.44}$$

We can substitute (5.10) in (17.42) together with the fact that the last term of (17.42) is

$$\frac{|\mathbf{g}|^2}{c^2} = \frac{|\mathbf{p}|^2}{p^{0 2}}\left(12\frac{|\mathbf{p}_{G}|}{|\mathbf{p}|}\frac{(p^0)}{(p^0_{G})}\cos\phi + \frac{|\mathbf{p}_{G}|^2}{|\mathbf{p}|^2}\frac{(p^0)^2}{(p^0_{G})^2}\right). \tag{17.45}$$

As the gas particles are non relativistic, we can consider the quantity $|\mathbf{p}_{G}|/p^0_{G}$ to be a small correction and then, neglect all the terms of order higher than 2. Because the partial derivatives of h do not depend neither on ϕ, nor on $|\mathbf{p}_{G}|$, we can directly integrate (17.42) over ϕ and $|\mathbf{p}_{G}|$. Substitution of the Maxwell distribution function $f^{(0)}_{G}$ and with the use of $p^0_{G} \approx m_{G}c$, we obtain

$$\frac{n_{G}}{(2\pi kT m_{G})^{\frac{3}{2}}}\int_0^\infty e^{-\frac{|\mathbf{p}_G|^2}{2kTm_G}}|\mathbf{p}_{G}|^2 d|\mathbf{p}_{G}| = \frac{n_{G}}{4\pi}, \tag{17.46}$$

$$\frac{n_{G}}{(2\pi kT m_{G})^{\frac{3}{2}}}\int_0^\infty e^{-\frac{|\mathbf{p}_G|^2}{2kTm_G}}|\mathbf{p}_{G}|^4 d|\mathbf{p}_{G}| = \frac{3n_{G}}{4\pi}m_{G}kT. \tag{17.47}$$

So we are now in position to give an explicit form for $\mathcal{I}$. This is Eq. (17.15).

References

1. G. Chacón-Acosta and G. M. Kremer, Phys. Rev. **E 76**, 021201 (2007).
2. F. Jüttner, Ann. Physik und Chemie **34**, 856-882 (1911); F. Jüttner, Zeitschr. Phys. **47**, 542-566 (1928).
3. A. G. Walker, Proc. Edinburgh Math. Soc. **2**, 238 (1934-36).
4. A. Lichnerowicz and R. Marrot, Compt. Rend. Acad. Sci. **210**, 759 París (1940). R. Marrot, J. Math Pures et Appl. **25**, 93 (1946).

[c]In[10,45] for the study of the spatial diffusion in a relativistic Lorentz gas, Kox uses the leading term in (17.43) and (5.10) to approximate the invariant flux.

5. J. L. Synge, *The Relativistic Gas.* North-Holland Publishing Company, Amsterdam (1957).
6. W. Israel, *Relativistic Kinetic Theory of a simple gas.* J. Math. Phys. **4**, 1163-1181 (1963).
7. N. A. Chernikov, *The Relativistic Gas in the Gravitational Field.* Acta Phys. Pol. **23** 629-645 (1963).
8. J. Ehlers, in *Proceedings of the International School of Physics, Enrico Fermi,* Ed. R. K. Sach, Academic Press, pp. 1-70 (1971).
9. J. M. Stewart, *Non-equilibrium relativistic kinetic theory.* Lecture Notes in Physics vol.10 Springer, Heidelberg (1971).
10. S. R. de Groot, W. A. van Leewen and Ch. G. van Weert, *Relativistic Kinetic Theory.* North-Holland, Amsterdam (1980).
11. C. Cercignani and G. M. Kremer, *The Relativistic Boltzmann Equation: Theory and Applications.* Progress in Mathematical Physics Vol. 22 Birkhäuser Verlag, Basel Botson Berlin (2002).
12. L. P. Horwitz, S. Shashoua and W. Schieve, Phys. A **161**, 300-338 (1989).
13. J. Dunkel and P. Hänggi, Phys. Rev. E **71**, 016124 (2005).
14. E. Lehmann, J. Math. Phys. **47**, 023303 (2006).
15. J. Dunkel, P. Talkner, P. Hänggi, New J. Phys. **9**, 144 (2007).
16. A. Sandoval-Villalbazo, A. L. García-Perciante and L. S. García-Colín, AIP Conf. Proc. **1122**, 388 (2009).
17. D. Cubero, J. Casado-Pascual, J. Dunkel, P. Talkner and P. Hänggi, Phys. Rev. Lett. **99**, 170601 (2007).
18. A. Montakhab, M. Ghodrat, and M. Barati, Phys. Rev. E **79**, 031124 (2009).
19. F. Peano, M. Marti, L. O. Silva and G. Coppa, Phys. Rev. E **79**, 025701(R) (2009).
20. S. Chandrasekar, *An Introduction to the Study of Stellar Structure.* Dover Publications Inc. (1958).
21. E. W. Kolb, M. S. Turner, *The Early Universe.* Paperback Ed. Westview Press (1994).
22. J. Bernstein, *Kinetic Theory in the Expanding Universe.* Cambridge University Press, Paperback Ed. (2004).
23. R. A. Sunyaev, Y. B. Zeldovich, Comm. Astrophys. and Space Phys. **4**, 173-178 (1972).
24. H. van Hees, V. Greco, and R. Rapp, Phys. Rev. C **73**, 034913 (2006); P. Huovinen and D. Molnar, Phys. Rev. C **79**, 014906 (2009).
25. G. L. Sewell, J. Phys. A: Math. Theor. **41** 382003 (2008).
26. A. L. García-Perciante, A. Sandoval-Villalbazo and L. S. García-Colín, Physica A **387**, 5073-5079 (2008).
27. A. Bret, L. Gremillet, D. Bénisti and E. Lefebvre, Phys. Rev. Lett. **100** 205008 (2008).
28. B. Cleuren, K. Willaert, A. Engel and C Van den Broeck, Phys. Rev. E **77** 022103 (2008).
29. C. Marle, C. R. Acad. Sc. Paris **260**, 6539 (1965).
30. J. L. Anderson and H. R. Witting, Physica **74**, 489 (1974).
31. P. Bhatnagar, E. Gross and M. Krook, Phys. Rev. **94**, 511 (1954).

32. E. Bertschinger, Albert Einstein Century International Conference. AIP Conf. Proc. **861**, 97-105 (2006).
33. R. Kuiper and G. Wolschin, Ann. Phys. (Leipzig) **16**, No. 1, 67 77 (2007).
34. N. Itoh, Y. Kohyama, S. Nozawa, Astrophys. J. **502**, 7-15 (1998).
35. J. Dunkel and P. Hänggi, *Relativistic Brownian Motion*, Phys. Rep. (2009)
36. F. Debbasch, Albert Einstein Century International Conference. AIP Conf. Proc. **861**, 488-493 (2006).
37. F. Dowker, J. Henson and R. D. Sorkin, Mod. Phys. Lett. **A19** 1829-1840, (2004); L. Philpott, F. Dowker and R. D. Sorkin, Phys. Rev. D **79**, 124047 (2009).
38. B. L. Hu and E. Verdaguer, Liv. Rev. Rel. **11**, 3 (2008). (http://www.livingreviews.org/lrr-2008-3).
39. H. A. Lorentz, Proc. Amst. Acad. **7**, 438, 585, 684 (1905).
40. S. Chapman and T. G. Cowling, *The Mathematical Theory of Non-Uniform Gases*. 3th ed. Cambridege University Press, Cambridge (1970).
41. K. Andersen and K. E. Shuler, J. Chem. Phys. **40**, 633 (1964).
42. H. Oser, K. E. Shuler and G. H. Weiss, J. Chem. Phys. **41**, 2661 (1964).
43. K. D. Knierim, M. Waldman and E. A. Mason, J. Chem Phys. **77**, 943 (1982).
44. D. C. Kelly, Phys. Fluids **12**, 799 (1969).
45. A. J. Kox, Phys. Lett **A 53**, 173 (1975); A. J. Kox, Physica **A 84**, 603-882 (1976).
46. D. Mosher, Phys. Fluids **18**, 846 (1979).
47. N. G. van Kampen, Physica **43** 244 (1969).
48. F. Debbasch, J. P. Rivet and W. A. van Leeuwen, Physica A **301** 181 (2001).
49. C. W. Misner, K. S. Thorne and J. A. Wheeler, *Gravitation*. W. H. Freeman and Company New York (1973).
50. C. S. Wang Chang and G. E. Uhlenbeck, in *Studies in Statistical Mechanics Vol. IV*, ed. J. de Boer and G.E. Uhlenbeck, 76-100 (1970).
51. M. S. Green, J. Chem. Phys. **19**, 1036-1046 (1951).
52. C. F. Lo, Europhys. Lett. **39**, 263-267 (1997).
53. V. M. Kenkre and M. N. Kuperman, Phys. Rev. **E 67**, 051921 (2003).
54. A. S. Cukrowski, A. L. Kawczński, J. Popielawskki, W. Stiller and R. Schmidt, Chem. Phys. **159**, 39 (1992).
55. K. Zabrocki, S. Tatur, S. Trimper, R. Mahnke, Phys. Lett. **A 359**, 349356 (2006); M. Schulz, S. Trimper and K. Zabrocki, J. Phys. A: Math. Theor. **40** 3369-3378 (2007).
56. G. Chacón-Acosta, L. Dagdug and H. A. Morales-Técotl, `arXiv:0910.1625` (2009).

Chapter 18

An Antecedence Principle Analysis of the Linearized Relativistic Navier-Stokes Equations

A. Sandoval-Villalbazo

Depto. de Fisica y Matematicas, Universidad Iberoamericana, Prolongacion Paseo de la Reforma 880, Mexico D. F. 01210, Mexico

A.L. Garcia-Perciante

Depto. de Matematicas Aplicadas y Sistemas, Universidad Autonoma Metropolitana-Cuajimalpa, Artificios 40 Mexico D.F 01120, Mexico

It is shown by means of a simple analysis that the linearized system of transport equations for a relativistic, simple fluid at rest obeys the *antecedence principle*, which is often referred to as causality principle. This task is accomplished by examining the roots of the dispersion relation for such a system. This result is important for recent experiments performed in relativistic heavy ion colliers, since it suggests that the Israel-Stewart like formalisms may be unnecessary in order to describe relativistic fluids.

Contents

18.1. Introduction

In 1940 C. Eckart published three papers entitled *The thermodynamics of irreversible processes*,[1] the third one addressing the problem of a relativistic simple fluid. In that paper, Eckart proceeded following the basic ideas of classical irreversible thermodynamics,[2] except for the fact that he introduced relativistic terms in the energy-momentum tensor. As part of his phenomenological approach, he proposed constitutive equations with relativistic corrections. Since Eckart's theory apparently leads to results that violate causality and involves undesirable unstable modes,[3] it has been patched up in several ways using formalisms introduced by Israel and coworkers[456] and sometimes using extended irreversible thermodynamics[7].[8] Recently, it has been shown that the unphysical behavior of the unstable modes is due to the coupling between heat an acceleration proposed by Eckart.[9] Indeed, it has been shown that such a relation is not sustained by kinetic theory.[10]

The so-called *causality problem* of heat conduction, which should be more precisely stated as *antecedence problem*, remains still a controversial issue which suggests the need of extended theories. However, as is shown in the following sections, relativistic classical linear irreversible thermodynamics, as obtained from relativistic kinetic theory features both stability for the equilibrium state[10] and satisfaction of the antecedence principle.

To accomplish this task we divide the rest of the paper as follows. In Sec. 18.2 we recall the Navier-Stokes equations for a simple relativistic fluid[11] and introduce the appropriate constitutive equation for the heat flux. The linearized set of transport equations is thoroughly analyzed in Sec. 18.3. Conclusions and final remarks are included in Sec. 18.4.

18.2. Transport Equations for the Relativistic Simple Fluid

The starting point are the balance equations for a relativistic fluid which are obtained from the conservation of the particle density flow

$$N^{\nu} = nu^{\nu} \tag{18.1}$$

and the energy-momentum tensor which, following Eckart[1] reads

$$T^{\mu}_{\nu} = \frac{n\varepsilon}{c^2} u^{\mu} u_{\nu} + p h^{\mu}_{\nu} + \pi^{\mu}_{\nu} + \frac{1}{c^2} q^{\mu} u_{\nu} + \frac{1}{c^2} u^{\mu} q_{\nu} \tag{18.2}$$

In Eqs. (18.1) and (18.2), n is the particle number density, u^{ν} the hydrodynamic velocity four vector, ε the internal energy per particle, c the speed of light, p the hydrostatic pressure and $h^{\mu}_{\nu} = \delta^{\mu}_{\nu} + u^{\mu} u_{\nu}/c^2$ the spatial projector. The dissipative fluxes are the Navier tensor π^{μ}_{ν} and the heat flux q^{ν} . The conservation equations $N^{\nu}_{;\nu} = 0$ and $T^{\mu}_{\nu;\mu} = 0$ for the quantities defined above yield the Navier-Stokes equations for the relativistic simple fluid namely,

$$\dot{n} + n\theta = 0 \tag{18.3}$$

$$\left(\frac{n\varepsilon}{c^2}+\frac{p}{c^2}\right)\dot{u}_\nu+\left(\frac{n\dot{\varepsilon}}{c^2}+\frac{p}{c^2}\theta\right)u_\nu+p_{,\mu}h^\mu_\nu+\pi^\mu_{\nu;\mu}$$
$$+\frac{1}{c^2}\left(q^\mu_{;\mu}u_\nu+q^\mu u_{\nu;\mu}+\theta q_\nu+u^\mu q_{\nu;\mu}\right)=0 \qquad (18.4)$$

$$nC_n\dot{T}+\left(\frac{T\beta}{\kappa_T}\right)\theta+u^\nu_{;\mu}\pi^\mu_\nu+q^\mu_{;\mu}+\frac{1}{c^2}\dot{u}^\nu q_\nu=0 \qquad (18.5)$$

where κ_T is the isothermal compressibility, β the thermal expansion coefficient and C_n the heat capacity at constant particle density. In order to close the system of equations, constitutive relations for the heat flux and Navier tensor must be introduced. The equation for the heat flux has been recently established by means relativistic kinetic theory and reads

$$q^\ell=-L_T\frac{T^{,\ell}}{T}+L_n\frac{n^{,\ell}}{n} \qquad (18.6)$$

where L_T and L_n are transport coefficients.[10] A detailed discussion on this equations can be found elsewhere.[10] The equations for π^μ_ν in Eq. (18.2) are well-known namely,

$$(\pi)^s=-2\eta\left(\nabla\vec{u}\right)^s \qquad (18.7)$$

$$tr\left(\pi\right)=-\xi\nabla\cdot\vec{u} \qquad (18.8)$$

where $(\pi)^s$ is the symmetric and traceless part of π^μ_ν and $tr\,(\pi)$ its trace. The transport coefficients in Eqs. (18.7) and (18.8) are the shear and bulk viscosities respectively.

18.3. Linearized Relativistic Hydrodynamics

In order to linearize the set of equations (18.3-18.5) we consider $n=n_0+\delta n$, $T=T_0+\delta T$ and $u^\nu=\delta u^\nu$ where naught subscripts denote equilibrium quantities and the δ prefix indicates small perturbations around it. With this hypothesis, the linearized transport equations for a simple, relativistic fluid in the absence of external fields are

$$\delta\dot{n}+n_0\delta\theta=0 \qquad (18.9)$$

$$\frac{1}{c^2}\left(n_0\varepsilon_0+p_0\right)\delta\dot{u}_\nu+\frac{1}{n\kappa_T}\delta n_{,\nu}+\frac{\beta}{\kappa_T}\delta T_{,\nu}$$
$$-\zeta\delta\theta_{,\nu}-2\eta\left(\delta\tau^\mu_\nu\right)_{;\mu}-\frac{L_T}{c^2}\delta\dot{T}_{,\nu}-\frac{L_n}{c^2}\delta\dot{n}_{,\nu}=0 \qquad (18.10)$$

$$nC_n\delta\dot{T}+\left(\frac{T_0\beta}{\kappa_T}\right)\delta\theta-\left(L_T\delta T^{,k}+L_n\delta n^{,k}\right)_{;k}=0 \tag{18.11}$$

where we have defined $\theta = u^{\nu}_{;\nu}$. It is crucial at this point to make the following observation. The so-called causality violation of the transport equations to first order in the gradients, given by linear irreversible thermodynamics, can be easily spotted by observing that, considering $u^{\ell}=0$ and linearizing, Eq. (18.5) leads to a parabolic equation for T. This clearly admits arbitrary propagation speeds for the corresponding signals. However, the hypothesis of a fluid at rest or the fact that calculations can be performed in the comoving frame should not be translated into a vanishing hydrodynamic velocity, but in $u^{\ell}_0=0$ as considered above. That is, δu^{ν} should not vanish even for the fluid at rest or in the comoving frame; *only the mean or equilibrium velocity can be zero.* This fact has already been pointed out in the analysis of the linearized relativistic Euler regime.[12]

In order to analyze the dynamics given the system of equations (18.9-18.11), we start by calculating the divergence of Eq. (18.10). The transverse mode is then uncoupled from the system and a set of three scalar differential equations for δn, $\delta\theta$ and δT is obtained. A Fourier-Laplace transform is then performed, leading to a system of algebraic equations depending on the time and space variables, s and q respectively, whose associated determinant reads

$$\begin{vmatrix} s & n_0 & 0 \\ -\frac{1}{n_0\kappa_T}q^2+\frac{L_n}{c^2}sq^2 & \tilde{\rho}_0 s+Aq^2 & \frac{L_T}{c^2}q^2 s-\frac{\beta}{\kappa_T}q^2 \\ \frac{L_n}{n_0c_n}q^2 & \frac{T_0\beta}{n_0c_n\kappa_T} & s+\frac{L_TT}{n_0c_n}q^2 \end{vmatrix}=0 \tag{18.12}$$

where for convenience we have introduced the following notation[11]

$$\tilde{\rho}_0=\frac{1}{c^2}\left(n_0\varepsilon_0+p_0\right) \tag{18.13}$$

$$A=\zeta+4\eta/3 \tag{18.14}$$

The dispersion relation is thus given by

$$s^3+d_2s^2q^2+s\left(d_3q^4+d_4q^2\right)+d_5q^4=0 \tag{18.15}$$

where the coefficients d_2 to d_5 have been specified in an earlier work.[11] The physical interpretation of the three roots of Eq. (18.15) is well known. The dynamics of the perturbations in the fluid are characterized by a strictly dissipative component which decays in time depending on the value of the real root while the other wave-like component propagates at a speed given by the imaginary parts of the conjugate roots, damped by a coefficient which depends on a Stokes-Kirchhoff like factor. Moreover, a plot of the dynamic structure factor as a function of s for a fixed $\vec{q}$, will feature three peaks. In this work, we are interested in the location of the symmetric Brillouin peaks,[13] which are given by the imaginary part of the conjugate roots, that is $\omega = \pm\sqrt{d_4}q$. Thus, in this case

$$\omega = \pm\sqrt{\frac{\gamma}{\kappa_T \tilde{\rho}_0}}q \tag{18.16}$$

such that, the distance between the peaks, i. e. the speed of propagation of the wave-like component of the fluctuations, and the origin is bounded by

$$c\sqrt{\frac{\gamma}{\kappa_T\left(n_0\varepsilon_0 + p_0\right)}} \tag{18.17}$$

Notice that, in the non-relativistic case, the fluctuations propagate at the speed of sound, i.e.

$$c_s^2 = \frac{\gamma}{\kappa_T \rho_0} \tag{18.18}$$

As an example, for an ideal gas $\gamma = 5/3$ and $\kappa_T = 1/p$, such that

$$c_s^2 = \frac{5}{3}\frac{kT}{m} \tag{18.19}$$

which is clearly unbounded and can be increasingly large for high temperatures. However, the speed of propagation in the relativistic calculation reads

$$c_R^2 = \frac{\gamma}{\kappa_T\left(n_0\varepsilon_0 + p_0\right)}c^2 \tag{18.20}$$

The internal energy for an ideal relativistic gas can be easily calculated through kinetic theory with a Juttner distribution function. This calculation can been found elsewhere[14] and yields

$$\varepsilon_0 = mc^2\left[3z + \frac{\mathcal{K}_1\left(\frac{1}{z}\right)}{\mathcal{K}_2\left(\frac{1}{z}\right)}\right] \tag{18.21}$$

where $z = \frac{kT}{mc^2}$ is the relativistic parameter and $\mathcal{K}_n\left(\frac{1}{z}\right)$ are the modified Bessel function of the second kind. Using this expression we now obtain

$$c_R^2 = \left\{\frac{5}{3}\frac{z}{\left[3z + \frac{\mathcal{K}_1\left(\frac{1}{z}\right)}{\mathcal{K}_2\left(\frac{1}{z}\right)}\right] + 1}\right\}c^2 \tag{18.22}$$

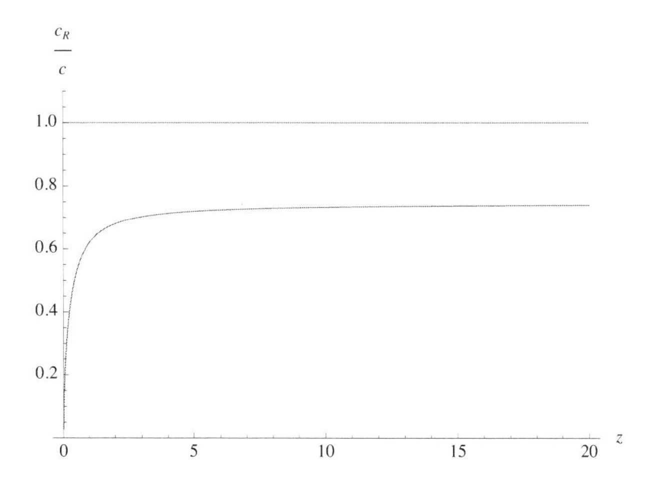

Fig. 18.1. The ratio of the speed of propagation to the speed of light, as a function of the relativistic parameter z.

As can be seen in Fig. 18.1, the expression in curly brackets never exceeds the unity which finally shows that the propagation speed for signals in the transport equations for the relativistic fluid does not violate the antecedence principle.

18.4. Final Remarks

The transport equations derived from Eckart's modified theory show, in its linear version, no problems regarding stability and causality features. This fact implies that there is no real motivation to introduce the so-called second order theories, which introduce non-fundamental adjustable parameters. A simple first order formalism is desirable to describe the fluids which are formed in RHIC type experiments.[15]

In the non-linear case the relativistic Navier-Stokes equations here employed are, by far, more complex. In this context, very little can be said regarding the problems of stability and causality with the techniques included in this paper. It is desirable to perform further work in this direction.

References

1. C. Eckart, The thermodynamics of irreversible processes I: the simple fluid. Phys. Rev. **58**, 267 (1940); II: fluid mixtures; ibid **58**, 919 (1940). III: The relativistic theory of the simple fluid; ibid **58**, 919 (1940).

2. S.R. de Groot, Thermodynamics of irreversible processes (North Holland publishing Co. Amsterdam 1952) Chap. XI is a very valuable reference for these purposes (1970).
3. W.A. Hiscock and L. Lindblom; Generic instabilities in first order dissipative relativistic fluids, Phys. Rev. D **31** 725 (1985).
4. W. Israel, Nonstationary Irreversible Thermodynamics: a Causal Relativistic Theory, Ann. Phys (N.Y.) **100** 310 (1976) and references therein.
5. W. Israel, Thermodynamics of relativistic systems, Physica **106A** 209 (1981).
6. W. Israel and J.M. Stewart, progress in relativistic thermodynamics and electrodynamics of continuous media in *General relativity and gravitation* Vol 2; A. Held, editor (Plenum press, N.Y. 1980).
7. W. Zimdahl, Bulk Viscous Cosmology; Phys. Rev. D **53**, 5483 (1996).
8. D. Pavon, D. Jou and J. Casas-Vazquez, Ann. Inst. Henri Poincare **A36,** 79 (1982).
9. A. L. Garcia-Perciante, L. S. Garcia-Colin and A. Sandoval-Villalbazo, Gen. Rel. and Grav. **41**, 1654 (2009).
10. A. Sandoval-Villalbazo, A. L. Garcia-Perciante and L. S. Garcia-Colin, Physica A **388**, 3765 (2009).
11. A. L. Garcia-Perciante, L. S. Garcia-Colin and A. Sandoval-Villalbazo, Phys. Rev. E **79,** 066310 (2009).
12. A. Sandoval-Villalbazo and D. Brun, arXiv:0905.2627 (2009).
13. B. J. Berne and R. Pecora; *Dynamic light scattering*, Dover, NY (2003).
14. C. Cercignani y G. Medeiros Kremer; *The relativistic Boltzmann equation: theory and applications*, Birkhuser, Berlin (2002).
15. C. Y. Wong, Phys. Rev. C **78**, 054902 (2008).

Chapter 19

Complete Classification of Geodesic Motion in Fast Kerr and Kerr–(anti-)de Sitter Space–Times

Eva Hackmann, Claus Lämmerzahl

ZARM, University of Bremen, Am Fallturm, 28359 Bremen, Germany

Alfredo Macias

Departamento de Física, Universidad Autónoma Metropolitana–Iztapalapa, Apartado Postal 55–534, C.P. 09340, México, D.F., México

The complete structure of geodesics in fast Kerr and Kerr–(anti-)de Sitter space–times are analyzed and classified. The analysis is based on the determination of the zeros of the polynomials underlying the equation of motion for the altitudinal and the radial coordinates.

Dedicated to Leopoldo García–Colín on the occasion of his 80th birthday.

Contents

19.1. Introduction and Motivation

The gravitational field can only be explored by the observation of the motion of test objects. In the most simplest case these test objects are point particles and light rays. From very simple and basic physically motivated postulates it can be shown[1,2] that point particles and light rays obey the geodesic equation of motion

$$0 = \frac{d^2x^\mu}{d\tau^2} + \left\{ {}^{\;\mu}_{\rho\sigma} \right\} \frac{dx^\rho}{d\tau}\frac{dx^\sigma}{d\tau}\,, \tag{19.1}$$

where $d\tau^2 = g_{\mu\nu}dx^\mu dx^\nu$ is the proper time along the geodesics and

$$\left\{ {}^{\;\mu}_{\rho\sigma} \right\} = \frac{1}{2}g^{\mu\nu}\left(\partial_\rho g_{\sigma\nu} + \partial_\sigma g_{\rho\nu} - \partial_\nu g_{\rho\sigma}\right). \tag{19.2}$$

The space–time metric $g_{\mu\nu}$ plays the role of the gravitational field which also defines the measured distance between two neighboring space–time events.

The space–time metric $g_{\mu\nu}$ is determined by the Einstein field equations

$$R_{\mu\nu} - \frac{1}{2}Rg_{\mu\nu} + \Lambda g_{\mu\nu} = \kappa T_{\mu\nu}\,, \tag{19.3}$$

where Λ is a cosmological constant. With the value of $|\Lambda| \lesssim 10^{-46}\,\mathrm{km}^{-23}$ all observations within sub–galactic as well as cosmological scales can be described. Deviations of the observations from this theory on intermediate scales is currently resolved through the assumption of the presence of dark matter.

The geodesic equation can be solved analytically only in the presence of symmetries. For spherical symmetry the complete solution of the geodesic equation in Schwarzschild space–time has been given by Hagihara[4] in terms of the Weierstraß $\wp$–function, and the solutions of the geodesic equation in Reissner–Nordström space–times are are also based on the Weierstraß $\wp$ function.[5] Recently we were able to find solutions of the geodesic equation in Schwarzschild–de Sitter and also Reissner–Nordström–de Sitter space–times.[6–8] The analytic solutions of the geodesic equation in axially symmetric Kerr space–time have been discussed in[5] and has been recently generalized to Kerr–(anti-)de Sitter space–times,[9] NUT–de Sitter,[10] and generally to solutions of the geodesic equation in Plebański–Demiański space–times without acceleration.[11] The latter result is of special interest: it states that all geodesic equation for which the Hamilon–Jacobi equation is separable which is the case in all Petrov type D space–times,[12] which are exhausted by all Plebański–Demiański space–times without acceleration, can indeed be analytically integrated.[11,13]

In this contribution we continue the discussion of the general analytic solution of the geodesic equation in Kerr–(anti-)de Sitter space–time found in.[9] While in[9] we discussed the complete set of analytic solutions in slow Kerr–(anti-)de Sitter space–times (the case with black hole), we will discuss here the fast Kerr–(anti-)de Sitter case (naked singularity). After summarizing the first steps of the procedure of how to solve the geodesic equation in all Kerr–(anti-)de Sitter space–times we present all possible solutions for the fast Kerr–(anti-)de Sitter case. Due to the cosmic censorship postulate and due to the lack of any observational hint, the fast Kerr–(anti-)de Sitter is very unlikely to be realized on the macroscopic level in nature. We nevertheless think that it is worth to discuss the solutions of the geodesic equation in these space–times since one may speculate that in the quantum realm classically forbidden solutions may have some importance.

19.2. Kerr–(anti-)de Sitter Space–Time

The Kerr–(anti-)de Sitter space–times are given by the metric

$$d\tau^2 = \frac{\Delta_r}{\chi^2\rho^2}\left(dt - a\sin^2\theta d\phi\right)^2 - \frac{\rho^2}{\Delta_r}dr^2 - \frac{\Delta_\theta \sin^2\theta}{\chi^2\rho^2}(adt - (r^2+a^2)d\phi)^2 - \frac{\rho^2}{\Delta_\theta}d\theta^2\,, \tag{19.4}$$

where

$$\Delta_r = \left(1 - \frac{\Lambda}{3}r^2\right)(r^2+a^2) - 2Mr\,, \tag{19.5}$$

$$\Delta_\theta = 1 + \frac{a^2\Lambda}{3}\cos^2\theta\,, \tag{19.6}$$

$$\chi = 1 + \frac{a^2\Lambda}{3}\,, \tag{19.7}$$

$$\rho^2 = r^2 + a^2\cos^2\theta \tag{19.8}$$

(in units where $c = G = 1$). This Boyer–Lindquist form of the Kerr–(anti)-de Sitter metric describes an axially symmetric and stationary vacuum solution of the Einstein equation and is characterized by $\bar{M} = 2M$ related to the mass M of the gravitating body, the angular momentum per mass $a = J/M$, and the cosmological constant Λ . Note that this metric has coordinate singularities on the axes $\theta = 0, \pi$ and on the horizons $\Delta_r = 0$. The only real singularity of the black hole is located at $\rho^2 = 0$, i.e. at simultaneously $r = 0$ and $\theta = \frac{\pi}{2}$ assuming $a \neq 0$.

Analogously to the situation in Kerr space–time, we classify this form of the metric according to the number of (disconnected) regions where $\Delta_r > 0$. This depends on the three parameters $\bar{M}$, a and Λ. We speak of 'slow' Kerr–(anti-)de Sitter if there are two such regions and of 'fast' Kerr–(anti-)de Sitter if there is one region where $\Delta_r > 0$. In the first case we have a black hole space–time, and in the second case a white hole (naked singularity) space–time classically forbidden by the cosmic censorship conjecture. The limiting case where two regions are connected by a zero Δ_r is called 'extreme' Kerr–(anti-)de Sitter. Other cases are not possible, as can be seen by a comparison of the coefficients in $\Delta_r = -\frac{\Lambda}{3}r^4 + (1 - \frac{\Lambda}{3}a^2)r^2 - \bar{M}r + a^2 = -\frac{\Lambda}{3}\prod_{i=1}^{4}(r - r_i)$ where r_i denote the zeros of Δ_r. Fig. 19.1 shows the modification of regions of slow, fast, and extreme Kerr–(anti-)de Sitter with varying Λ.

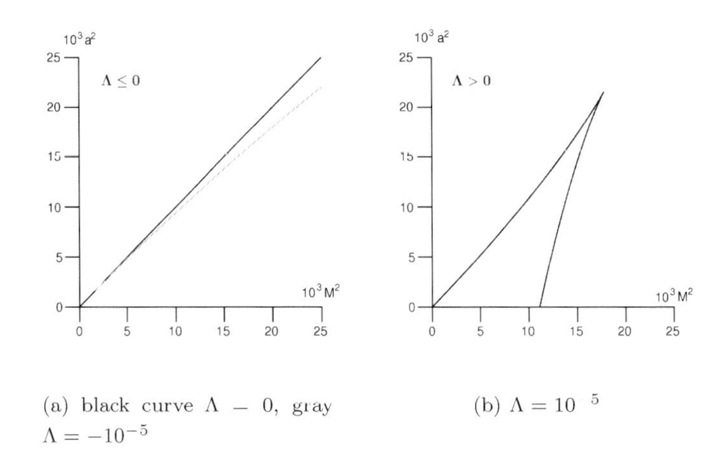

(a) black curve $\Lambda = 0$, gray $\Lambda = -10^{-5}$

(b) $\Lambda = 10^{-5}$

Fig. 19.1. Regions of slow and fast Kerr–(anti-)de Sitter (KdS) for different values of Λ. 19.1(a): below the line we have slow KdS and fast above 19.1(b): the region bounded by the two curves corresponds to slow KdS, outside to fast.

19.3. The Geodesic Equation

For a geodesic motion in an axially symmetric space-time we have four constants of motion. The first constant is given by the normalization condition $\delta = g_{\mu\nu}\dot{x}^\mu\dot{x}^\nu$ with $\delta = 1$ for timelike and $\delta = 0$ for lightlike geodesics. The next two constants of motion correspond to the energy per unit mass E and the angular momentum per unit mass in z direction L_z and are given by the generalized momenta p_t and p_ϕ

$$p_t = g_{tt}\dot{t} + g_{t\phi}\dot{\phi} =: E\,, \tag{19.9}$$

$$-p_\phi = -g_{\phi\phi}\dot{\phi} - g_{t\phi}\dot{t} =: L_z\,, \tag{19.10}$$

where the dot denotes a derivative with respect to the proper time τ. In addition, a fourth constant of motion can be obtained by separating of the Hamilton-Jacobi equation

$$2\frac{\partial S}{\partial \tau} = g^{ij}\frac{\partial S}{\partial x^i}\frac{\partial S}{\partial x^j} \tag{19.11}$$

using the ansatz

$$S = \frac{1}{2}\delta\tau - Et + L_z\phi + S_r(r) + S_\theta(\theta)\,. \tag{19.12}$$

If we insert this ansatz into (19.11) we obtain

$$\begin{aligned}\delta a^2\cos^2\theta + \Delta_\theta\left(\frac{\partial S_\theta}{\partial\theta}\right)^2 &+ \frac{\chi^2}{\Delta_\theta\sin^2\theta}\left(aE\sin^2\theta - L_z\right)^2 \\ &= -\delta r^2 - \Delta_r\left(\frac{\partial S_r}{\partial r}\right)^2 + \frac{\chi^2}{\Delta_r}\left((r^2+a^2)E - aL_z\right)^2,\end{aligned} \tag{19.13}$$

where each side depends on r reps. θ only. This means that each side is equal to a constant K, the famous Carter constant.[14]

From the separation ansatz (19.12) we derive the equations of motion

$$\rho^4\dot{r}^2 = \chi^2(\bar{M}^2\mathbb{P})^2 - \Delta_r(\delta r^2 + K)\,, \tag{19.14}$$

$$\rho^4\dot{\theta}^2 = \Delta_\theta(K - a^2\delta\cos^2\theta) - \frac{\chi^2\bar{M}\mathbb{T}^2}{\sin^2\theta}\,, \tag{19.15}$$

$$\frac{\rho^2}{\chi^2}\dot{\phi} = \frac{a}{\Delta_r}\bar{M}^2\mathbb{P} - \frac{\bar{M}\mathbb{T}}{\Delta_\theta\sin^2\theta}\,, \tag{19.16}$$

$$\frac{\rho^2}{\chi^2}\dot{t} = \frac{r^2+a^2}{\Delta_r}\bar{M}^2\mathbb{P} - \frac{a\bar{M}\mathbb{T}}{\Delta_\theta}\,. \tag{19.17}$$

where

$$\bar{M}^2\mathbb{P} = (r^2+a^2)E - aL_z \tag{19.18}$$

$$\bar{M}\mathbb{T} = aE\sin^2\theta - L_z\,. \tag{19.19}$$

The equations for $\dot{r}$ and $\dot{\theta}$ are still coupled through $\rho^2 = r^2 + a^2\cos^2\theta$. This difficulty can be overcome by introducing the Mino time λ[15] which is related to the proper time τ by $\frac{d\tau}{d\lambda} = \rho^2$. For simplicity, we rescale the parameters appearing in eqs. (19.14)-(19.17) such that they are dimensionless. Thus, we introduce

$$\bar{r} = \frac{r}{\bar{M}}, \quad \bar{a} = \frac{a}{\bar{M}}, \quad \bar{\Lambda} = \frac{1}{3}\Lambda\bar{M}^2, \quad \bar{L}_z = \frac{L_z}{\bar{M}}, \quad \bar{K} = \frac{K}{\bar{M}^2}\,, \tag{19.20}$$

and accordingly

$$\begin{aligned}\Delta_{\bar{r}} &= \left(1 - \bar{\Lambda}\bar{r}^2\right)\left(\bar{r}^2 + \bar{a}^2\right) - \bar{r}\,, \quad (\Delta_r = \bar{M}^2\Delta_{\bar{r}})\\ \Delta_\theta &= 1 + \bar{a}^2\bar{\Lambda}\cos^2\theta\,,\\ \chi &= 1 + \bar{a}^2\bar{\Lambda}\,.\end{aligned} \tag{19.21}$$

Then the equations (19.14)-(19.17) decouple and read

$$\left(\frac{d\bar{r}}{d\lambda}\right)^2 = R(\bar{r}) = \chi^2\mathbb{P}^2 - \Delta_{\bar{r}}(\delta\bar{r}^2 + \bar{K})\,, \tag{19.22}$$

$$\left(\frac{d\theta}{d\lambda}\right)^2 = \Theta(\theta) = \Delta_\theta(\bar{K} - \delta\bar{a}^2\cos^2\theta) - \frac{\chi^2\mathbb{T}^2}{\sin^2\theta}\,, \tag{19.23}$$

$$\frac{1}{\chi^2}\frac{d\phi}{d\lambda} = \frac{\bar{a}}{\Delta_{\bar{r}}}\mathbb{P} - \frac{\mathbb{T}}{\Delta_\theta\sin^2\theta}\,, \tag{19.24}$$

$$\frac{1}{\bar{M}\chi^2}\frac{dt}{d\lambda} = \frac{\bar{r}^2 + \bar{a}^2}{\Delta_{\bar{r}}}\mathbb{P} - \frac{\bar{a}}{\Delta_\theta}\mathbb{T}\,, \tag{19.25}$$

where, as before,

$$\mathbb{P} = (\bar{r}^2 + \bar{a}^2)E - \bar{a}\bar{L}_z\,, \tag{19.26}$$

$$\mathbb{T} = \bar{a}E\sin^2\theta - \bar{L}_z\,. \tag{19.27}$$

In the following we will analyze the structure of possible orbits dependent on the space–time parameters $\bar{a}$, $\bar{\Lambda}$ and the particle parameters $\delta, E, \bar{L}_z, \bar{K}$. The major point in this analysis is that (19.22) and (19.23) imply $R(\bar{r}) \geq 0$ and $\Theta(\theta) \geq 0$ as a necessary condition for the existence of a geodesic. Although we concentrate here on timelike geodesics light can be treated in the same manner and is in general easier to deal with. As a complete discussion of orbits in slow Kerr–(anti-)de Sitter was already carried through in,[11] here we will discuss orbits in fast Kerr–(anti-)de Sitter.

19.4. Timelike Orbits in Fast Kerr–(anti-)de Sitter Space–Times

A geodesic in a Kerr–(anti-)de Sitter space–time can only exist if the two conditions $R(\bar{r}) \geq 0$ and $\Theta(\theta) \geq 0$ are both fulfilled. Before studying the consequences of this two conditions let us consider some geometric features of geodesics connected with the Carter constant K.

As $\Theta(\theta)$ is symmetric with respect to $\theta = \frac{\pi}{2}$ the latitudinal motion is symmetric with respect to the equatorial plane. Therefore, a geodesic lies entirely in the equatorial plane if and only if $\theta = \frac{\pi}{2}$ is a multiple zero of

Θ. As $\Theta\left(\frac{\pi}{2}\right) = 0$ iff the modified Carter constant $Q = K - \chi^2(aE - L_z)$ vanishes and $\frac{d\Theta}{d\theta}\left(\frac{\pi}{2}\right) = 0$ for all values of $\bar{M}$, $\bar{a}$, E, $\bar{L}_z$ and $\bar{K}$, a geodesic lies entirely in the equatorial plane if and only if $Q = 0$.

In a Kerr and Kerr–(anti-)de Sitter space–time the singularity is located at simultaneously $\bar{r} = 0$ and $\theta = \frac{\pi}{2}$. Thus, it is possible that $\bar{r} = 0$ can be reached (and crossed) without falling into the singularity. The value $\bar{r} = 0$ is allowed iff

$$0 \leq R(0) = \chi^2(\bar{a}^2 - \bar{a}\mathcal{D})^2 - \bar{a}^2\kappa = -\bar{a}^2\frac{Q}{E^2 r_S^2}\,. \tag{19.28}$$

This implies that $\bar{r} = 0$ can be reached if and only if $Q \leq 0$.

We study now the two conditions $R(\bar{r}) \geq 0$ and $\Theta(\theta) \geq 0$ separately.

19.4.1. *Types of latitudinal motion*

Geodesics can take an angle θ if and only if $\Theta(\theta) \geq 0$. Thus, we want to determine which values of $\bar{a}$, $\bar{\Lambda}$, E, $\bar{L}_z$, $\bar{K}$, and $\theta \in [0, \pi]$ result in positive $\Theta(\theta)$. For simplicity, we substitution $\nu = \cos^2\theta$ giving

$$\Theta(\nu) = (1 + \bar{a}^2\bar{\Lambda}\nu)(\kappa - \delta_2\bar{a}^2\nu) - \chi^2\left(\bar{a}^2(1-\nu) - 2\bar{a}\mathcal{D} + \frac{\mathcal{D}^2}{1-\nu}\right). \tag{19.29}$$

Assume now that for a given set of parameters there exists a certain number of zeros of $\Theta(\nu)$ in $[0, 1]$. If we vary the parameters, the position of the zeros varies and the number of real zeros in $[0, 1]$ can change only if (i) a zero crosses 0 or 1 or (ii) two zeros merge. Let us consider case (i). 0 is a zero iff

$$\Theta(\nu = 0) = \kappa - \chi^2(\bar{a} - \mathcal{D})^2 = 0 \tag{19.30}$$

or

$$\bar{L}_z = \bar{a}E \pm \frac{\sqrt{\bar{K}}}{\chi}\,. \tag{19.31}$$

As $\nu = 1$ is in general a pole of $\Theta(\nu)$ it is a necessary condition for 1 being a zero of $\Theta(\nu)$ that this pole becomes a removable singularity. From (19.29) it follows that this is the case for $\mathcal{D} = 0$ or, equivalently, $\bar{L}_z = 0$. Under this assumption we obtain

$$\Theta(\nu = 1) = (1 + \bar{a}^2\bar{\Lambda})(\kappa - \delta_2\bar{a}^2) \qquad \text{for } \bar{L}_z = 0\,. \tag{19.32}$$

If we additionally assume that $\Lambda > -\frac{3}{a^2}$ and, therefore, $(1 + \bar{a}^2\bar{\Lambda}) > 0$ we can conclude that $\Theta(\nu = 1) = 0$ iff $\bar{L}_z = 0$ and $\kappa = \delta_2\bar{a}^2$. Summarized, $\bar{L}_z = \bar{a}E \pm \frac{\sqrt{\bar{K}}}{\chi}$ and simultaneously $\bar{L}_z = 0$ and $\kappa = \delta_2\bar{a}^2$ (assuming

$\Lambda > -\frac{3}{a^2}$) give us boundaries of the θ motion. Since from observation the cosmological constant has a small positive value, the condition $\Lambda > -\frac{3}{a^2}$ is always fulfilled.

Now let us consider case (ii). If we exclude the coordinate singularities $\theta = 0, \pi$ or $\nu = 1$ the zeros of $\Theta(\nu)$ are given by the zeros of

$$\Theta_\nu = (1-\nu)(1+\bar{a}^2\bar{\Lambda}\nu)(\kappa - \delta_2\bar{a}^2\nu) - \chi^2\left(\bar{a} - \mathcal{D} - \bar{a}\nu\right)^2 , \qquad (19.33)$$

which is in general a polynomial of degree 3. Then two zeros coincide at $x \in [0, 1)$ iff

$$\Theta_\nu = (\nu - x)^2(a_1\nu + a_0) \qquad (19.34)$$

for some real constants a_1, a_0. By a comparison of coefficients we can solve this equation for $\bar{L}_z(x)$ and $E^2(x)$ dependent the remaining parameters $\bar{a}, \bar{\Lambda}$, and $\bar{K}$. This parametric representation of values of $\bar{L}_z$ and E^2 again correspond to boundary cases of the θ motion. Let us additionally consider the conditions for $\nu = 1$ being a double zero. With $\bar{L}_z = 0$ and $\kappa = \delta_2\bar{a}^2$ (assuming $\Lambda > -\frac{3}{a^2}$) it follows

$$\left.\frac{d\Theta(\nu)}{d\nu}\right|_{\nu=1} = \bar{a}^2\chi(\chi - \delta_2) , \qquad (19.35)$$

which is zero for $\chi = \delta_2$ or, equivalently, $E^2 = \chi^{-1}$.

For given parameters of the space time $\bar{a}$ and $\bar{\Lambda}$, we can use this information to analyze the θ motion of all possible geodesics in this space-time. As a typical example for timelike geodesics in fast Kerr–(anti-)de Sitter consider Fig. 19.2, where the curves divide the half plane into four regions (a)-(d) which correspond to different arrangement of zeros in $[0, 1]$. A geodesic motion is only possible in regions (b) and (d) because in all other regions Θ_ν is negative for all $\nu \in [0, 1]$. Note that for the special case of $\kappa = \delta_2\bar{a}^2$ (assuming $\Lambda > -\frac{3}{a^2}$), strictly speaking, regions (b) and (d) are divided by $\bar{L}_z = 0$. However, in each region we have for $\bar{L}_z > 0$ and $\bar{L}_z < 0$ the same number of zeros in $[0, 1]$ and, thus, the same type of motion. (More precisely, near $\bar{L}_z = 0$ a zero $\nu_0 < 1$ of $\Theta(\nu)$ approaches 1, but does not cross it.) Therefore, in each region we put the parts above and below $\bar{L}_z = 0$ together. The arrangement of zeros in the two regions (b) and (d) correspond to the following different types of motion in θ direction (cp. Fig. 19.3)

- Region (b): Θ_ν has one real zero $\nu_{\max}$ in $[0, 1]$ with $\Theta_\nu \geq 0$ for $\nu \in [0, \nu_{\max}]$, i.e. θ oscillates around the equatorial plane $\theta = \frac{\pi}{2}$.

- Region (d): Θ_ν has two real zeros $\nu_{\min}, \nu_{\max}$ in $[0,1]$ with $\Theta_\nu \geq 0$ for $\nu \in [\nu_{\min}, \nu_{\max}]$, i.e. θ oscillates between $\arccos(\pm\sqrt{\nu_{\min}})$ and $\arccos(\pm\sqrt{\nu_{\max}})$.

The boundaries of region (b) are given by $\bar{L}_z = \bar{a}E \pm \frac{\sqrt{\bar{K}}}{\chi}$ and, therefore, the regions gets larger if $\frac{\sqrt{\bar{K}}}{\chi}$ grows, i.e. if $\bar{K}$ grows or $\bar{a}^2\bar{\Lambda}$ gets smaller. A change of $\bar{a}$ in addition causes region (b) to shift up or down. The dependence of region (d) on the parameters $\bar{K}, \bar{a}$ and $\bar{\Lambda}$ is much more involved. The upper boundary of (d) is also the lower boundary of region (b). The lower boundary is given in a complicated parametric form which makes it apparently impossible to determine an explicit connection between the form of region (d) and the parameters. However, the point where the upper and lower boundaries of region (d) touch each other is where 0 is a double zero of Θ_ν, which is given by $x = 0$ in $E(x)$ and $\bar{L}_z(x)$ from (19.34),

$$E(0) = \frac{1}{2}\frac{\bar{a}^2 + \bar{K} - \bar{a}^2\bar{\Lambda}\bar{K}}{\bar{a}\sqrt{\bar{K}}\chi}, \quad \bar{L}_z(0) = \frac{1}{2}\frac{\bar{a}^2 - \bar{K}\chi}{\sqrt{\bar{K}}\chi}. \tag{19.36}$$

The regions (b) and (d) are characterized in an easy way in terms of the modified Carter constant Q. As in region (d) $\Theta_\nu(0) < 0$ it follows that this region corresponds to $Q < 0$ because of $\Theta_\nu(0) = \frac{Q}{E^2 r_S^2}$. In the same way we can conclude that region (b), where $\Theta_\nu(0) > 0$, corresponds to $Q > 0$.

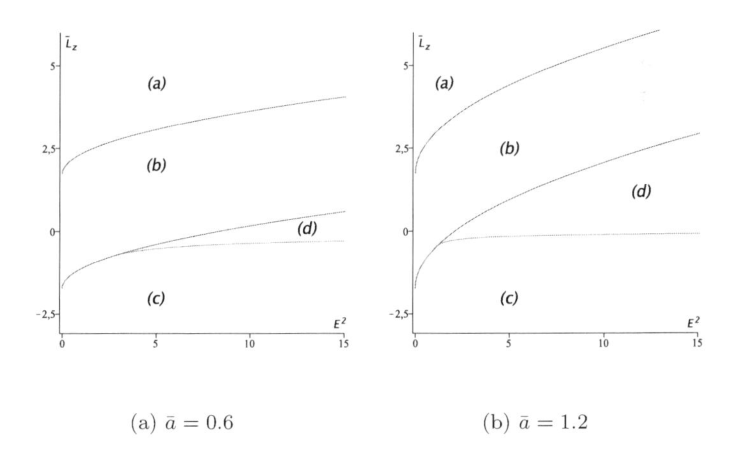

(a) $\bar{a} = 0.6$ (b) $\bar{a} = 1.2$

Fig. 19.2. Typical regions of different types of θ-motion in fast Kerr–(anti-)de Sitter for $\bar{M} = 2$, $\bar{\Lambda} = 10^{-5}$, and $\bar{K} = 3$.

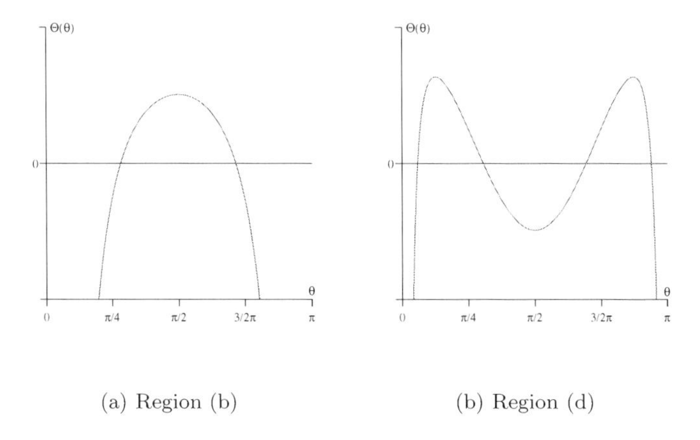

(a) Region (b) (b) Region (d)

Fig. 19.3. Examples of $\Theta(\theta)$ in the two regions (b) and (d). Allowed values of θ are given by $\Theta(\theta) \geq 0$.

19.4.2. *Types of radial motion*

A geodesic can take a radial coordinate $\bar{r}$ if and only if $R(\bar{r}) \geq 0$. The zeros of R are extremal values of $\bar{r}(\gamma)$ and determine the type of geodesic. The polynomial R is in general of degree six in $\bar{r}$ and, therefore, has six possibly complex zeros of which the real zeros are of interest for the type of motion. As a Kerr–(anti-)de Sitter space–time has no singularity in $\bar{r} = 0, \theta \neq \frac{\pi}{2}$, we also consider negative $\bar{r}$ as valid. As discussed at the beginning of the section, $\bar{r} = 0$ can only be crossed iff $Q < 0$, which corresponds to region (d) of the θ motion due to $R(0) = -\bar{a}\Theta_\nu(0)$. In region (b) of the θ motion where $Q > 0$ holds a transition from positive to negative $\bar{r}$ is not possible.

To clarify the discussion we introduce some types of orbits.[16]

- Flyby orbit: $\bar{r}$ starts from $\pm\infty$, then approaches a periapsis $\bar{r} = r_0$ and back to $\pm\infty$.
- Bound orbit: $\bar{r}$ oscillates between to extremal $r_1 \leq \bar{r} \leq r_2$ with $-\infty < r_1 < r_2 < \infty$.
- Transit orbit: $\bar{r}$ starts from $\pm\infty$ and goes to $\mp\infty$ crossing $\bar{r} = 0$.

All other types of orbits are exceptional and treated separately. They are either connected with the ring singularity $\rho^2 = 0$ or with the appearance of multiple zeros in R, which simplifies the structure of the differential equation (19.22) considerably. Examples for the latter type are homoclinic orbits, cp.[17] As large negative $\bar{r}$ correspond to negative mass of the black or white hole,[18] we will assign the attribute 'crossover' to flyby or bound orbits which pass from positive to negative $\bar{r}$ or vice versa. (By definition, a transit orbit is always a crossover orbit.) Therefore, in region (d) of the θ motion there exists a crossover orbit, whereas all orbits located in region (b) of the θ motion do not cross $\bar{r} = 0$.

For a given set of parameters we have a certain number of real zeros of R. If we vary the parameters this number can change only if two zeros merge to one. This happens at $\bar{r} = x$ iff

$$R = (\bar{r} - x)^2(a_4\bar{r}^4 + a_3\bar{r}^3 + a_2\bar{r}^2 + a_1\bar{r} + a_0) \tag{19.37}$$

for some real constants a_i. By a comparison of coefficients we can solve the resulting 7 equations for $E^2(x)$ and $\bar{L}_z(x)$ depending on the remaining parameters $\bar{a}$, $\bar{\Lambda}$, and $\bar{K}$. A typical result in fast Kerr–(anti-)de Sitter space–times for small Λ including the results of the foregoing subsection is shown in Figs. 19.6, 19.8 and 19.10. An analysis of the influence of each of the parameters $\bar{a}$, $\bar{\Lambda}$, and $\bar{K}$ is not done easily due to the complexity of the expressions for E^2 and $\bar{L}_z$. Examples of $R(\bar{r})$ for different numbers of real zeros are given in Fig. 19.4.

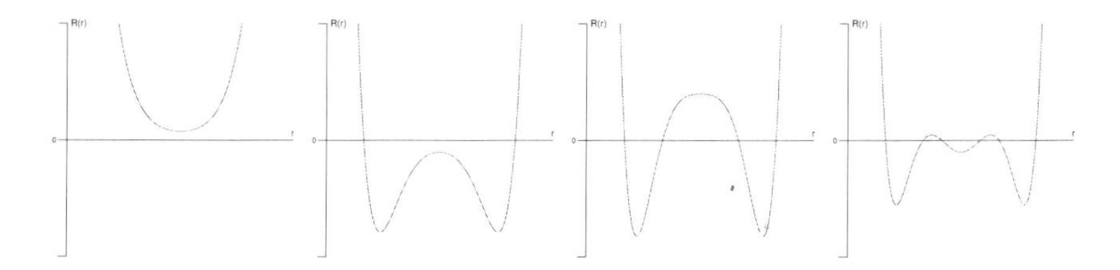

Fig. 19.4. Typical examples of $R(\bar{r})$ for $\Lambda > 0$ and for different numbers of real zeros. Allowed values of $\bar{r}$ are given by $R(\bar{r}) \geq 0$.

For comparison, let us first study the situation in fast Kerr space–time with $\Lambda = 0$.

Case $\Lambda = 0$. In this case there are 5 different regions of types of $\bar{r}$ motion with up to 3 additional regions dependent on the combination of $\bar{a}$ and $\bar{K}$. In Fig. 19.5 some $(E^2, \bar{L}_z)$-diagrams are shown for different values of $\bar{a}$ and $\bar{K}$. The 5 basic regions, which are present for every combination of $\bar{a}$ and $\bar{K}$ are given by (here we always assume $r_i < r_{i+1}$)

- Region (I): all zeros of R are complex and $R(\bar{r}) \geq 0$ for all $\bar{r}$. Possible orbit types: transit orbit.

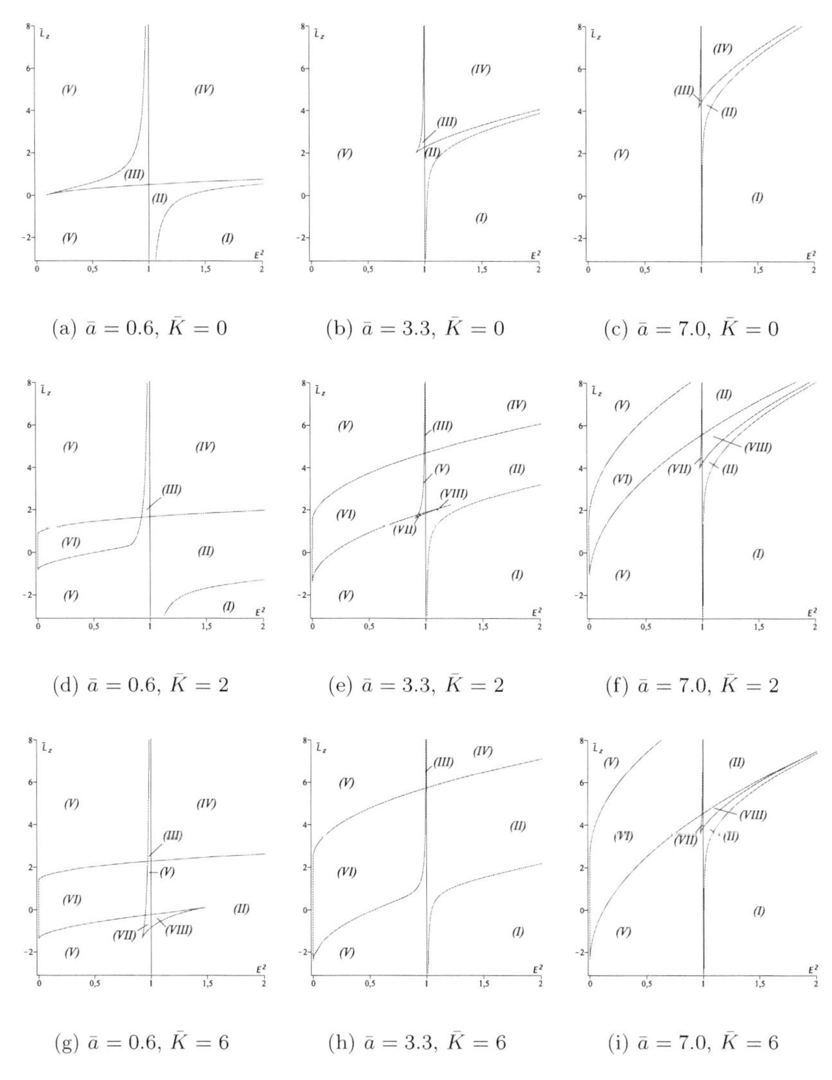

(a) $\bar{a} = 0.6$, $\bar{K} = 0$ (b) $\bar{a} = 3.3$, $\bar{K} = 0$ (c) $\bar{a} = 7.0$, $\bar{K} = 0$

(d) $\bar{a} = 0.6$, $\bar{K} = 2$ (e) $\bar{a} = 3.3$, $\bar{K} = 2$ (f) $\bar{a} = 7.0$, $\bar{K} = 2$

(g) $\bar{a} = 0.6$, $\bar{K} = 6$ (h) $\bar{a} = 3.3$, $\bar{K} = 6$ (i) $\bar{a} = 7.0$, $\bar{K} = 6$

Fig. 19.5. Different regions of types of $\bar{r}$ motion in fast Kerr space–time with $\Lambda = 0$ and different $\bar{a}$ and $\bar{K}$.

- Region (II): R has two real zeros r_1, r_2 and $R(\bar{r}) \geq 0$ for $\bar{r} \leq r_1$ and $r_2 \leq \bar{r}$. Possible orbit types: two flyby orbits, one to $+\infty$ and one to $-\infty$.
- Region (III): all four zeros r_i, $1 \leq i \leq 4$, of R are real and $R(\bar{r}) \geq 0$ for $r_{2k-1} \leq \bar{r} \leq r_{2k}$, $k = 1, 2$. Possible orbit types: two different bound orbits.

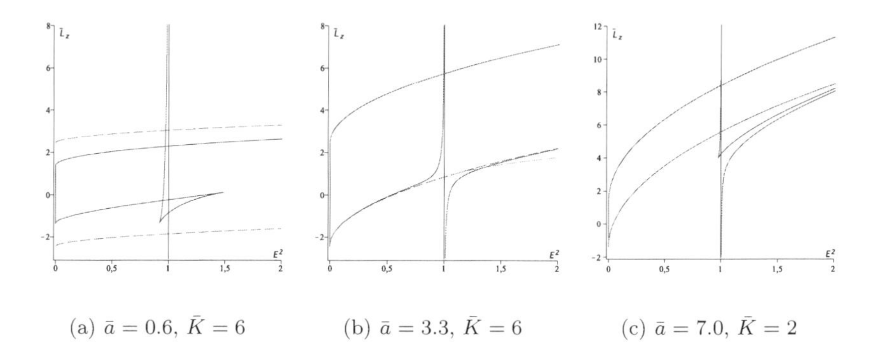

(a) $\bar{a} = 0.6$, $\bar{K} = 6$ (b) $\bar{a} = 3.3$, $\bar{K} = 6$ (c) $\bar{a} = 7.0$, $\bar{K} = 2$

Fig. 19.6. Different regions of $\bar{r}$ and θ motion in fast Kerr space–time with $\Lambda = 0$. The solid lines represent regions of $\bar{r}$ motion whereas dashed lines represents regions of θ motion. In (b) the upper and in (c) the upper two 'horizontal' lines border $\bar{r}$ as well as θ regions. Compared to Fig. 19.5 note the rescaled axis in (c).

- Region (IV): again all four zeros of R are real but $R(\bar{r}) \geq 0$ for $\bar{r} \leq r_1$, $r_2 \leq \bar{r} \leq r_3$, and $r_4 \leq \bar{r}$. Possible orbit types: two flyby orbits, one to each of $\pm\infty$ and a bound orbit.
- Region (V): R has two real zeros r_1, r_2 and $R(\bar{r}) \geq 0$ for $r_1 \leq \bar{r} \leq r_2$. Possible orbit types: a bound orbit.

The additional regions, which may disappear for certain parameter combinations, are

- Region (VI): all zeros of R are complex but $R(\bar{r}) < 0$ for all $\bar{r}$. Here a geodesic motion is not possible.
- Region (VII): all zeros of R are real and $R(\bar{r}) \geq 0$ for $r_{2k-1} \leq \bar{r} \leq r_{2k}$, $k = 1, 2$. Possible orbit types: two different bound orbits.
- Region (VIII): again all four zeros of R are real but $R(\bar{r}) \geq 0$ for $\bar{r} \leq r_1$, $r_2 \leq \bar{r} \leq r_3$ and $r_4 \leq \bar{r}$. Possible orbit types: two flyby orbits and a bound orbit.

Let us also analyze where we have crossover orbits. As region (I) can only contain a transit orbit, which is by definition a crossover orbit, it can only intersect region (d). Region (II) may intersect both (b) and (d) but all other regions contain only region (b) of the θ motion and, therefore do not have any crossover orbits. The combination of results for the latitudinal and radial motion for chosen combinations of $\bar{a}$ and $\bar{K}$ can be seen in Fig. 19.6. The results of this paragraph together with the numbers of positive and negative zeros for each region are summarized in Tab. 19.1.

Table 19.1. Orbit types for $\Lambda = 0$.
The + and − columns give the number of positive and negative real zeros of the polynomial R. Here the thick lines represent the range of orbits. Turning points are shown by thick dots. The small vertical line denotes $\bar{r} = 0$.

region	−	+	range of $\bar{r}$	types of orbits
Id	0	0		transit
IIb	1	1	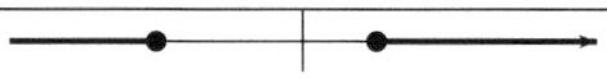	2x flyby
IId	2	0		flyby, crossover flyby
IIIb	0	4	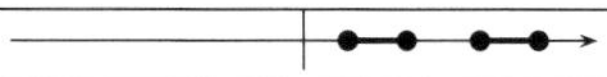	2x bound
IVb	1	3	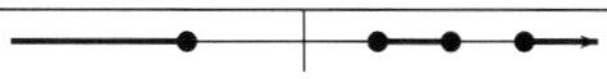	2x flyby, bound
Vb	0	2	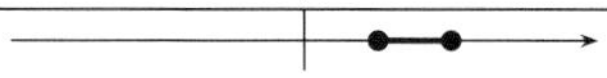	bound
VIIb	0	4	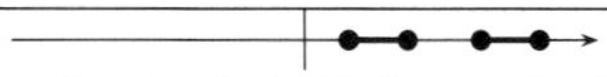	2x bound
VIIIb	1	3	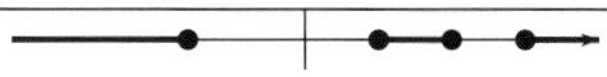	2x flyby, bound

Case $\mathbf{\Lambda > 0}$. Let us analyze now which regions changes compared to the case $\Lambda = 0$. At first, we recognize that region (V) of $\Lambda = 0$ merges with region (IV). The vertical line at $E^2 = 1$ for $\bar{\Lambda} = 0$ shifts to the left and has a maximum or even vanishes. In the latter case region (III) of $\bar{\Lambda} = 0$ also vanishes. As for $\bar{\Lambda} = 0$ we have basic regions which appear for every combination of $\bar{a}$ and $\bar{K}$ and additional regions. However, for $\bar{\Lambda} > 0$ there are only 3 basic regions, namely (I), (II), and (IV). The regions (III), (VI), (VII), and (VIII) are additional.

A comparison of the possible orbit types for $\bar{\Lambda} > 0$ with the one for $\bar{\Lambda} = 0$ shows that in regions (I), (II), and (VIII) there are no differences. However, these regions are deformed and a pair of parameters $(E^2, \bar{L}_z)$ located in region (I) or (II) for $\bar{\Lambda} > 0$ may be located in a different region for $\bar{\Lambda} = 0$. Let us consider regions (III), (IV), (VI), and (VII).

- Region (III) and (VII): all six zeros r_i of R are real and $R(\bar{r}) \geq 0$ for $\bar{r} \leq r_1, r_6 \leq \bar{r}$ and $r_{2k} \leq \bar{r} \leq r_{2k+1}$ for $k = 1, 2$. Possible orbit types: two flyby orbits, one to each of $\pm\infty$, and two different bound orbits.
- Region (IV) (and (VIII)): R has four real zeros and $R(\bar{r}) \geq 0$ for $\bar{r} \leq r_1, r_2 \leq \bar{r} \leq r_3, r_4 \leq \bar{r}$. Possible orbit types: two flyby orbits, one to each of $\pm\infty$ and a bound orbit.

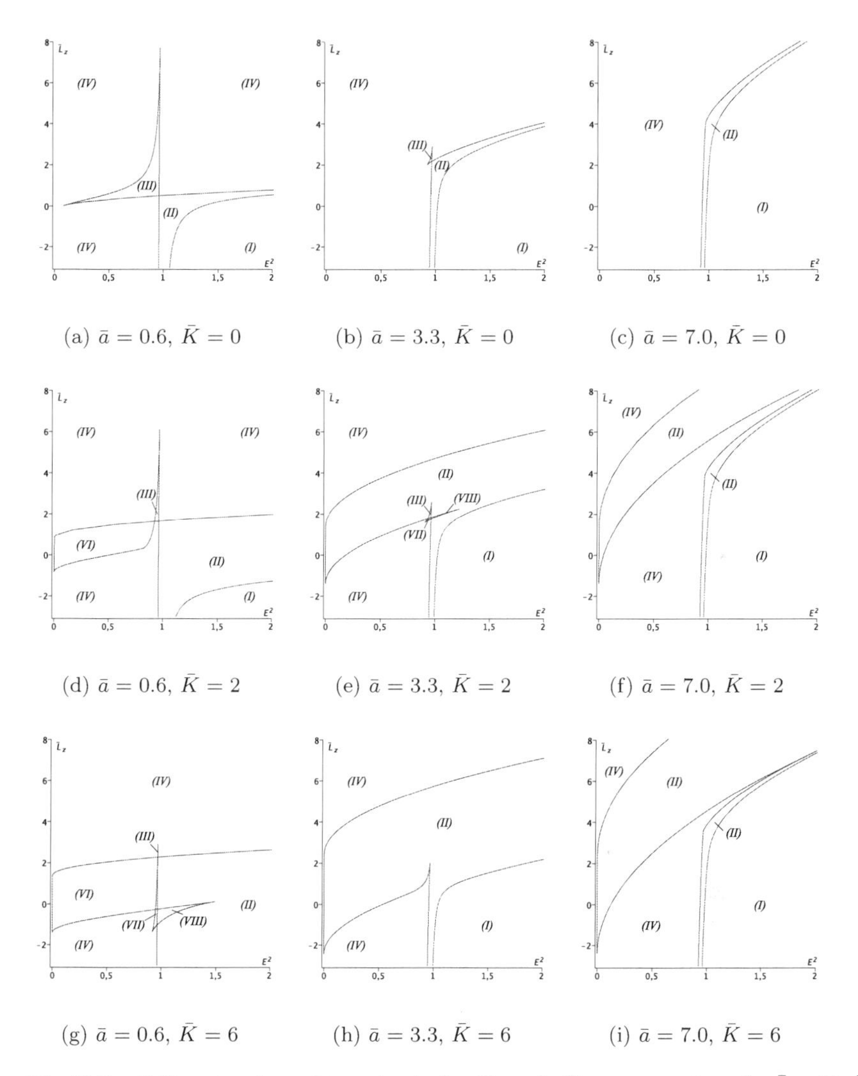

(a) $\bar{a} = 0.6,\ \bar{K} = 0$ (b) $\bar{a} = 3.3,\ \bar{K} = 0$ (c) $\bar{a} = 7.0,\ \bar{K} = 0$

(d) $\bar{a} = 0.6,\ \bar{K} = 2$ (e) $\bar{a} = 3.3,\ \bar{K} = 2$ (f) $\bar{a} = 7.0,\ \bar{K} = 2$

(g) $\bar{a} = 0.6,\ \bar{K} = 6$ (h) $\bar{a} = 3.3,\ \bar{K} = 6$ (i) $\bar{a} = 7.0,\ \bar{K} = 6$

Fig. 19.7. Different regions of $\bar{r}$ motion in fast Kerr–de Sitter space–times for $\bar{\Lambda} = 10^{-5}$ and different $\bar{a}$ and $\bar{K}$. Compared to $\bar{\Lambda} = 4$ region (V) vanishes completely.

- Region (VI): R has two real zeros and $R(\bar{r}) \geq 0$ for $\bar{r} < r_1$ and $r_2 < \bar{r}$. Possible orbit types: two flyby orbits, one to $+\infty$ and one to $-\infty$.

Here an additional dramatic change compared to $\bar{\Lambda} = 0$ becomes obvious: now a geodesic motion is possible in region (VI). Concerning crossover orbits regions (III), (IV), (VI), (VII), and (VIII) only contain region (b) of

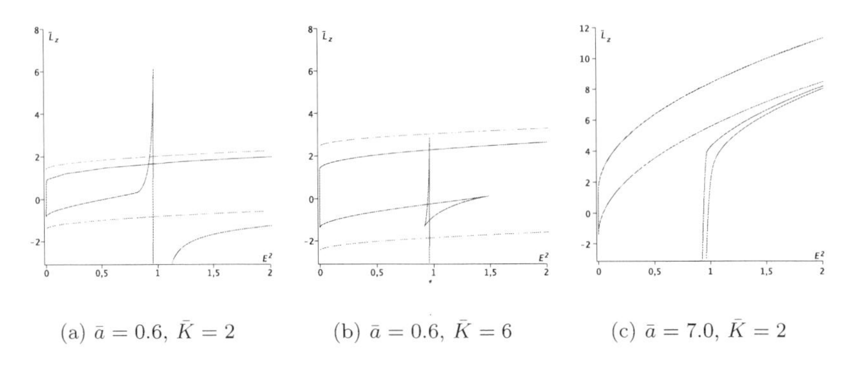

(a) $\bar{a} = 0.6$, $\bar{K} = 2$ (b) $\bar{a} = 0.6$, $\bar{K} = 6$ (c) $\bar{a} = 7.0$, $\bar{K} = 2$

Fig. 19.8. Different regions of $\bar{r}$ and θ motion in fast Kerr–de Sitter space–time with $\bar{\Lambda} = 10^{-5}$. The solid lines represent regions of $\bar{r}$ motion whereas dashed lines represents regions of the θ motion. In (c) the upper two 'horizontal' lines border $\bar{r}$ as well as θ regions. Compared to Fig. 19.5 note the rescaled axis in (c).

θ motion, analogously to $\bar{\Lambda} = 0$. This implies that there are no crossover orbits possible. Region (I) can only intersect region (d) because only transit orbits are possible. The remaining region (II) is the only one which intersects regions (b) and (d).

We conclude that for $E^2 > 1$ the types of orbits are not noticeable changed, whereas for $E^2 \leq 1$ there are significant changes. In the former region (V) (for $\bar{\Lambda} = 0$), which is now in region (IV), and in regions (III) and (VII) we have two additional flyby orbits which are not present for $\bar{\Lambda} = 0$. In a small vertical stripe left of $E^2 = 1$ there are even orbits which are bound for $\bar{\Lambda} = 0$ but reaching infinity for $\bar{\Lambda} > 0$. In particular, it is independent of the value of E if a geodesic may reach infinity as expected from the repulsive cosmological force related to $\bar{\Lambda} > 0$. Furthermore, a geodesic motion is now possible in region (VI), which was forbidden for $\bar{\Lambda} = 0$.

Note that for huge $\bar{\Lambda}$ the separation in regions (I) to (VIII) is no longer possible because the repulsive cosmological force becomes so strong that bound orbits become disintegrated. In this case we have only two regions, one with two real zeros corresponding to two flyby orbits and one with only complex zeros corresponding to a transit orbit.

All orbit types for small $\bar{\Lambda} > 0$ are summarized in Tab. 19.2.

Case $\Lambda < 0$. Here region (V) from $\Lambda = 0$ merges with region (I) and the vertical line at $E^2 = 1$ for $\bar{\Lambda} = 0$ shifts to the right and has a minimum.

Table 19.2. Orbit types for small $\Lambda > 0$.
For the description of the $+$, $-$ and range of $\bar{r}$ columns see Tab. 19.1.

region	$-$	$+$	range of $\bar{r}$	types of orbits
Id	0	0		transit
IIb	1	1	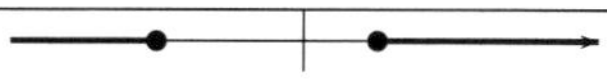	2x flyby
IId	2	0		flyby, crossover flyby
IIIb	1	5		2x flyby, 2x bound
IVb	1	3		2x flyby, bound
VIb	1	1	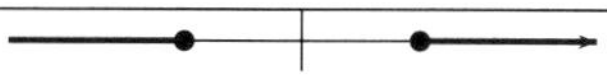	2x flyby
VIIb	1	5		2x flyby, 2x bound
VIIIb	1	3	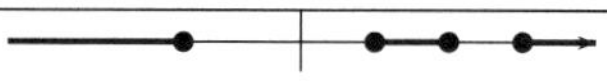	2x flyby, bound

Compared to the situation for $\Lambda = 0$ the possible orbit types in regions (III) and (VI) do not change, but a set of parameters located there may be located in a different region for $\Lambda = 0$. Let us examine the remaining regions.

- Region (I): R has two real zeros $r_1 < r_2$ and $R(\bar{r}) \geq 0$ for $r_1 \leq \bar{r} \leq r_2$. Possible orbit types: bound orbit.
- Region (II),(III),and (VII): R has four real zeros with $R(\bar{r}) \geq 0$ for $r_{2k-1} \leq \bar{r} \leq r_{2k}$, $k = 1, 2$. Possible orbit types: two different bound orbits.
- Region (IV) and (VIII): all six zeros of R are real and $R(\bar{r}) \geq 0$ for $r_{2k-1} \leq \bar{r} \leq r_{2k}$, $k = 1, 2, 3$. Possible orbit types: three different bound orbits.

Let us combine these results with the analysis of the latitudinal motion. The only regions which can intersect region (d) of the θ motion and which, therefore, allow crossover orbits are regions (I) and (II). All other regions only are located in region (b) of the θ motion where crossover orbits are not possible.

Summarizing, the types of orbits significantly change if $E^2 > 1$. The transit orbit in region (I) for $\Lambda = 0$ is transformed to a bound orbit for $\Lambda < 0$ as well as the flyby orbits in regions (II), (IV), and (VIII). Although

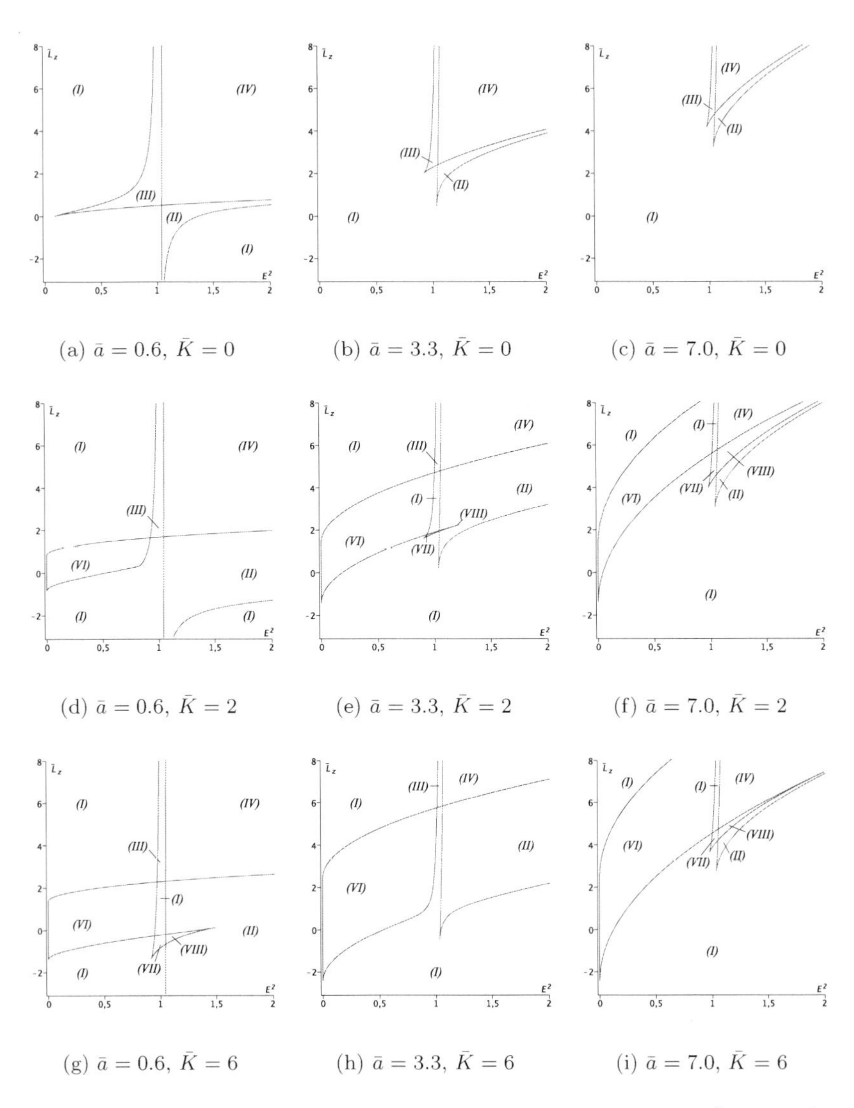

(a) $\bar{a} = 0.6$, $\bar{K} = 0$ (b) $\bar{a} = 3.3$, $\bar{K} = 0$ (c) $\bar{a} = 7.0$, $\bar{K} = 0$

(d) $\bar{a} = 0.6$, $\bar{K} = 2$ (e) $\bar{a} = 3.3$, $\bar{K} = 2$ (f) $\bar{a} = 7.0$, $\bar{K} = 2$

(g) $\bar{a} = 0.6$, $\bar{K} = 6$ (h) $\bar{a} = 3.3$, $\bar{K} = 6$ (i) $\bar{a} = 7.0$, $\bar{K} = 6$

Fig. 19.9. Different regions of $\bar{r}$ motion in fast Kerr–anti-de Sitter for $\bar{\Lambda} = -10^{-5}$ and different $\bar{a}$ and $\bar{K}$. Compared to $\bar{\Lambda} = 0$ region (V) vanishes completely.

region (V) for $\Lambda = 0$ merges with region (I) for $\Lambda < 0$, the types of orbits do not change there. In general, because of $R \to -\infty$ if $\bar{r} \to \pm\infty$ we can not have orbits reaching $\bar{r} = \pm\infty$ at all as expected due to the attractive cosmological force related to $\Lambda < 0$.

All orbit types for $\Lambda < 0$ are summarized in Tab. 19.3.

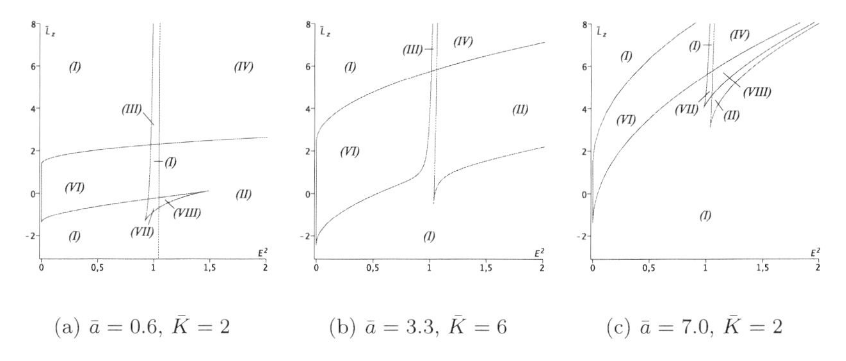

(a) $\bar{a} = 0.6$, $\bar{K} = 2$ (b) $\bar{a} = 3.3$, $\bar{K} = 6$ (c) $\bar{a} = 7.0$, $\bar{K} = 2$

Fig. 19.10. Different regions of $\bar{r}$ and θ motion in fast Kerr–anti-de Sitter space–times with $\bar{\Lambda} = -10^{-5}$. The solid lines represent regions of $\bar{r}$ motion whereas dashed lines represents regions of θ motion. In (b) the upper and in (c) the upper two 'horizontal' lines border $\bar{r}$ as well as θ regions. Compared to Fig. 19.5 note the rescaled axis in (c).

Table 19.3. Orbit types for $\Lambda < 0$.
For the description of the +, − and range of $\bar{r}$ columns see Tab. 19.1.

region	−	+	range of $\bar{r}$	types of orbits
Ib	0	2		bound
Id	1	1		crossover bound
IIb	2	2		2x bound
IId	3	1		bound crossover bound
IIIb	0	4		2x bound
IVb	2	4		3x bound
VIIb	0	4		2x bound
VIIIb	2	4		3x bound

19.5. Summary

In this paper we classified possible types of geodesic motion in fast Kerr and Kerr–(anti-)de Sitter space–times by an analysis of the zeros of the polynomials underlying the θ and $\bar{r}$ motion. We discussed the influence of a non–vanishing cosmological constant on the orbit types and found several differences.

The same method of classifying geodesic motion has already been applied to slow Kerr and Kerr–(anti-)de Sitter space–times in.[9] This method should also be applied to extreme Kerr and Kerr–(anti-)de Sitter space–times in order to obtain a exhaustive discussion of all types of geodesic motion for all possible Kerr and Kerr–(anti-)de Sitter space–times. It should also be possible to apply this method of classification to more general axially symmetric space–times as the Plebański–Demiański space–time.[13] It will also be interesting to apply the presented method to higher dimensional stationary axially symmetric space–times like the Myers–Perry solutions.

Acknowledgments

We are grateful to H. Dullin, W. Fischer, and P. Richter for helpful discussions. E.H. thanks the German Research Foundation DFG for financial support, and C.L. and A.M. for support from the joint German–Mexican DFG–CONACyT program.

References

1. J. Ehlers, F. Pirani, and A. Schild. The geometry of free fall and light propagation. In ed. L. O'Raifeartaigh, *General Relativity, Papers in Honour of J.L. Synge*, p. 63. Clarendon Press, Oxford, (1972).
2. J. Ehlers. Survey of general relativity theory. In ed. W. Israel, *Relativity, Astrophysics and Cosmology*, p. 1. Reidel, Dordrecht, (1973).
3. J. Dunkley et al., Five-Year Wilkinson Microwave Anisotropy Probe (WMAP) Observations: Likelihoods and Parameters from the WMAP data, *Astrophys. J. Suppl.* **180**, 306, (2009).
4. Y. Hagihara, Theory of relativistic trajectories in a gravitational field of Schwarzschild, *Japan. J. Astron. Geophys.* **8**, 67, (1931).
5. S. Chandrasekhar, *The Mathematical Theory of Black Holes.* (Oxford University Press, Oxford, 1983).
6. E. Hackmann and C. Lämmerzahl, Complete analytic solution of the geodesic equation in Schwarzschild–(anti) de Sitter space–times, *Phys. Rev. Lett.* **100**, 171101–1, (2008).
7. E. Hackmann and C. Lämmerzahl, Geodesic equation in Schwarzschild–(anti-)de Sitter space–times: Analytical solutions and applications, *Phys. Rev.* **D 78**, 024035, (2008).
8. E. Hackmann, V. Kagramanova, J. Kunz, and C. Lämmerzahl, Analytic solutions of the geodesic equations in higher dimensional static spherically symmetric space–times, *Phys. Rev.* **D 78**, 124018, (2008).
9. E. Hackmann, V. Kagramanova, J. Kunz, and C. Lämmerzahl, Analytic solution of the geodesic equation in Kerr–(anti-)de Sitter space–times, *submitted.* (2009).
10. E. Hackmann, V. Kagramanova, J. Kunz, and C. Lämmerzahl, Analytic solution of the geodesic equation in NUT-de Sitter space-time, *in preparation.*

11. E. Hackmann, V. Kagramanova, J. Kunz, and C. Lämmerzahl, Analytic solution of the geodesic equation in axially symmetric space–times, *submitted.* (2009).
12. M. Demiański and M. Francaviglia, Type-D space-times with a Killing tensor, *J. Phys. A: Math. Gen.* **14**, 173, (1981).
13. E. Hackmann, V. Kagramanova, J. Kunz, and C. Lämmerzahl, Analytic solution of the geodesic equation in Plebanski–Demianski space-time in terms of hyperelliptic functions, *in preparation.* (2009).
14. B. Carter, Global structure of the Kerr family of gravitational fields, *Phys. Rev.* **174, 5**, 1559, (1968).
15. Y. Mino, Perturbative approach to an orbital evolution around a supermassive black hole, *Phys. Rev.* **D 67**, 084027, (2003).
16. B. O'Neill, *The Geometry of Black holes.* (A K Peters, Wellesly, Massasuchetts, 1995).
17. J. Levin and G. Perez-Giz, Homoclinic orbits around spinning black holes. I. Exact solution for the Kerr separatrix, *Phys. Rev.* **D 79**, 124013, (2009).
18. S. W. Hawking and G. Ellis, *The large scale structure of space-time.* (Cambridge Univ. P., Cambridge, 1973).

Chapter 20

A Shortcut to the Derivation of 2 + 1 Cyclic Symmetric Stationary Solutions

Alberto A. García

Departamento de Física,
Centro de Investigación y de Estudios Avanzados del IPN,
Apdo. Postal 14-740, 07000 México DF, México
Department of Physics, University of California, Davis, CA 95616, USA
aagarcia@fis.cinvestav.mx

From a general cyclic symmetric stationary metric of the 2 + 1–gravity a shortcut in the derivation of the three families of cyclic solutions is given. There arise two branches for the BTZ black hole solution, and additionally the Coussaert-Henneaux solution and the cyclic $SO(2) \times SO(2)$ cosmological metric.

Contents

20.1. Introduction

During the last two decades three–dimensional gravity has received some attention, in particular, in topics such as: black hole physics, search of exact solutions, quantization of fields coupled to gravity, cosmology, topological aspects, and others. This interest in part has been motivated by the discovery, in 1992, of the $2+1$ stationary circularly symmetric black hole solution by Bañados, Teitelboim and Zanelli [1]–the BTZ black hole– see also[2–4] , which possesses certain features inherent to $3+1$ black holes. On the other hand, it is believed that $2+1$ gravity may provide new insights towards a better understanding of the physics of $3+1$ gravity. In the framework of exact solutions in $2+1$ gravity the list of references on the topic is extremely vast; one finds works on point masses, cosmological and perfect fluid solutions, dilaton and string fields, and on electromagnetic fields coupled to gravity, among others.

The purpose of this contribution is to provide a breakthrough in the derivation of the cyclic symmetric solutions to the Anti–de Sitter–Einstein $2+1$ gravity. There exist three families: the cyclic symmetric stationary BTZ class, the Coussaert-Henneaux metric, cyclic $SO(2)\times SO(2)$ cosmological, and the static AdS class with its sub-branches: the one–parametric AdS solution, the AdS black hole, the AdS naked singularity metric.

20.2. Cyclic Symmetric Stationary Solutions in Canonical Coordinates

Within (2+1)–dimensional Einstein theory, the general form of a stationary cyclic symmetric line element– allowing for a timelike Killing vector $\partial_{\mathbf{t}}$ and a spacelike Killing vector field ∂_ϕ– can be given as

$$\mathbf{g} = -\frac{F(r)}{H(r)}\mathbf{dt}^2 + \frac{\mathbf{dr}^2}{F(r)} + H(r)(\mathbf{d}\phi + W(r)\mathbf{dt})^{\mathbf{2}}. \tag{20.1}$$

For this metric, the Einstein equations $E_\mu{}^\nu := G_\mu{}^\nu - \delta_\mu{}^\nu/l^2 = 8\pi\, T_\mu{}^\nu$, with negative cosmological constant, in the vacuum case $T_\mu{}^\nu = 0$ yield:

$$E_\phi{}^t = \frac{1}{2}\frac{d}{dr}(H^2\frac{dW}{dr}) = 0 \rightarrow \frac{dW}{dr} = \frac{J}{H^2}, \tag{20.2}$$

$$E_t{}^t + E_\phi{}^\phi + 2\,E_r{}^r = \frac{1}{2}\left(\frac{d^2F}{dr^2} - \frac{8}{l^2}\right) = 0 \rightarrow F(r) = \frac{4}{l^2}r^2 + F_1 r + F_0, \tag{20.3}$$

$$E_t{}^t - E_r{}^r - W\,E_\phi{}^t = \frac{F}{2H}\frac{d^2H}{dr^2} = 0 \rightarrow H(r) = H_1 r + H_0. \tag{20.4}$$

The integration of (20.2), assuming $H_1 \neq 0$, gives

$$W(r) = -\frac{J}{H_1}\frac{1}{H(r)} + W_0 = -\frac{J}{H_1}\frac{1}{H_1 r + H_0} + W_0, \tag{20.5}$$

without loss of generality one can set $W_0 = 0$. Moreover, the substitution of the above structural functions into the remaining Einstein equations gives rise to a condition on the integration constants, which can be solved, for instance, for F_0

$$F_0 = \frac{H_0}{H_1}F_1 + \frac{J^2}{H_1^2} - \frac{4}{l^2}\frac{H_0^2}{H_1^2}. \tag{20.6}$$

Therefore the metric tensor components become

$$\begin{aligned} g_{t\,t} &= -\frac{1}{H_1}(\frac{4}{l^2}r + F_1 - 4\frac{H_0}{l^2\,H_1}),\\ 1/g_{r\,r} &= \frac{4}{l^2}r^2 + F_1\,r + \frac{J^2}{H_1^2} + \frac{H_0}{H_1}(F_1 - 4\frac{H_0}{l^2\,H_1}),\\ g_{t\,\phi} &= -\frac{J}{H_1},\; g_{\phi\,\phi} = H_1\,r + H_0, \end{aligned} \tag{20.7}$$

20.3. BTZ Solution in Canonical R and Polar ρ–Coordinates

Accomplishing in (20.1), with structural functions determined by (20.3)–(20.5) and constraint (20.6), the linear transformations and re-parameterizations

$$t = T\sqrt{H_1},\ \phi = \Phi/\sqrt{H_1},\ r = R - H_0/H_1,\ J = J_0\,H_1, F_1 = \frac{8}{l^2}\frac{H_0}{H_1} - M, \tag{20.8}$$

one arrives at the BTZ solution in the form

$$\begin{aligned} \mathbf{g} &= -\frac{F(R)}{R}\mathbf{dT}^2 + \frac{\mathbf{dR}^2}{F(R)} + R\left(\mathbf{d\Phi} - \frac{J_0}{R}\mathbf{dT}\right)^2,\\ F(R) &= \frac{4}{l^2}R^2 - MR + J_0^2. \end{aligned} \tag{20.9}$$

Notice that this metric representation is quite close to the one reported by Clement;[9] by doing in (20.9) the coordinate transformations and constant rescaling

$$R = \rho + \frac{l^2}{4}M_{Cl}, T = t/\sqrt{2}, \Phi = \sqrt{2}\theta, M = 2M_{Cl}, J_0 = J/2$$

one arrives,modulo signature, just at Clement's ($Cl.20$).

A further transformation

$$T \to t,\ \Phi \to 2\phi,\ R \to \rho^2/4,\ 4F(\rho)/\rho^2 \to f(\rho), \tag{20.10}$$

brings (20.9) to the standard form

$$\begin{aligned} \mathbf{g} &= -f(\rho)\mathbf{dt}^2 + \frac{\mathbf{d}\rho^2}{f(\rho)} + \rho^2\left(\mathbf{d}\phi - 2\frac{J_0}{\rho^2}\mathbf{dt}\right)^2, \\ f(\rho) &= \frac{\rho^2}{l^2} - M + 4\frac{J_0^2}{\rho^2}. \end{aligned} \tag{20.11}$$

20.4. BTZ Solution Counterpart in Canonical R and Polar ρ–Coordinates

On the other hand, accomplishing in (20.1) the following linear transformations and re-parametrization

$$\begin{aligned} t &= T\frac{\sqrt{H_1}l}{2}, \quad \phi = \frac{2}{l\sqrt{H_1}}\Phi, \quad r = R - \frac{l^2}{4}(F_1 - 4\frac{H_0}{l^2 H_1}), \\ J &= J_0\, H_1, \quad F_1 = \frac{8}{l^2}\frac{H_0}{H_1} - M, \end{aligned} \tag{20.12}$$

one arrives at the BTZ solution in the form

$$\begin{aligned} \mathbf{g} &= -R\left(\mathbf{d\,T} + \frac{J_0}{R}\mathbf{d\Phi}\right)^2 + \frac{\mathbf{d\,R}^2}{F(R)} + \frac{F(R)}{R}\mathbf{d\Phi}^2, \\ &= -\frac{F(R)}{4R/l^2+M}\mathbf{dT}^2 + (4R/l^2+M)\left(\mathbf{d\Phi} - 2\frac{J_0}{4R/l^2+M}\mathbf{dT}\right)^2 + \frac{\mathbf{dR}^2}{F(R)}, \\ F(R) &= \frac{4}{l^2}R^2 + MR + J_0^2. \end{aligned} \tag{20.13}$$

The transformations

$$T \to 2t,\ \Phi \to \phi,\ R \to \rho^2/4,\ 4F(\rho)/\rho^2 \to f(\rho), \tag{20.14}$$

brings (20.13) to form

$$\begin{aligned} \mathbf{g} &= -\rho^2\left(\mathbf{dt} + 2\frac{J_0}{\rho^2}\mathbf{d}\phi\right)^2 + \frac{\mathbf{d}\rho^2}{f(\rho)} + f(\rho)\mathbf{d}\phi^2 \\ &= -\rho^2\,\frac{f(\rho)}{\rho^2/l^2 + M}\mathbf{dt}^2 + \frac{\mathbf{d}\rho^2}{f(\rho)} + (\rho^2/l^2 + M)[\mathbf{d}\phi - 2\frac{J_0}{\rho^2/l^2 + M}\,\mathbf{dt}]^2, \\ f(\rho) &= \frac{\rho^2}{l^2} + M + 4\frac{J_0^2}{\rho^2}. \end{aligned} \tag{20.15}$$

It becomes apparent that this metric form is just another real cut of the metric (20.11) when subjecting it to the complex transformations $t \to i\,\phi$ and $\phi \to i\,t$. Moreover, by accomplishing in the above metric the transformation of the radial coordinate

$$t \to t/l,\ \rho \to l\sqrt{\rho^2/l^2 - M},\ \phi \to l\,\phi \tag{20.16}$$

one gets the standard BTZ metric (20.11). In this representation, the coordinate ϕ loses its interpretation of circular angular coordinate; the same fact takes place in the stationary solution below.

20.5. Stationary Anti–de Sitter Solution with $H(r) = J\,l/2$; Coussaert–Henneaux and Cosmological Metrics

There exists the branch of solutions with constant $H(r) = H_0 = J\,l/2$, thus the integration of (20.2) yields $W(r) = 4r/(Jl^2)$, and therefore the resulting metric can be given as

$$\begin{aligned} \mathbf{g} &= -2\frac{F(r)}{J\,l}\mathbf{dt}^2 + \frac{\mathbf{dr}^2}{F(r)} + \frac{J\,l}{2}\left(\mathbf{d}\phi + \frac{4\,r}{J\,l^2}\mathbf{dt}\right)^2, \\ F(r) &= \frac{4}{l^2}r^2 + F_1\,r + F_0, \end{aligned} \tag{20.17}$$

which, by changing $t/\sqrt{Jl/2} \to t$, $\phi\sqrt{Jl/2} \to \phi$, can be brought to the form

$$\begin{aligned} \mathbf{g} &= -F(r)\mathbf{dt}^2 + \frac{\mathbf{dr}^2}{F(r)} + \left(\mathbf{d}\phi + \frac{2\,r}{l}\mathbf{dt}\right)^2, \\ F(r) &= \frac{4}{l^2}r^2 + F_1\,r + F_0 = \frac{4}{l^2}[(r + \frac{l^2}{8}F_1)^2 \pm A_{\pm}^2], \\ A_{\pm}^2 &:= \pm\frac{l^2}{4}(F_0 - \frac{l^2}{16}F_1^2). \end{aligned} \tag{20.18}$$

It is worthwhile to point out that this metric representation is close to the metric reported by Clement[9] ; by accomplishing in (20.18) the coordinate transformations and constant identifications

$$R = \rho - \frac{l^2}{8}F_1, t = \frac{l}{2A_{\pm}}t_{Cl}, \phi = \frac{2A_{\pm}}{l}\theta + \frac{l^2}{8}F_1, b := \frac{4A_{\pm}^2}{l^2}$$

one arrives,modulo signature, at Clement's (Cl.21).

Coussaert–Henneaux cyclic metric

Subjecting (20.18) to the transformations

$$t = \frac{l^2}{4\,A_+}\tilde{t}, \quad r = -\frac{l^2}{2}F_1 + A_+\,\sinh\tilde{r}, \quad \phi = -\frac{l^3\,F_1}{4\,A_+}\,\tilde{t} + \frac{l}{2}\tilde{\phi},$$

$$A_+ := \frac{l}{2}\sqrt{F_0 - \frac{l^2}{16}\,F_1^2}, F(r) \to \frac{4}{l^2}\,A_+^2\,\cosh^2\tilde{r} \tag{20.19}$$

one gets the standard Coussaert–Hennaux metric[8] , see also,[10]

$$\mathbf{g} = \frac{l^2}{4}\left[-\mathbf{d}\tilde{\mathbf{t}}^2 + 2\,\sinh\tilde{r}\mathbf{d}\tilde{\mathbf{t}}\,\mathbf{d}\tilde{\phi} + \mathbf{d}\tilde{\phi}^2 + \mathbf{d}\tilde{\mathbf{r}}^2\right]. \tag{20.20}$$

The time–dependent $SO(2) \times SO(2)$ cyclic metric

Moreover, if one were allowing the structural function $F(r)$ to range negative values, which takes place in the case of $A_-^2 = -\frac{l^2}{4}(F_0 - l^2\,F_1^2)$, lower negative sign in $F(r)$ from (20.18), subjecting (20.18) to the transformations

$$t = \frac{l^2}{4\,A_-}\theta, \; r = -\frac{l^2}{8}F_1 + A_-\,\sin\tau, \; \phi = \frac{l^3\,F_1}{16\,A_-}\,\theta + \frac{l}{2}\Phi,$$

$$A_- := \frac{l}{2}\sqrt{\frac{l^2}{16}\,F_1^2 - F_0}, F(r) \to -\frac{4}{l^2}\,A^2\,\cos^2\tau, \tag{20.21}$$

one arrives at a time–dependent (τ–coordinate) spacetime

$$\mathbf{g} = \frac{l^2}{4}\left[\mathbf{d}\theta^2 + \mathbf{d}\mathbf{\Phi}^2 + 2\,\sin\tau\mathbf{d}\theta\,\mathbf{d}\mathbf{\Phi} - \mathbf{d}\tau^2\right]. \tag{20.22}$$

This spacetime occurs to be the self–dual spacetime with isometry $SO(2) \times SO(2)$, see equation (34) of [10] and details therein.

20.6. Quasilocal Mass, Energy and Angular Momentum

To evaluate quasilocal mass, energy, and angular momentum of spacetimes with asymptotic different from the flat one, in particular the anti-de Sitter, one uses the quasilocal formalism developed in[5,6] .

For a stationary cyclic symmetric metric of the form

$$\mathbf{g} = -N(\rho)^2\,\mathbf{d}\,\mathbf{T}^2 + \frac{\mathbf{d}\rho^2}{L(\rho)^2} + K(\rho)^2\left[\mathbf{d}\mathbf{\Phi} + W(\rho)\mathbf{d}\,\mathbf{T}\right]^2, \tag{20.23}$$

the surface energy density is given by

$$\epsilon(\rho) = -\frac{L(\rho)}{\pi\, K(\rho)}\frac{d}{d\rho}K(\rho) - \epsilon_0, \tag{20.24}$$

where ϵ_0 is the reference energy density, which in the case of solutions with negative cosmological constant $\Lambda = -1/l^2$ corresponds to the density of the anti–de–Sitter spacetime, namely $\epsilon_0 = -\frac{1}{\pi\rho}\sqrt{1+\frac{\rho^2}{l^2}}$. The momentum density is determined from

$$j(\rho) = \frac{K(\rho)^2\, L(\rho)}{2\,\pi\, N(\rho)}\frac{d}{d\rho}W(\rho). \tag{20.25}$$

The integral momentum $J(\rho)$, global energy $E(\rho)$, and the integral mass $M(\rho)$ are correspondingly given by

$$\begin{aligned} J(\rho) &= 2\pi\, K(\rho)\, j(\rho), \\ E(\rho) &= 2\pi\, K(\rho)\, \epsilon(\rho), \\ M(\rho) &= E(\rho)\, K(\rho) - W(\rho)\, J(\rho). \end{aligned} \tag{20.26}$$

The evaluation of these physical quantities for the studied classes of solutions is given in the corresponding forthcoming subsections.

20.7. Mass, Energy and Momentum of the BTZ Solution

The energy–momentum characterization of the metric (20.9) is given by the following expressions

$$\begin{aligned} j(R) &= \frac{J_0}{2\pi\sqrt{R}},\ J(R) = J_0, \\ \epsilon(R,\epsilon_0) &= -\frac{1}{2\pi\, R}\sqrt{F(R,M,J_0)} - \epsilon_0, \\ E(R,\epsilon_0) &= -\frac{\sqrt{F(R,M,J_0)}}{\sqrt{R}} - 2\pi\epsilon_0\sqrt{R},\ E(R) \approx -\frac{2}{l}\sqrt{R} + M\frac{l}{4\sqrt{R}}, \\ M(R,\epsilon_0) &= -\frac{4R}{l^2} + M - 2\pi\epsilon_0\sqrt{F(R,M,J_0)},\ M(R) \approx -\frac{4}{l^2}R + M, \\ F(R,M,J_0) &:= \frac{4}{l^2}R^2 - MR + J_0^2. \end{aligned} \tag{20.27}$$

While these characteristics in coordinates $\{t,\rho,\phi\}$ amount to

$$\begin{aligned} j_\phi &= \frac{2J_0}{\pi\rho},\ j_\phi \approx \frac{2J_0}{\pi\rho}, \\ J_\phi &= 4J_0,\ J_\phi \approx 4J_0, \end{aligned} \tag{20.28a}$$

$$\epsilon(\rho, M, \epsilon_0) = -\frac{\sqrt{f(\rho, M, J_0)}}{\pi\rho} - \epsilon_0, \ \epsilon(\rho, M, 0) \approx -\frac{1}{\pi l} + \frac{l}{2\pi}\frac{M}{\rho^2},$$
$$\epsilon(\rho, M, \epsilon_0(M_0)) \approx \frac{l}{2\pi}\frac{M - M_0}{\rho^2}, \tag{20.28b}$$

$$E(\rho, M, \epsilon_0) = -2\sqrt{f(\rho, M, J_0)} - 2\pi\rho\epsilon_0,$$
$$E(\rho, M, 0) \approx -2\frac{\rho}{l} + \frac{l}{\rho}M, E(\rho, M, \epsilon_0(M_0)) \approx \frac{l}{\rho}(M - M_0), \tag{20.28c}$$

$$M(\rho, M, \epsilon_0) = -2\frac{\rho^2}{l^2} + 2M - 2\pi\rho\,\epsilon_0\sqrt{f(\rho, M, J_0)},$$
$$M(\rho, M, 0) = -2\frac{\rho^2}{l^2} + 2M, M(\rho, M, \epsilon_0(M_0)) \approx M - M_0, \tag{20.28d}$$

$$f(\rho, M, J_0) = \frac{\rho^2}{l^2} - M + 4\frac{J_0^2}{\rho^2},$$
$$\epsilon_0(M_0) := -\frac{\sqrt{\rho^2/l^2 - M_0}}{\pi\,\rho}, \ \epsilon_0(M_0) \approx -\frac{1}{\pi\, l} + \frac{l\, M_0}{2\pi\,\rho^2}. \tag{20.28e}$$

20.8. Mass, Energy and Momentum of the BTZ Solution Counterpart

Using the second representation of the metric (20.13),the quasi local energy–momentum characterization of the this metric is given by

$$\epsilon(R, \epsilon_0) = -\frac{2}{\pi\, l^2}\frac{\sqrt{F(R, M, J_0)}}{(4R/l^2 + M)} - \epsilon_0, \epsilon(R) \approx -\frac{1}{\pi\, l} + \frac{l\, M}{8\,\pi\, R},$$
$$j(R) = \frac{2J_0}{2\pi\sqrt{4R/l^2 + M}}, \ j(R) \approx -\frac{J}{\pi l R}\sqrt{R}, \ J(R) = \frac{4}{l^2}J_0,$$
$$E(R, \epsilon_0) = -\frac{4}{l^2}\frac{\sqrt{F(R, M, J_0)}}{\sqrt{4R/l^2 + M}} - 2\pi\epsilon_0\sqrt{4R/l^2 + M}, \ E(R) \approx -\frac{4}{l^2}\sqrt{R},$$
$$M(R, \epsilon_0) = -\frac{4R}{l} - 2\pi\epsilon_0\sqrt{F(R, M, J_0)}, \ M(R) \approx -\frac{4}{l^2}R,$$
$$F(R, M, J_0) := \frac{4}{l^2}R^2 + MR + J_0^2. \tag{20.29}$$

As far as to the energy-momentum characteristics are concerned, for the second form of the metric (20.15), one has

$$\begin{aligned} j_\phi &= -\frac{2J_0}{l^2\,\pi\sqrt{\rho^2/\,l^2+M}},\ j_\phi \approx -\frac{2J_0}{l\,\pi\rho}, \\ J_\phi &= -\frac{4J_0}{l^2},\ J_\phi \approx -\frac{4J_0}{l^2}, \end{aligned} \tag{20.30a}$$

$$\begin{aligned} \epsilon(\rho,M,\epsilon_0) &= -\frac{\rho}{l^2\pi}\frac{\sqrt{f(\rho,M,J_0)}}{\rho^2/\,l^2+M}-\epsilon_0, \\ \epsilon(\rho,M,0) &\approx -\frac{1}{\pi l}+\frac{l}{2\pi}\frac{M}{\rho^2}, \end{aligned} \tag{20.30b}$$

$$\begin{aligned} E(\rho,M,\epsilon_0) &= -\frac{2}{l^2}\,\rho\frac{\sqrt{f(\rho,M,J_0)}}{\sqrt{\rho^2/\,l^2+M}}-2\pi\,\epsilon_0\sqrt{\rho^2/\,l^2+M}, \\ E(\rho,M,0) &\approx -2\frac{\rho}{l^2},\ E(\rho,M,\epsilon_0(M_0))\approx\frac{1}{\rho}(M-M_0), \end{aligned} \tag{20.30c}$$

$$\begin{aligned} M(\rho,M,\epsilon_0) &= -\frac{2}{l^2}\rho^2-2\pi\,\rho\,\epsilon_0\sqrt{f(\rho,M,J_0)}, \\ M(\rho,M,0) &= -2\frac{\rho^2}{l^2},\,M(\rho,M,\epsilon_0(M_0))\approx M-M_0, \end{aligned} \tag{20.30d}$$

$$\begin{aligned} f(\rho,M,J_0) &= \frac{\rho^2}{l^2}+M+4\frac{J_0^2}{\rho^2}, \\ \epsilon_0(M_0) &:= -\frac{\rho}{\pi\,l^2\sqrt{\rho^2/\,l^2+M_0}}, \\ \epsilon_0(M_0) &\approx -\frac{1}{\pi\,l}+\frac{l\,M_0}{2\pi\,\rho^2}. \end{aligned} \tag{20.30e}$$

Mass, energy and momentum of the Coussaert–Henneaux solution

If one were interested in the quasi local energy–momentum characterization of the the Coussaert–Henneaux metric (20.18) one would arrive at

$$\begin{aligned} j(r) &= \frac{1}{\pi\,l},\ J(r)=\frac{2}{l}, \\ \epsilon(r,\epsilon_0) &= -\epsilon_0,\ \epsilon(r,0)=0,\ E(r,\epsilon_0)=-2\pi\epsilon_0,\ E(r,0)=0, \\ M(r,\epsilon_0) &= -\frac{4r}{l^2}-2\pi\epsilon_0,\ M(r,0)=-\frac{4r}{l^2}. \end{aligned} \tag{20.31}$$

20.9. Concluding Remarks

In the framework of the (2+1)–dimensional Einstein theory with cosmological constant different families of exact solutions for cyclic symmetric stationary (static) metrics have been derived. Specific branches of solutions in the general case are determined via a straightforward integration. In this systematic approach all known cyclic symmetric solutions are properly identified.

Acknowledgments

This work is dedicated to Leopoldo García–Colín on the occasion of his eightieth birthday. This work has been partially supported by UC MEXUS–CONACyT for Mexican and UC Faculty Fellowships 09012007, and grants CONACyT 57195, and 82443. The author thanks S. Carlip and acknowledges the hospitality of the Physics Department, University of California, Davis, at which most of the related work has been accomplished.

A.1. Static AdS Black Hole in Canonical Coordinates

Working with the canonical metric (20.1) the Einstein equations in the static case, $W(r) = 0$, simplify considerably. One can obtain the anti–de Sitter static black hole solution from the BTZ metric (20.9 simply by setting $J = 0$, or following step by step the integration process exhibited in the BTZ case. By setting $J_0 = 0$ in the metric (20.9), one arrives at the anti–de Sitter static black hole solution in the form

$$\mathbf{g} = -\frac{F(R)}{R}\mathbf{dT}^2 + \frac{\mathbf{dR}^2}{F(R)} + R\,\mathbf{d\Phi}^2,$$

$$F(R, M) = \frac{4}{l^2}R^2 - M\,R, \tag{A.1}$$

The quasi local energy and mass of this metric are given by the corresponding quantities of (20.27) except that now $F(R, M, J_0) \rightarrow F(R, M, 0) = F(R, M)$.

Explicitly, the quasi local energy and mass for the above metric are given as

$$\begin{aligned}
F(R) &= \frac{4}{l^2}R^2 - M\,R,\\
\epsilon(R,\epsilon_0) &= -\frac{1}{4\pi R}\sqrt{F(R)} - \epsilon_0,\\
E(R,\epsilon_0) &= -\frac{\sqrt{F(R)}}{\sqrt{R}} - 2\pi\epsilon_0\sqrt{R},\ E(R) \approx -\frac{2}{l}\sqrt{R} + M\,\frac{l}{4\sqrt{R}},\\
M(R,\epsilon_0) &= -\frac{4R}{l^2} + M - 2\pi\epsilon_0\sqrt{F(R)},\ M(R) \approx -\frac{4}{l^2}R + M\,. \qquad \text{(A.2)}
\end{aligned}$$

In polar coordinates the corresponding static metric is given by metric (20.11) with $J_0 = 0$, namely

$$\begin{aligned}
\mathbf{g} &= -f(\rho, M\,)\mathbf{dt}^2 + \frac{\mathbf{d}\rho^2}{f(\rho, M\,)} + \rho^2\,\mathbf{d}\phi^2,\\
f(\rho, M\,) &:= \frac{\rho^2}{l^2} - M\,. \qquad \text{(A.3)}
\end{aligned}$$

In this coordinates the mass–energy characteristics, introducing the energy density function $\epsilon(\rho, M_0)$,

$$\epsilon(\rho, M) := -\frac{1}{\pi\rho}\sqrt{f(\rho, M)} \qquad \text{(A.4)}$$

are given in the following form

$$\begin{aligned}
\epsilon(\rho, M\,,\epsilon_0) &= \epsilon(\rho, M\,) - \epsilon_0,\\
E(\rho, M\,,\epsilon_0) &= 2\pi\rho\left[\epsilon(\rho, M\,) - \epsilon_0\right],\\
M(\rho, M\,,\epsilon_0) &= 2\pi\,\rho\sqrt{f(\rho, M\,)}\left[\epsilon(\rho, M\,) - \epsilon_0\right], \qquad \text{(A.5)}
\end{aligned}$$

they explicitly amount to

$$\begin{aligned}
\epsilon(\rho, M\,,\epsilon_0) &= -\frac{1}{\pi\rho}\sqrt{f(\rho, M\,)} - \epsilon_0,\ \epsilon(\rho, M\,,0) \approx -\frac{1}{\pi l} + \frac{l}{2\pi}\frac{M}{\rho^2},\\
E(\rho, M\,,\epsilon_0) &= -2\sqrt{f(\rho, M\,)} - 2\pi\,\rho\,\epsilon_0,\ E(\rho, M\,,0) \approx -2\frac{\rho}{l} + \frac{l}{\rho}M\,,\\
M(\rho, M\,,\epsilon_0) &= -2f(\rho, M\,) - 2\pi\rho\sqrt{f(\rho, M\,)}\;\epsilon_0,\\
M(\rho, M\,,0)) &\approx -2\frac{\rho^2}{l^2} + 2M. \qquad \text{(A.6)}
\end{aligned}$$

Choosing the base energy density; AdS spacetime

For $M = -1$ one obtains the standard anti–de Sitter solution in polar coordinates

$$\begin{aligned} \mathbf{g} &= -f(\rho)\mathbf{dt}^2 + \frac{\mathbf{d}\rho^2}{f(\rho)} + \rho^2\,\mathbf{d}\phi^2, \\ f(\rho) &= \frac{\rho^2}{l^2} + 1, \end{aligned} \tag{A.7}$$

with energy density

$$\begin{aligned} \epsilon(\rho,-1,0) = \epsilon_{AdS}(\rho) &:= -\frac{1}{\pi\rho}\sqrt{\frac{\rho^2}{l^2}+1},\ \epsilon_{AdS}(\rho) \approx -\frac{1}{\pi l} - \frac{l}{2\pi\,\rho^2}, \\ \epsilon(\rho,-1,\epsilon_0) &= \epsilon_{AdS}(\rho) - \epsilon_0, \end{aligned} \tag{A.8}$$

therefore, for the referential energy density choice $\epsilon_0 = \epsilon_{AdS}(\rho)$ the mass–energy functions of the the anti–de Sitter become zero :

$$\epsilon(\rho,-1,\epsilon_{AdS}) = 0,\ E(\rho,-1,\epsilon_{AdS}(\rho)) = 0, M(\rho,-1,\epsilon_{AdS}(\rho)) = 0.$$

This is one of the plausible choices for the base energy density of the anti–de Sitter spacetime, see[6] in the paragraph below Eq. (4.12). Another possibility takes place for the naked singularity solution with $M = 0$,

$$\begin{aligned} \mathbf{g} &= -f(\rho,0)\mathbf{dt}^2 + \frac{\mathbf{d}\rho^2}{f(\rho,0)} + \rho^2\,\mathbf{d}\phi^2, \\ f(\rho,0) &= \frac{\rho^2}{l^2}, \end{aligned} \tag{A.9}$$

with energy characteristics

$$\begin{aligned} \epsilon(\rho,0) &:= -\frac{1}{\pi\,l},\ \epsilon(\rho,0,\epsilon_0) = -\frac{1}{\pi\,l} - \epsilon_0, \\ E(\rho,0,\epsilon_0) &= 2\pi\,\rho\,\epsilon(\rho,0,\epsilon_0),\ M(\rho,0,\epsilon_0) = \frac{\rho}{l}\,E(\rho,0,\epsilon_0). \end{aligned} \tag{A.10}$$

For the particular referential energy density $\epsilon_0 = -\frac{1}{\pi\,l}$, all energy characteristics of this naked anti–de Sitter solution vanish,

$$\epsilon(\rho,0,\epsilon_0 = -\frac{1}{\pi\,l}) = 0, E(\rho,0,\epsilon_0 = -\frac{1}{\pi\,l}) = 0, M(\rho,0,\epsilon_0 = -\frac{1}{\pi\,l}) = 0,$$

in agreement with,[6] Eq. (4.12).

Limits of the mass–energy functions of the anti de Sitter black hole

For the static black hole metric (A.3), with M parameter, the energy characteristics at infinity, $\rho \to \infty$, become:
for $\epsilon_0 = -\frac{1}{\pi l}$

$$\begin{aligned}\epsilon\left(\rho, M, \epsilon_0 = -\frac{1}{\pi l}\right) &\approx \frac{M\,l}{2\pi\rho^2},\\ E\left(\rho, M, \epsilon_0 = -\frac{1}{\pi l}\right) &\approx \frac{M\,l}{\rho},\\ M\left(\rho, M, \epsilon_0 \approx -\frac{1}{\pi l}\right) &\approx M\,,\end{aligned} \tag{A.11}$$

while for $\epsilon_0 = \epsilon_{AdS}$ they are

$$\begin{aligned}\epsilon(\rho, M, \epsilon_{AdS}) &\approx \frac{l}{2\pi\rho^2}(1 + M),\\ E(\rho, M, \epsilon_{AdS}) &\approx \frac{l}{\rho}(1 + M),\\ M(\rho, M, \epsilon_{AdS}) &\approx 1 + M\,.\end{aligned} \tag{A.12}$$

These results can be gathered in a more compact form by introducing the base energy density equipped with the discrete parameter $m = -1$, and 0,

$$\epsilon_0(\rho, m) =:= -\frac{1}{\pi\rho}\sqrt{\frac{\rho^2}{l^2} - m},\ \epsilon_0(\rho, m) \approx -\frac{1}{\pi l} + \frac{l\,m}{2\pi\,\rho^2}, \tag{A.13}$$

as

$$\begin{aligned}\epsilon(\rho, M, \epsilon_0(\rho, m)) &\approx \frac{l}{2\pi\rho^2}(M - m),\\ E(\rho, M, \epsilon_0(\rho, m)) &\approx \frac{l}{\rho}(M - m),\\ M(\rho, M, \epsilon_0(\rho, m)) &\approx M - m.\end{aligned} \tag{A.14}$$

A.2. Static AdS Solution Counterpart in Canonical Coordinates

Setting $J_0 = 0$ in the canonical metric (20.13) one gets

$$\begin{aligned}\mathbf{g} &= -R\mathbf{d}\,\mathbf{T}^2 + \frac{\mathbf{d}\,\mathbf{R}^2}{F(R, M)} + (\frac{4}{l^2}R + M)\mathbf{d}\mathbf{\Phi}^2,\\ F(R) &= \frac{4}{l^2}R^2 + M\,R.\end{aligned} \tag{A.15}$$

The mass–energy characteristics are given by expressions (20.27), where instead of $F(R, M, J_0)$ has to be replaced $F(R, M, 0) = F(R, M)$. The corresponding static metric is obtained from (20.15) by equating $J_0 = 0$, this yields

$$\mathbf{g} = -\rho^2 \mathbf{d}\,\mathbf{t}^2 + \frac{\mathbf{d}\,\rho^2}{(\rho^2/l^2 + M)} + (\rho^2/l^2 + M)\mathbf{d}\phi^2. \tag{A.16}$$

Defining the base energy density function $\epsilon(\rho, M)$,

$$\epsilon(\rho, M) := -\frac{\rho}{\pi\, l^2}\frac{1}{\sqrt{\rho^2/l^2 + M}},\ \epsilon(\rho, M) \approx -\frac{1}{\pi l} + \frac{l}{2\pi}\frac{M}{\rho^2}. \tag{A.17}$$

In this coordinates the energy characteristics are given in the following form

$$\begin{aligned} \epsilon(\rho, M\,, \epsilon_0) &= \epsilon(\rho, M) - \epsilon_0, \\ E(\rho, M\,, \epsilon_0) &= 2\pi\sqrt{f(\rho, M)}\left[\epsilon(\rho, M\,) - \epsilon_0\right], \\ M(\rho, M\,, \epsilon_0) &= 2\pi\, f(\rho, M\,)\left[\epsilon(\rho, M\,) - \epsilon_0\right], \end{aligned} \tag{A.18}$$

or explicitly

$$\begin{aligned} \epsilon(\rho, M, \epsilon_0) &= -\frac{\rho}{\pi l^2}\frac{1}{\sqrt{\rho^2/l^2 + M}} - \epsilon_0,\ \epsilon(\rho, M, 0) \approx -\frac{1}{\pi l} + \frac{l}{2\pi}\frac{M}{\rho^2}, \\ E(\rho, M, \epsilon_0) &= -2\frac{\rho}{l^2} - 2\pi\,\epsilon_0\sqrt{\rho^2/l^2 + M},\ E(\rho, M, 0) \approx -2\frac{\rho}{l^2}, \\ M(\rho, M, \epsilon_0) &= -2\frac{\rho^2}{l^2} - 2\pi\rho\,\epsilon_0\sqrt{\rho^2/l^2 + M}, \\ M(\rho, M, 0) &\approx -2\frac{\rho^2}{l^2}. \end{aligned} \tag{A.19}$$

One could define the base energy density as the one corresponding to the black hole limit of the anti–de Sitter spacetime or that one associated with the naked singularity; both cases can be handled by introducing the discrete parameter $m = 0, -1$, and defining the base energy density function $\epsilon(\rho, m)$,

$$\epsilon_0(\rho, m) := -\frac{\rho}{\pi\, l^2}\frac{1}{\sqrt{\rho^2/l^2 + m}},\ \epsilon_0(\rho, m) \approx -\frac{1}{\pi l} + \frac{l}{2\pi}\frac{m}{\rho^2}. \tag{A.20}$$

For such choice of the base energy density, the mass–energy characteristics of the proper anti–de Sitter solution counterpart, as it should be, vanish: $\epsilon(\rho, m, \epsilon_0(\rho, m)) = 0$, $E(\rho, m, \epsilon_0(\rho, m)) = 0$, $M(\rho, m, \epsilon_0(\rho, m)) = 0$.

The limit at spatial infinity $\rho \to \infty$ of the energy quantities are

$$\begin{aligned} \epsilon(\rho, M, \epsilon_0(\rho, m)) &\approx \frac{l}{2\pi\rho^2}(M-m) \\ E(\rho, M, \epsilon_0(\rho, m)) &\approx \frac{1}{\rho}(M-m), \\ M(\rho, M, \epsilon_0(\rho, m)) &\approx M-m. \end{aligned} \tag{A.21}$$

By accomplishing in metric (A.15) the transformations (20.16) one gets the anti de Sitter black hole metric (A.3). The presence of powers of l in these transformations explains the appearance of different powers in the limits at spatial infinity $\rho \to \infty$ of the energy–momentum characteristics for the BTZ and AdS black hole solutions compared with the corresponding quantities of their solution counterparts.

References

1. M. Bañados, C. Teitelboim and J. Zanelli, The black hole in three–dimensional spacetime, *Phys. Rev. Lett.* **69**, (1992) 1849 [hep-th/9204099].
2. M. Bañados, M. Henneaux, C. Teitelboim and J. Zanelli, Geometry of the 2+1 black hole, *Phys. Rev.* **D 48**, (1993) 1506 [gr-qc/9302012].
3. D. Cangemi, M. Leblanc, R.B. Mann, Gauge formulation of the spinning black hole in (2+1)-dimensional anti–de Sitter space, *Phys. Rev.* **D 48**, (1993) 3606.
4. S. Carlip, The (2+1)-dimensional black hole, *Class. Quantum Grav.* **12**, (1995) 2853.
5. J.D. Brown and J.W. York, Jr. Quasilocal energy and conserved charges derived from the gravitational action, *Phys. Rev.* **D 47** 1407 (1993).
6. J.D. Brown, J. Creighton, and R.B. Mann, Temperature, energy, and heat capacity of asymptotically anti–de Sitter black holes, *Phys. Rev.* **D 50** 6384 (1994).
7. A.A. García, Three–dimensional stationary cyclic symmetric Einstein–Maxwell solutions, *Annals of Phys.* **324**, 2004–2050 (2009).
8. O. Coussaert and M. Henneaux, in *The Black Hole, 25 Years Later*, Ed. C. Teitelboim and J. Zanelli (World Scientific, Singapore 1998) [arXiv: hepth/9407181].
9. G. Clement, Classical solutions in three–dimensional cosmological gravity, *Phys. Rev.* **D 49** 5131(1994).
10. E. Ayon–Beato, C. Martinez, and J. Zanelli. Birkhoffs Theorem for Three–Dimensional AdS Gravity, *Phys. Rev.* **D 70** 044027 (2004).

Chapter 21

Averaging and Back–Reaction in Spherically Symmetric Dust Models

Roberto A. Sussman
Instituto de Ciencias Nucleares, UNAM, México D.F. 04510, México
sussman@nucleares.unam.mx

Quasi–local (QL) scalar variables are introduced for spherically symmetric LTB models in order to examine the "back–reaction" term $\mathcal{Q}$ in the context of Buchert's scalar averaging formalism. Since QL scalars defined as functionals become weighed averages that generalize the standard proper volume averages, we are able to provide rigorous proof that back–reaction is positive for (i) all LTB models with negative and asymptotically negative spatial curvature, and (ii) models with positive curvature decaying to zero asymptotically in the radial direction. Qualitative, but robust, arguments are given in order to prove that generic LTB models exist, either with clump or void profiles, for which an "effective" acceleration associated with Buchert's formalism can mimic the effects of dark energy.

Contents

21.1. Introduction

Recent observations apparently reveal that the universe is spatially flat and is undergoing an accelerated expansion. To account for these observations, a large variety of theoretical and empiric models have molded a dominant theoretical paradigm: the "concordance" model, based on the assumption that cosmic dynamics appears to be dominated by an elusive source ("dark energy") that behaves as a cosmological constant or as a fluid with negative pressure (see[1] for a review).

The concordance model also assumes that inhomogeneities play a minimal role in cosmic dynamics at scales over 100 Mpc, and thus can be adequately dealt with in terms of linear perturbations on a FLRW background. This assumption, and the existence of dark energy, has been challenged from various angles.[2] In trying to account with supernovae observations, numerous articles (see[2] for a review, see also[3]) show that cosmic acceleration can be reproduced simply by considering large scale inhomogeneities in photon trajectories within the so–called "homogeneity scale" (100-300 Mpc). From a theoretical point of view, it has been argued that inhomogeneity implies that observations from distant high redshift sources must be understood in terms of averaged quantities,[3–8] which in homogeneous conditions would be trivially identical with local quantities. Thus, non–linear spatial gradients of the Hubble expansion scalar and quasi–local effective energies would have an important effect in the interpretation of these observations.[8]

The spherically symmetric LTB dust models[9–11] have often been used to test the effects of inhomogeneity in cosmic observations, as well as the issue of back–reaction in the context of Buchert's spatial averaging.[12,13] In the present article we examine the dynamical equations of there models in terms of suitably defined quasi–local variables and of spatially averaged scalars.[14,15] We show that the back–reaction term is the difference between squared fluctuations of the expansion Hubble scalar, which in turn are related to spatial gradients of its average and quasi–local equivalent. Necessary conditions for a positive "effective" accelerations, mimicking the effect of dark energy, follow from comparing these fluctuations, and are satisfied as long as spatial curvature is negative in a sufficiently large averaging domain (even if smaller domains contain bound structures). Since these conditions are compatible with the observed void dominated structure, it is highly likely that such "effective" acceleration could be observed. However, further steps in this direction require a more elaborate numerical study of these models (see[16]).

21.2. LTB Dust Models in the "Fluid Flow" Description

Spherically symmetric inhomogeneous dust sources are usually described by the well known Lemaître–Tolman–Bondi metric and energy–momentum tensor in a comoving frame[9–11]

$$\mathrm{d}s^2 = -c^2\mathrm{d}t^2 + \frac{R'^2}{\mathcal{F}^2}\mathrm{d}r^2 + R^2\left(\mathrm{d}\theta^2 + \sin^2\theta\mathrm{d}\phi^2\right). \tag{21.1}$$

$$T^{ab} = \rho\, u^a u^b, \tag{21.2}$$

where $R = R(ct, r)$, $R' = \partial R/\partial r$, $\mathcal{F} = \mathcal{F}(r)$, $u^a = \delta^a_0$ and $\rho(t,r)$ is the rest matter–energy density. The field equations $G^{ab} = \kappa T^{ab}$ (with $\kappa = 8\pi G/c^2$) for (21.1) and (21.2) reduce to

$$\dot{R}^2 = \frac{2M}{R} + \mathcal{F}^2 - 1, \tag{21.3}$$

$$2M' = \kappa\,\rho\,R^2 R', \tag{21.4}$$

where $M = M(r)$ and $\dot{R} = u^a\nabla_a R$. The sign of $\mathcal{F}^2 - 1$ determines the zeroes of $\dot{R}$ and thus classifies LTB models in terms of the following kinematic classes:

$$\begin{aligned}
\mathcal{F} &= 1, && \text{parabolic models}\\
\mathcal{F} &\geq 1, && \text{hyperbolic models}\\
0 < \mathcal{F} &\leq 1, && \text{“open” elliptic models}\\
-1 \leq \mathcal{F} &\leq 1, && \text{“closed” elliptic models}
\end{aligned} \tag{21.5}$$

where by “open” or “closed” models we mean the cases where the $\mathcal{T}^{(3)}$ are, respectively, topologically equivalent to $\mathbf{R}^3$ and $\mathbf{S}^3$. Regularity conditions[10,11] require $\mathcal{F}' \geq 0$ for all r in hyperbolic models, while in open elliptic models $\mathcal{F}'$ can be, either negative for all $r > 0$, or it can have a zero $\mathcal{F}'(y) = 0$, so that $\mathcal{F}' \leq 0$ for $0 \leq x \leq y$ and $\mathcal{F}' > 0$ for $x > y$.

Besides ρ and R given above, other covariant objects of LTB spacetimes are the expansion scalar Θ, the Ricci scalar ${}^3\mathcal{R}$ of the hypersurfaces $\mathcal{T}^{(3)}$ orthogonal to u^a, the shear tensor σ_{ab} and the electric Weyl tensor E^{ab}:

$$\Theta = \tilde{\nabla}_a u^a = \frac{2\dot{R}}{R} + \frac{\dot{R}'}{R'}, \quad {}^3\mathcal{R} = \frac{2[(1-\mathcal{F})\,R]'}{R^2 R'}, \tag{21.6}$$

$$\sigma_{ab} = \tilde{\nabla}_{(a}u_{b)} - (\Theta/3)h_{ab} = \Sigma\,\Xi^{ab}, \quad E^{ab} = u_c u_d C^{abcd} = \mathcal{E}\,\Xi^{ab}, \tag{21.7}$$

where $h_{ab} = u_a u_b - g_{ab}$, $\tilde{\nabla}_a = h^b_a\nabla_b$, and C^{abcd} is the Weyl tensor, $\Xi^{ab} = h^{ab} - 3\eta^a\eta^b$ with $\eta^a = \sqrt{h^{rr}}\delta^a_r$ being the unit radial vector orthogonal to u^a. The scalars $\mathcal{E}$ and Σ in (21.7) are

$$\Sigma = \frac{1}{3}\left[\frac{\dot{R}}{R} - \frac{\dot{R}'}{R'}\right], \qquad \mathcal{E} = -\frac{\kappa}{6}\rho + \frac{M}{R^3}. \tag{21.8}$$

The dynamics of LTB spacetimes can be fully characterized by the local covariant scalars $\{\mu, \Theta, \Sigma, \mathcal{E}, {}^3\mathcal{R}\}$. Given the covariant "1+3" slicing afforded by u^a, the evolution of the models can be completely determined by a "fluid flow" description of scalar evolution equations for these scalars (as in[17]):

$$\dot{\Theta} = -\frac{\Theta^2}{3} - \frac{\kappa}{2}\rho - 6\,\Sigma^2, \tag{21.9}$$

$$\dot{\rho} = -\rho\,\Theta, \tag{21.10}$$

$$\dot{\Sigma} = -\frac{2\Theta}{3}\Sigma + \Sigma^2 - \mathcal{E}, \tag{21.11}$$

$$\dot{\mathcal{E}} = -\frac{\kappa}{2}\rho\,\Sigma - 3\,\mathcal{E}\left(\frac{\Theta}{3} + \Sigma\right), \tag{21.12}$$

together with the spacelike constraints

$$\left(\Sigma + \frac{\Theta}{3}\right)' + 3\,\Sigma\,\frac{R'}{R} = 0, \qquad \frac{\kappa}{6}\rho' + \mathcal{E}\,' + 3\,\mathcal{E}\,\frac{R'}{R} = 0, \tag{21.13}$$

and the Friedman equation (or "Hamiltonian" constraint)

$$\left(\frac{\Theta}{3}\right)^2 = \frac{\kappa}{3}\,\mu - \frac{{}^3\mathcal{R}}{6} + \Sigma^2, \tag{21.14}$$

The solutions of system (21.9)–(21.14) are equivalent to the solution of the field plus conservation equations $\nabla_b T^{ab} = 0$.

21.3. Proper Volume Average: Buchert's Formalism

The time slicing defined by u^a defines as the space slices the hypersurfaces $\mathcal{T}^{(3)}$ marked by constant t, whose metric is $h_{ab} = u_a u_b + g_{ab}$ and their proper volume element is

$$\mathrm{d}\mathcal{V}_p = \sqrt{\det(h_{ab})}\,\mathrm{d}r\,\mathrm{d}\theta\,\mathrm{d}\phi = \mathcal{F}^{-1}\,R^2\,R'\,\sin\theta\,\mathrm{d}r\,\mathrm{d}\theta\,\mathrm{d}\phi. \tag{21.15}$$

Consider spherical comoving regions of the form

$$\mathcal{D}[r] = \vartheta[r] \times \mathbf{S}^2 \subset \mathcal{T}^{(3)}, \qquad \vartheta[r] \equiv \{x \in \mathbf{R} \,|\, 0 \leq x \leq r\} \tag{21.16}$$

where $\mathbf{S}^2$ is the unit 2–sphere and $x = 0$ marks a symmetry center. We introduce now the following definitions:

> Let $X(\mathcal{D}[r])$ be the set of all smooth integrable scalar functions in $\mathcal{D}$, then

Definition 1. For every $A \in X(\mathcal{D}[r])$, the p-map is defined as

$$\mathcal{J}_p : X(\mathcal{D}[r]) \to X(\mathcal{D}[r]),\ \ A_p = \mathcal{J}_p(A) = \frac{\int_{\mathcal{D}[r]} A\,\mathrm{d}\mathcal{V}_p}{\int_{\mathcal{D}[r]} \mathrm{d}\mathcal{V}_p} = \frac{\int_0^r A\mathcal{F}^{-1}R^2R'\mathrm{d}x}{\int_0^r \mathcal{F}^{-1}R^2R'\mathrm{d}x}. \tag{21.17}$$

where $\int_0^r ..\,\mathrm{d}x = \int_{x=0}^{x=r} ..\,\mathrm{d}x$. The scalar functions $A_p : \mathcal{D} \to \mathrm{R}$ that are images of $\mathcal{J}_p$ will be denoted by "p–functions". In particular, we will call A_p the p–dual of A.

Definition 2. For every $A \in X(\mathcal{D})$, the proper volume average is the functional

$$\langle\ \rangle_p[r] : X(\mathcal{D}) \to \mathbf{R}, \qquad \langle A\rangle_p[r] = A_p(r). \tag{21.18}$$

The real number $\langle A\rangle_p[r]$, associated to the full domain $\mathcal{D}[r]$, will be denoted by the proper volume average of A on $\mathcal{D}[r]$. In order to simplify notation, we will drop the "$[r]$" and $_p$ symbols and express (21.18) simply as $\langle A\rangle$, as it is clear that the average is evaluated as a proper volume integral and it depends on the domain boundary as a functional.

The proper volume average (as well as the p–functions) satisfy the following commutation rule:

$$\langle A\rangle^{\cdot} - \langle \dot{A}\rangle = \langle \Theta A\rangle - \langle\Theta\rangle\langle A\rangle, \tag{21.19}$$

It is important to emphasize that only the functionals $\langle A\rangle$ can be considered as average distributions, as they satisfy $\langle A\rangle\rangle = \langle A\rangle$, and thus

$$\langle (A - \langle A\rangle)^2\rangle = \langle A^2\rangle - \langle A\rangle^2, \tag{21.20}$$

$$\langle (A - \langle A\rangle)\,(B - \langle B\rangle)\rangle = \langle AB\rangle - \langle A\rangle\langle B\rangle, \tag{21.21}$$

which define variance and covariance moment definitions for continuous random variables. It is straightforward to verify that the functions A_p do not satisfy these equations. The relation between the p–functions A_p and the average $\langle A\rangle$ is illustrated by figure 1.

The well known evolution equations of Buchert's formalism[4] follow by applying the proper volume average functional (21.18) to both sides of the energy balance (21.10), Raychaudhuri (21.9) and Friedman (21.14) equations, and then using (21.19) and (21.20)–(21.21) to eliminate averages $\langle\dot{A}\rangle$ in terms of derivatives of averages $\langle A\rangle^{\cdot}$ and squares of averages as averages

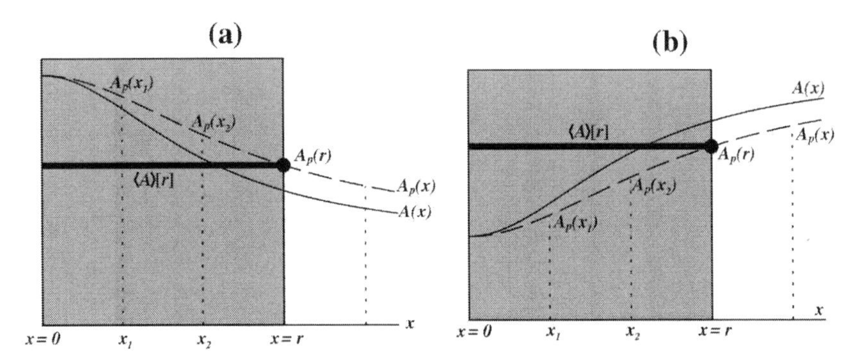

Fig. 21.1. **The difference between A_p and $\langle A\rangle$.** The figure displays the radial profile of a scalar function $A(x)$ (solid curve) along a regular hypersurface $\mathcal{T}^{(3)}(t)$, together with its dual p–function $A_p(x)$ (dotted curve) defined by (21.17). Panels (a) and (b) respectively display the cases when $A' \leq 0$ ("clump") and $A' \geq 0$ ("void"). The average functional (21.18) assigns the real number $\langle A\rangle[r]$ to the full domain (shaded area) marked by $\vartheta[r] = \{x \,|\, 0 \leq x \leq r\}$, whereas the function A_p varies along this domain. Hence, A_p and $\langle A\rangle$ are only equal at the domain boundary $x = r$, and so they satisfy the same differentiation rules locally, *i.e.* $\dot{A}_p(r) = \langle A\rangle^{\cdot}[r]$ and $A_p'(r) = \langle A\rangle'[r]$, but behave differently when integrated along the domain. Notice that, from (21.46) and (21.47), if $A' \leq 0$ in all $\vartheta[r]$ then $A - \langle A\rangle \leq 0$ and the opposite situation occurs if $A' \geq 0$. This figure also applies for the quasi–local functions and averages.

of squares. The averaged forms of these equations are:

$$\langle \dot{\rho} + \rho\Theta\rangle = \langle\rho\rangle^{\cdot} + \langle\rho\rangle\langle\Theta\rangle = 0, \tag{21.22}$$

$$\langle\Theta\rangle^{\cdot} + \frac{\langle\Theta\rangle^2}{3} = -\frac{\kappa}{2}\langle\rho\rangle + \mathcal{Q}, \tag{21.23}$$

$$\frac{\langle\Theta\rangle^2}{9} - \frac{\kappa}{3}\langle\rho\rangle = -\frac{\langle {}^3\mathcal{R}\rangle + \mathcal{Q}}{6}, \tag{21.24}$$

where the kinematic "back–reaction" term, $\mathcal{Q}$, is given by

$$\mathcal{Q}[r] \equiv \frac{2}{3}\,\langle(\Theta - \langle\Theta\rangle)^2\rangle - 6\,\langle\Sigma^2\rangle. \tag{21.25}$$

Equation (21.22) simply expresses the compatibility between the averaging (21.18) and the conservation of rest mass, but (21.23) and (21.24) lead to an interesting re–interpretation of the dynamics because of the presence of $\mathcal{Q}$. This follows by re–writing these equations as

$$\langle\Theta\rangle^{\cdot} + \frac{\langle\Theta\rangle^2}{3} = -\frac{\kappa}{2}\left[\rho_{\text{eff}} + 3\,P_{\text{eff}}\right], \tag{21.26}$$

$$\frac{\langle\Theta\rangle^2}{9} = \frac{\kappa}{3}\,\rho_{\text{eff}}, \tag{21.27}$$

where the "effective" density and pressure are

$$\kappa\,\rho_{\text{eff}} \equiv \kappa\,\langle\rho\rangle - \frac{\langle {}^3\mathcal{R}\rangle + \mathcal{Q}}{2}, \qquad \kappa\,P_{\text{eff}} \equiv \frac{\langle {}^3\mathcal{R}\rangle}{6} - \frac{\mathcal{Q}}{2}, \tag{21.28}$$

The compatibility condition between (21.23), (21.24) and (21.25) is given by the following relation between $\dot{\mathcal{Q}}$ and $\langle {}^3\mathcal{R}\rangle^{\cdot}$ (as in[4]):

$$\dot{\mathcal{Q}} + 2\langle\Theta\rangle\,\mathcal{Q} + \frac{2}{3}\langle\Theta\rangle\,\langle {}^3\mathcal{R}\rangle + \langle {}^3\mathcal{R}\rangle^{\cdot} = 0, \tag{21.29}$$

which is equivalent to the compatibility between the time derivative of (21.14), equations (21.9)–(21.10), the commutation rule (21.19) and the variance (21.20) for Θ and ${}^3\mathcal{R}$. From (21.26), the condition for an "effective" cosmic acceleration mimicking dark energy is

$$\frac{\kappa}{2}\,[\rho_{\text{eff}} + 3\,P_{\text{eff}}] = \frac{\kappa}{2}\,\langle\rho\rangle - \mathcal{Q} < 0, \tag{21.30}$$

which, apparently, could be possible to fulfill for a sufficiently large and positive back reaction $\mathcal{Q}$. We will evaluate this condition for spherically symmetric LTB dust solutions (see Ref. 16 and Refs. 12 and 13 for previous work on this).

21.4. Quasi–Local (QL) Variables

The Misner–Sharp quasi–local mass–energy function, $\mathcal{M}$, is a well known invariant in spherically symmetric spacetimes.[18,19] For LTB dust models (21.1)–(21.2) it satisfies the equations

$$2\mathcal{M}' = \kappa\rho\,R^2R', \qquad 2\dot{\mathcal{M}} = 0, \qquad \Rightarrow \mathcal{M} = M(r) \tag{21.31}$$

where M is the function appearing in the field equation (21.4). Comparing (21.31) and (21.4) suggest obtaining an integral expression for $\mathcal{M}$ that can be related to R and $\dot{R}$. This integral along the $\mathcal{T}^{(3)}$ exists and is bounded if we consider an integration domain of the form (21.16) containing a symmetry center.[19] Since $\mathcal{M}(ct, 0) = 0$ for all t, we integrate both sides of (21.4) and also (21.31). This allows us to define a scalar ρ_q as

$$\frac{\kappa}{3}\rho_q \equiv \frac{2\mathcal{M}}{R^3} = \frac{2M}{R^3} = \frac{\kappa}{3}\,\frac{\int_0^r \rho R^2 R' \mathrm{d}x}{\int_0^r R^2 R' \mathrm{d}x}, \tag{21.32}$$

This integral definition of ρ_q, which is related to ρ and to the quasi–local mass–energy function, $\mathcal{M}$, motivates us to generalize it to other scalars by means of the following:

Definition 3: Quasi–local (QL) scalar map. Let $X(\mathcal{D})$ be the set of all smooth integrable scalar functions in $\mathcal{D}$. For every $A \in X(\mathcal{D})$, the quasi–local map is defined as

$$\mathcal{J}_q : X(\mathcal{D}) \to X(\mathcal{D}),\; A_q = \mathcal{J}_q(A) = \frac{\int_{\mathcal{D}[r]} A\,\mathcal{F}\,\mathrm{d}\mathcal{V}_p}{\int_{\mathcal{D}[r]} \mathcal{F}\,\mathrm{d}\mathcal{V}_p} = \frac{\int_0^r AR^2R'\mathrm{d}x}{\int_0^r R^2R'\mathrm{d}x}. \tag{21.33}$$

The scalar functions $A_q : \mathcal{D} \to \mathrm{R}$ that are images of $\mathcal{J}_q$ will be denoted by "quasi–local" (QL) scalars. In particular, we will call A_q the QL dual of A. Notice that a quasi–local average can also be defined by means of a functional with the correspondence rule (21.33), but we will not need it in this article.

Applying the map (21.33) to the scalars Θ and ${}^3\mathcal{R}$ in (21.6) we obtain

$$\Theta_q = \frac{3\dot{R}}{R}, \qquad {}^3\mathcal{R}_q = \frac{6(1-\mathcal{F}^2)}{R^2}. \tag{21.34}$$

Applying now (21.33) to ρ, comparing with (21.3) and (21.4), and using (21.34), these two field equations yield

$$\left(\frac{\Theta_q}{3}\right)^2 = \frac{\kappa}{3}\rho_q - \frac{{}^3\mathcal{R}_q}{6}, \tag{21.35}$$

$$\dot{\Theta}_q = -\frac{\Theta_q^2}{3} - \frac{\kappa}{2}\rho_q. \tag{21.36}$$

$$\dot{\rho}_q = -\rho_q\,\Theta_q. \tag{21.37}$$

which are identical to the Friedman, Raychaudhuri and energy balance equations for dust FLRW cosmologies, but given among QL scalars (notice that we are using here the QL functions, not QL averages). By applying (21.33) to (21.8) the remaining covariant scalars $\{\Sigma,\, \mathcal{E}\}$ can be expressed as deviations or fluctuations of ρ and Θ with respect to their QL duals:

$$\Sigma = -\frac{1}{3}\left[\Theta - \Theta_q\right], \qquad \mathcal{E} = -\frac{\kappa}{6}\left[\rho - \rho_q\right]. \tag{21.38}$$

21.5. Sufficient Conditions for a Positive Back–Reaction and an Effective Acceleration

Equation (21.30) provides the relation between back–reaction ($\mathcal{Q}$) and the an effective acceleration mimicking dark matter. A necessary (but not sufficient) condition for the existence of this acceleration in a given comoving domain $\mathcal{D}[r]$ of the form (21.16) is evidently

$$\frac{3}{2}\mathcal{Q} = \langle \tilde{\mathcal{C}}(x,r) \rangle \geq 0, \qquad \tilde{\mathcal{C}}(x,r) \equiv \left(\Theta(x) - \langle \Theta \rangle [r]\right)^2 - \left(\Theta(x) - \Theta_q(x)\right)^2, \tag{21.39}$$

where we have used (21.38) to eliminate Σ^2 in terms of $\Theta - \Theta_q$. Testing the fulfillment of this condition from the integral definitions (21.17) and (21.33) is very difficult without resorting to numerical methods. However, for every domain $\mathcal{D}[r]$ and every scalar we have $A(x) \geq 0 \ \forall x \in \vartheta[r] \ \Rightarrow \ \langle A \rangle [r] \geq 0$ (though the converse is not necessarily true). Hence, a sufficient condition for the fulfillment of (21.39) in a given domain (21.16) is simply

$$\tilde{\mathcal{C}}(x,r) \geq 0. \tag{21.40}$$

Moreover, this condition is still too difficult to handle because of the dependence of $\tilde{\mathcal{C}}$ on points inside ($x < r$) and in the boundary ($x = r$) of the domain. Fortunately, by means of the following lemma we can find a condition equivalent to (21.40) that depends only on the domain boundary and is applicable to any domain (r variable).

> Lemma 1: $\langle \mathcal{P} \rangle = 0$ in every domain $\vartheta[r]$ for $\mathcal{P} = \mathcal{P}(x,r)$ given by
>
> $$\mathcal{P}(x,r) = \left[\Theta(x) - \langle \Theta \rangle [r]\right]^2 - \left[\Theta(x) - \Theta_p(x)\right]^2, \tag{21.41}$$
>
> Proof. Expanding (21.41) and applying (21.17) and (21.18) we obtain with the help of (21.20)
>
> $$\langle \mathcal{P} \rangle [r] = -\langle \Theta \rangle [r]^2 + \frac{1}{\mathcal{V}_p(r)} \int_0^r \left[2\Theta\Theta_p - \Theta_p^2\right] \mathcal{V}_p' \, dx. \tag{21.42}$$
>
> Inserting $\Theta = \dot{\mathcal{V}}_p'/\mathcal{V}_p'$ and $\Theta_p = \dot{\mathcal{V}}_p/\mathcal{V}_p$ in the integrand above, and bearing in mind that $\langle \Theta \rangle$ and Θ_p coincide at the domain boundary $x = r$, leads to the desired result:
>
> $$\langle \mathcal{P} \rangle [r] = -\langle \Theta \rangle^2 [r] + \frac{1}{\mathcal{V}_p(r)} \int_0^r [\dot{\mathcal{V}}_p^2/\mathcal{V}_p]' \, dx = -\langle \Theta \rangle^2 [r] + \Theta_p^2(r) = 0. \tag{21.43}$$
>
> An analogous result follows for the quasi–local average acting on a scalar like $\mathcal{P}$ with $\langle \, \rangle_q$ and Θ_q instead of $\langle \, \rangle$ and Θ_p.

As a consequence of this lema, we have $\langle (\Theta(x) - \langle \Theta \rangle [r])^2 \rangle = \langle (\Theta(r) - \Theta_p(r))^2 \rangle$ and so the sufficient condition for $\mathcal{Q} \geq 0$ given by (21.40) can be rewritten now as

$$\mathcal{C}(r) \equiv \left(\Theta(r) - \Theta_p(r)\right)^2 - \left(\Theta(r) - \Theta_q(r)\right)^2 \geq 0, \tag{21.44}$$

holding for comoving domains of the form (21.16). This condition is domain dependent, in the sense that it may hold for some domains and not for others. Since the condition for an effective acceleration in (21.30) is given as the average of the scalar $(2/3)\tilde{\mathcal{C}}(x,r)-(\kappa/2)\rho(x)$, lemma 1 also provides the following sufficient condition for its fulfillment

$$\mathcal{A}_(r)\equiv\frac{2}{3}\mathcal{C}(r)-\frac{\kappa}{2}\rho(r)\geq 0. \tag{21.45}$$

where $\mathcal{C}(r)$ is given by (21.44). For the remaining of this paper we will examine conditions (21.44) and (21.45), finding out first the necessary restrictions on LTB models to comply with (21.44), and then exploring for these models the fulfillment of (21.45).

21.6. Probing the Sign of the Back–Reaction Term

In order to look at conditions (21.44) and (21.45), we need to explore the behavior of LTB scalars along radial rays of the hypersurfaces $\mathcal{T}^{(3)}$. For this purpose, the integral definitions (21.17) and (21.33) yield the following properties of p–functions and averages and their quasi–local analogues:

$$A'_p=\frac{3R'}{R}\frac{\mathcal{F}_p}{\mathcal{F}}\left[A-A_p\right],\qquad A'_q=\frac{3R'}{R}\left[A-A_q\right] \tag{21.46}$$

$$A(r)-A_p(r)=\frac{1}{\mathcal{V}_p(r)}\int_0^r A'(x)\,\mathcal{V}_p(x)\,\mathrm{d}x\,,$$

$$A(r)-A_q(r)=\frac{1}{\mathcal{V}_q(r)}\int_0^r A'(x)\,\mathcal{V}_q(x)\,\mathrm{d}x, \tag{21.47}$$

where $\mathcal{F}_p$ is the p–function associated to $\mathcal{F}(r)$ and $\mathcal{V}_q=\int\mathcal{F}\,\mathrm{d}\mathcal{V}_p=4\pi\int_0^r R^2R'\mathrm{d}x=(4\pi/3)R^3(r)$. Considering (21.47), condition (21.44) for $\mathcal{Q}\geq 0$ becomes

$$\mathcal{C}(r)=\Phi(r)\,\Psi(r)\geq 0, \tag{21.48}$$

with:

$$\Phi(r)\equiv\int_0^r\Theta'(x)\,\varphi(x,r)\,\mathrm{d}x,\qquad \Psi(r)\equiv\int_0^r\Theta'(x)\,\psi(x,r)\,\mathrm{d}x, \tag{21.49}$$

and with φ and ψ given by

$$\varphi(x,r)=\frac{\mathcal{V}_p(x)}{\mathcal{V}_p(r)}-\frac{\mathcal{V}_q(x)}{\mathcal{V}_q(r)}=\frac{\mathcal{V}_p(x)}{\mathcal{V}_p(r)}\left[1-\frac{\mathcal{F}_p(x)}{\mathcal{F}_p(r)}\right], \tag{21.50}$$

$$\psi(x,r)=\frac{\mathcal{V}_p(x)}{\mathcal{V}_p(r)}+\frac{\mathcal{V}_q(x)}{\mathcal{V}_q(r)}=\frac{\mathcal{V}_p(x)}{\mathcal{V}_p(r)}\left[1+\frac{\mathcal{F}_p(x)}{\mathcal{F}_p(r)}\right], \tag{21.51}$$

where we have used the relation $\mathcal{V}_q/\mathcal{V}_p = \mathcal{F}_p$, which follows directly from (21.17) and (21.33) and is valid for all domains.

The fulfillment of (21.48) clearly depends on the signs of φ and ψ (besides the sign of Θ') at all points in any given domain. This fulfillment might hold only in some domains and not in others. Since, by their definition, $\mathcal{V}_p(0) = \mathcal{V}_q(0) = 0$ and $\mathcal{F}(0) = 1$, (21.51) and (21.50) imply that for every domain we have $\psi(0,r) = 0$, $\psi(r,r) = 2$, whereas $\varphi(0,r) = \varphi(r,r) = 0$. Thus, as long as $\mathcal{F}$ and R' are non–negative for all $x \in \vartheta[r]$, the sign of ψ is non–negative, and so Ψ basically depends only on the sign of Θ'. On the other hand, the sign of φ is not determined, and so the sign of Φ requires more examination as it depends on both: the sign of Θ' and the ratio $\mathcal{F}_p(x)/\mathcal{F}_p(r)$ (which relates to the sign of $\mathcal{F}'$). Since $\mathcal{F}$ and R' can become negative in elliptic configurations where the $\mathcal{T}^{(3)}(t)$ have spherical topology, we will only consider in this paper "open" LTB models in which these slices are homeomorphic to $\mathbf{R}^3$. Since the sign of $\mathcal{C}$ depends on the sign of the product $\Phi\Psi$, we will need to obtain the conditions for both terms having the same sign.

In order to to probe condition (21.48), we consider the following sign relations that emerge from (21.46)–(21.47) and are valid for all $x \in \vartheta[r]$ in every domain and in every $\mathcal{T}^{(3)}$:

$$\mathcal{F}' \geq 0 \quad \Rightarrow \quad \mathcal{F} \geq \mathcal{F}_p \quad \Rightarrow \quad \mathcal{F}(r) \geq \mathcal{F}_p(x), \tag{21.52}$$

$$\mathcal{F}' \leq 0 \quad \Rightarrow \quad \mathcal{F} \leq \mathcal{F}_p \quad \Rightarrow \quad \mathcal{F}(r) \leq \mathcal{F}_p(x), \tag{21.53}$$

Considering (21.48)–(21.51) and the relations above, we have the following rigorous results:

Lemma 2: Let Θ' be a monotonous function in a domain (21.16) in a given $\mathcal{T}^{(3)}$, then :

$$\mathcal{Q}[r] = 0 \quad \text{if the domain belongs to a parabolic model,} \tag{21.54}$$

$$\mathcal{Q}[r] \geq 0 \quad \text{if the domain belongs to a hyperbolic model,} \tag{21.55}$$

$$\mathcal{Q}[r] \leq 0 \quad \text{if the domain belongs to an open elliptic model for which } \mathcal{F}' \leq 0, \tag{21.56}$$

The proof follows directly from (21.48)–(21.51). It is trivial for the parabolic case, since $\mathcal{F} = 1$ for every domain. For the hyperbolic and elliptic cases, having assumed a monotonous Θ implies that the sign of $\mathcal{C}$ only depends on the sign of $\varphi(x,r)$

in (21.50), but then regularity conditions[10,11] require $\mathcal{F}' \geq 0$ to hold for all r in hyperbolic models, while regularity admits (but does does not require) that $\mathcal{F}' \leq 0$ holds for all r in open elliptic models (we are not considering closed models, see[16] for the study of back–reaction in these models). Hence, we have $\mathcal{C} \geq 0$ and $\mathcal{C} \leq 0$, respectively, for the hyperbolic and elliptic cases, and the result follows.

Notice that (21.54) is domain independent (holds at every domain $\vartheta[r]$), whereas (21.55) necessarily becomes domain independent in those $\mathcal{T}^{(3)}$ of hyperbolic models for which Θ is monotonous for all r. For hypersurfaces $\mathcal{T}^{(3)}$ of open elliptic models in which Θ is monotonous for all r, then (21.55) may become domain dependent: it holds for all domains sufficiently close to the center (where $\mathcal{F}' \leq 0$) but may not hold if $\mathcal{F}'$ changes sign, though (21.55) is domain independent if $\mathcal{F}'$ does not change sign.[16]

The restriction that Θ be monotonous for all r does not hold, in general, for all $\mathcal{T}^{(3)}$ in any hyperbolic or open elliptic model.[20] However, the effect of Θ' changing sign simply makes the sign of $\mathcal{Q}$ domain dependent along any given $\mathcal{T}^{(3)}$. The same remark holds for open elliptic models for which $\mathcal{F}'$ changes sign:

Lemma 3: Consider the following two possible configurations along a given $\mathcal{T}^{(3)}$: (i) Θ' changes sign at a given $x = y \in \vartheta[r]$ in a hyperbolic model, and (ii) $\mathcal{F}'$ and Θ' respectively change sign at a $x = y_1$, $x = y_2$ inside $\vartheta[r]$ in an open elliptic model. For both cases there are always domains $\vartheta[r_0]$ such that $\mathcal{Q}[r] \geq 0$ holds for all $r > r_0$.

The proof in both cases is based on the fact that the zero of Θ' is fixed at any given $\mathcal{T}^{(3)}$, whereas the choice of domain is arbitrary. Consider the hyperbolic case: the zero of Θ' implies that Θ is monotonous in the full "external" range $r > x > y$. Hence if regularity conditions hold, r can take arbitrarily large values and we can always find sufficiently large domains such that $r \gg y$ holds, and thus we can use (21.52) to obtain the expected sign in the integration of Φ and Ψ in (21.48) along the range $y < x \leq r$. For sufficiently large domains the "external" contribution $y < x \leq r$ outweighs that of the inner range $0 \leq x \leq y$. The same situation occurs in the elliptic case, for which $\mathcal{F}'$ must pass from negative to positive at $x = y_1$. The zeroes of Θ' and $\mathcal{F}'$ imply that F and Θ are monotonous for $x > \max(y_1, y_2)$, and thus the same argument applies. The reader is advised to consult[16] for further detail on these proofs.

As a consequence of lemmas 2 and 3, the condition $\mathcal{Q} > 0$ is compatible with both, negative and positive spatial curvature (hyperbolic and open elliptic models), at least in their asymptotic radial range. We need now to find if the positive back–reaction is comparable to the average density.

21.7. Probing the Sign of the Effective Acceleration

In order to examine (21.45), we will need the following rigorous results concerning the radial profiles of scalars A along the $\mathcal{T}^{(3)}$:

> Lemma 4. If $R' > 0$ everywhere and $A(r) \to A_0 =$ constant as $r \to \infty$, then $A_p(r) \to A_0$ and $A_q(r) \to A_0$ as $r \to \infty$, while $A'_p(r) \to 0$ and $A'_q(r) \to 0$ in this limit.

> Proof. We consider the case when $A' \geq 0$ for sufficiently large values of r. The case when $A' \leq 0$ is analogous. If the limit of A as $r \to \infty$ is A_0, then for all $\epsilon > 0$ there exists $L(\epsilon)$ such that $A_0 - \epsilon < A(r) < A_0$ holds for all $r > L(\epsilon)$. Constraining A by means of this inequality in the definitions of A_p and A_q in (21.17) and (21.33) leads immediately to $A_0 - \epsilon < A_q(r) < A_0$ and $A_0 - \epsilon < A_p(r) < A_0$. Hence, the limit of A_p and A_q as $r \to \infty$ is also A_0. The limits $A'_p(r) \to 0$ and $A'_q(r) \to 0$ follow trivially.

> Lemma 5. If there is a zero of A' at $x = y$ and $R' > 0$ in $\vartheta[r]$, then for sufficiently large r there will be a zero of A'_p at $x = r_1 > y$ and a zero of A'_q at $x = r_2 > y$, with $A(r_1) = A_p(r_1)$ and $A(r_2) = A_q(r_2)$.

> Proof. Let A' pass from positive to negative at $x = y$. As x reaches y the first integral in (21.47) is still positive and so $A(y) > A_p(y)$, but for $y < x < r$ the integrand becomes negative, and so the contributions to the integral are increasingly negative. Since $\mathcal{V}_p(x)/\mathcal{V}_p(r)$ is increasing, if r is sufficiently large, then a value $x = r_1 > y$ is necessarily reached so that this integral vanishes (thus $A(r_1) = A_p(r_1)$). From (21.46), we have $A'_p(r_1) = 0$ and $A'_p < 0$ for $x > r_1$. An analogous situation occurs when A' passes from negative to positive. The proof is identical for A_q, but using the second integral in (21.47).

In order to apply these lemmas to the case $A = \mathcal{C}$ with $\mathcal{C}$ given by (21.44), we use (21.46) to rewrite this scalar in terms of the gradients of Θ_q and Θ_p

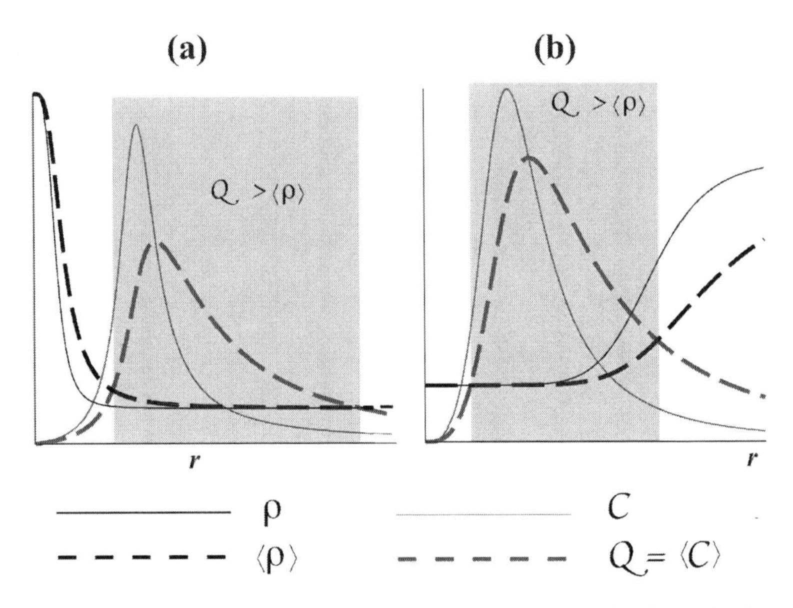

Fig. 21.2. **Scenarios with positive effective acceleration.** The figure displays, for a density clump (a) and a void (b), the radial profiles of $\rho(x)$ (black solid curve) and $\mathcal{C}(x)$ (red solid curve), compared with $\langle\rho\rangle$ (black dotted curve) and back–reaction $\mathcal{Q} \propto \langle\mathcal{C}\rangle$ (red dotted curve) along a given regular hypersurface $\mathcal{T}^{(3)}$ of either a hyperbolic or an open elliptic model. Notice an intermediate region (gray area) where density is low and nearly homogeneous, while the growth of back–reaction is driven by the gradients of spatial curvature and reaches a maximum, thus allowing for (21.45) to hold.

as

$$\mathcal{C} = \left(\frac{R}{3R'}\right)^2 \left[\left(\Theta'_p \frac{\mathcal{F}}{\mathcal{F}_p}\right)^2 - \Theta'^2_q\right] = \left(\frac{R}{3R'}\right)^2 \left[\Theta'_p \frac{\mathcal{F}}{\mathcal{F}_p} - \Theta'_q\right] \left[\Theta'_p \frac{\mathcal{F}}{\mathcal{F}_p} + \Theta'_q\right]. \tag{21.57}$$

Considering now LTB models (hyperbolic and open elliptic) in which the scalars $\{\Theta, \rho, {}^3\mathcal{R}\}$ tend to nonzero finite values $\{\Theta_0, \rho_0, {}^3\mathcal{R}_0\}$ as $r \to \infty$ (an asymptotic FLRW state in the radial direction), then Lemma 4 and (21.57) imply that $\mathcal{C} \to 0$, $\Theta - \Theta_p \to 0$ and $\Theta - \Theta_q \to 0$ as $r \to \infty$, and also Θ'_p and Θ'_q vanish in this limit. Since $\mathcal{C}(0) = 0$ follows from (21.47), then domains must exist for which $\mathcal{C}'$ must have a zero for some $x = y \in \vartheta[r]$. Lemma 5 implies then that domains must exist in which $\mathcal{Q}' = (2/3)\langle\mathcal{C}\rangle'$ has also a zero at $r_1 > y$, corresponding to a local maximum of $\mathcal{Q}$ where it reaches its maximal value in the domain (see figure 2).

The next step is to compare $\mathcal{C}$ and $\mathcal{Q}$ with ρ and $\langle\rho\rangle$ in order to test the fulfillment of (21.45). For this purpose we note that Θ, Θ_p, Θ_q are respectively related to ρ, ρ_p, ρ_q and ${}^3\mathcal{R}$, ${}^3\mathcal{R}_p$, ${}^3\mathcal{R}_q$ by the constraints (21.14), (21.24) and (21.35). Hence, the magnitude of the expansion gradients is closely connected with the magnitude of the gradients of the density and spatial curvature. Given the fact that $\mathcal{Q}(0) = 0$ and $\mathcal{Q} \to 0$ as $r \to \infty$, reaching a maximum in the intermediate range, the best possible situation in which (21.45) could hold is if $\mathcal{Q}$ reaches its maximum in the same region where ρ (and thus $\langle\rho\rangle$) has a low and almost constant value. In this scenario we would have a large intermediate region in which $\rho(r)/\rho(0) \ll 1$ and $\rho' \approx 0$ (and thus $\langle\rho\rangle[r]/\rho(0) \ll 1$ and $\langle\rho\rangle' \approx 0$ hold), so that the growth of $\mathcal{C}$ (and thus $\mathcal{Q}$) is driven by the gradients of the spatial curvature. The scenario described above is illustrated in figure 2 for a density clump and void profile.

21.8. Conclusion

We have introduced quasi–local variables and averages in LTB dust models. These scalar variables are very useful to discuss various theoretical issues concerning these models, such as the application of Buchert's scalar averaging formalism.[4,5] We have found in this paper analytic conditions for a positive back–reaction term $\mathcal{Q}$ and for an effective acceleration mimicking dark energy in these models. The present paper provides a quick summary of comprehensive articles[15,16,20] that are currently under revision, all of them dealing with different aspects of radial profiles of scalars and the issue of back–reaction in LTB models. We have chosen here the simplest boundary conditions (radial asymptotical homogeneity) to illustrate how generic LTB configurations with open topology (hyperbolic and elliptic) can fulfill the conditions for such an effective acceleration. More general conditions are examined in.[20] It is very likely that the astrophysical and cosmological effects of Buchert's formalism will require numerical methods applied to more "realistic" configurations not restricted by spherical symmetry and compatible with observations. The analytic study carried on here and in the associated references can certainly provide a useful guideline for this important task.

Acknowledgments

The author acknowledges financial support form grant PAPIIT–DGAPA IN–119309.

References

1. Copeland, E. J., Sami, M. and Tsujikawa, S. 2006 (*Preprint* arXiv:hep-th/0603057); Sahni, V. 2004 *Lect. Notes Phys.* **653** 141-180 (*Preprint* `arXiv:astro-ph/0403324v3`).
2. Celerièr, M. N. 2007, *New Advances in Physics* **1** 29 (*Preprint* `arXiv:astro-ph/0702416`).
3. Kolb, E. W., Matarrese, S., Notari, A. and Riotto, A. 2005, *Phys. Rev.* D **71** 023524 (*Preprint* `arXiv:hep-ph/0409038v2`); Marra, V., Kolb, E. W. and Matarrese, S. 2008, *Phys. Rev.* D **77** 023003; Marra, V, Kolb, E. W., Matarrese, S. and Riotto, A. 2007, *Phys. Rev.* D **76** 123004.
4. Buchert, T., 2000 *Gen. Rel. Grav.* **32** 105; Buchert, T., 2000, *Gen. Rel. Grav.* **32** 306-321; Buchert, T. 2001, *Gen. Rel. Grav.* **33** 1381-1405; Ellis, G. F. R. and Buchert, T. 2005, *Phys. Lett.* A **347** 38-46; Buchert, T. and Carfora, M. 2002, *Class. Quant. Grav.* **19** 6109-6145; Buchert, T. 2006, *Class. Quantum Grav.* 23 819; Buchert, T., Larena, J. and Alimi, J. M. 2006, *Class. Quantum Grav.* 23 6379.
5. Buchert, T. 2008, *Gen. Rel. Grav.* **40**, 467.
6. Zalaletdinov, R. M., *Averaging Problem in Cosmology and Macroscopic Gravity*, Online Proceedings of the Atlantic Regional Meeting on General Relativity and Gravitation, Fredericton, NB, Canada, May 2006, ed. R. J. McKellar. Preprint `arXiv:gr-qc/0701116`.
7. Coley, A. A. and Pelavas, N. 2007, Phys. Rev. D **75** 043506; Coley, A. A., Pelavas, N. and Zalaletdinov, R. M. 2005, *Phys. Rev. Lett.* **95** 151102.
8. Wiltshire, D. L. 2007, *New J. Phys.* **9** 377.
9. Krasinski, A. 1998, *Inhomogeneous Cosmological Models* (Cambridge University Press).
10. Matravers, D. R. and Humphreys, N. P. 2001, *Gen. Rel. Grav.* **33** 53152.
11. Sussman, R. A. and García–Trujillo, L. 2002, *Class. Quantum Grav.* **19** 2897-2925.
12. Moffat, J. W. 2006, *J. Cosmol. Astropart. Phys.* JCAP (2006) 001; Rasanen, S. 2006, *Class. Quant. Grav.* **23** 1823-1835; Kai, T., Kozaki, H., Nakao, K., Nambu, Y. and Yoo, C. 2007, *Prog. Theor. Phys.* **117** 229-240 (*Prepint* `arXiv:gr-qc/0605120v2`); Enqvist, K. and Mattsson, T. 2007, *JCAP* **0702** 019 (*Preprint* `arXiv:astro-ph/0609120v4`).
13. Paranjape, A. and Singh, T. P. 2006, *Class. Quant. Grav.* **23** 69556969.
14. Sussman, R. A. 2009, *Phys. Rev.* D **79** 025009 *Preprint* `arXiv:0801.3324v4 [gr-qc]`.
15. Sussman, R. A. 2008, Quasi–local variables, non–linear perturbations and back–reaction in spherically symmetric spacetimes *Preprint* `ArXiv:0809.3314v1 [gr-qc]`.
16. Sussman, R. A. 2008, Conditions for a back–reaction and "effective" acceleration in Lematre–Tolman–Bondi dust models *Preprint* `arXiv:0807.1145 [gr-qc]`.
17. Ellis, G. F. R. and Bruni, M. 1989, *Phys. Rev.* D **40** 1804; Ellis, G. F. R. and

van Elst, H. 1998, Cosmological Models (Cargèse Lectures 1998) *Preprint* `arXiv gr-qc/9812046 v4`.
18. Kodama, H. 1980, *Prog. Theor. Phys.* **63**; Szabados, L. B. 2004, *Living Rev. Relativity* **7** 4.
19. Hayward, S. A. 1996, *Phys. Rev.* D **53** 1938 (*Preprint* `ArXiv gr-qc/9408002`).
20. Sussman, R. A. 2009, "Radial asymptotics and profiles in regular LTB models". In preparation.

Author Index

Subject Index